AF902643

Road Survey and Design

Bin Yu • Jianchuan Cheng

# Road Survey and Design

东南大学出版社
SOUTHEAST UNIVERSITY PRESS

Springer

Bin Yu
Southeast University
Nanjing, China

Jianchuan Cheng
Professor of School of Transportation
Southeast University
Jiangsu, China

ISBN 978-981-97-4835-8 ISBN 978-981-97-4836-5 (eBook)
https://doi.org/10.1007/978-981-97-4836-5

Jointly published with Southeast University Press
The print editor is not for sale in China Mainland. Customer from China Mainland, please order this book from Southeast University Press

© The Editor(s) (if applicable) and The Author(s), under exclusive license to Springer Nature Singapore Pte Ltd. 2025

This work is subject to copyright. All rights are solely and exclusively licensed by the Publisher, whether the whole or part of the material is concerned, specifically the rights of translation, reprinting, reuse of illustrations, recitation, broadcasting, reproduction on microfilms or in any other physical way, and transmission or information storage and retrieval, electronic adaptation, computer software, or by similar or dissimilar methodology now known or hereafter developed.
The use of general descriptive names, registered names, trademarks, service marks, etc. in this publication does not imply, even in the absence of a specific statement, that such names are exempt from the relevant protective laws and regulations and therefore free for general use.
The publishers, the authors, and the editors are safe to assume that the advice and information in this book are believed to be true and accurate at the date of publication. Neither the publishers nor the authors or the editors give a warranty, express or implied, with respect to the material contained herein or for any errors or omissions that may have been made. The publishers remain neutral with regard to jurisdictional claims in published maps and institutional affiliations.

This Springer imprint is published by the registered company Springer Nature Singapore Pte Ltd.
The registered company address is: 152 Beach Road, #21-01/04 Gateway East, Singapore 189721, Singapore

If disposing of this product, please recycle the paper.

# Preface

"Road survey and design" is the core course of civil engineering (road and bridge major), highway and bridge engineering, traffic engineering and other majors. With the fast development of highway infrastructure construction and urbanization process in China, new concepts of planning and development, such as people orientation, safety first, environment-friendly and ecological protection, are merging. Moreover, the rapid development of road survey and design technologies, and the continuous updating of the related design standards and specifications of highway and urban roads all put forward requirements for the supplement and improvement of "road survey and design".

This book refines the traditional knowledge of road geometric design and focuses on the new concepts of road planning and design. It updates technologies of survey and design, introduces innovative methods of road geometric design, and is compiled by paying attention to the engineering practice. Meanwhile, videos and QR codes of key points are provided to facilitate the utilization of this book.

This book consists of eight chapters and is compiled by Bin Yu and Jianchuan Cheng. Shuyi Wang, Jinzhou Liu, Qi Liu, Wen Zhou, Shuying Qin, Qian Yang, Haotian Di, Tianheng Chen, Jiaqiang Zhang, Chuanyu Xiao, and Na Li participated in the compilation, illustration and partial proofreading of the book.

This book is compiled with reference to the relevant books, and the related standards and specifications of highways and urban roads. The engineering projects involved in the book are provided by China Design Group Co., Ltd. (formerly Jiangsu Provincial Transportation Planning and Design Institute) and China Merchants Chongqing Communications Technology Research & Design Institute Co., Ltd. We hereby express our sincere gratitude to the authors of the references and industry experts.

Subject to the capability of the editors, there might be deficiencies or inadequacies in the book. Please contact us for further revision and improvement.

**Bin YU**

**Nanjing, September 2022**

# Contents

# Chapter 1 Introduction

【**Objectives**: 1. Understand the characteristics of road transportation; 2. Be familiar with the history, status quo and planning of road development, and understand the requirements for road design (macro level: layout, grade coordination, etc.; micro level: safety, mobility, etc.); 3. Grasp the principles and methods of road classification and grading; 4. Comprehend the relationship between road class selection and design elements, design vehicle, design speed, traffic volume, capacity and other design control factors; 5. Master the main contents and objectives of the course and the relationship with related courses.】

## 1.1 Introduction to roadway transportation

Transportation is composed of railway, roadway, waterway, aviation and pipeline modes. Among them, roadway transportation plays an important role. On the premise of jointly promoting the development of national economy, the five transportation modes have clear functional divisions, and meanwhile complement each other to form a unified, integrated transportation system. Railway transportation undertakes the major burden of bulk passenger and cargo transportation and long-distance transportation. Water transportation shows its economic benefits due to its low freight costs. Air transportation is well-known for its competence in the rapid transportation of passengers and valuable emergency supplies. Pipeline transportation is mostly used for the transportation of liquid and gaseous goods such as oil and gas, etc.

Roadway transportation is characterized by its:

1) High degree of flexibility: vehicles can be collected and distributed quickly with limited time and quantity constraints;

2) Convenience of service: vehicles can go directly to the departure and arrival points of passenger and cargo transportation without intermediate transshipment, i.e. the so-called "door-to-door" service, which is also an

© The Author(s), under exclusive license to Springer Nature Singapore Pte Ltd. 2025
B. Yu and J. Cheng, *Road Survey and Design*,
https://doi.org/10.1007/978-981-97-4836-5_1

essential supplement to the other modes of transportation.

Roads in this book refer to highways and urban roads (streets). Highways refer to roads located in the suburb of the city, serving the long-distance traffic that connects cities. Urban roads refer to roads within the city, which are not only responsible for the daily traffic, but also provide space for the layout of the urban public pipelines, street greening, organization of buildings along the street and division of neighborhoods. And to a certain extent, they affect the sunshine, ventilation and architectural art of urban street buildings. Therefore, urban roads are an important part of urban municipal facilities.

## 1.2 Introduction to road development in China

The history of roads is also the development history of human society. Road development in China has a long history. The pottery unearthed in the Xia Dynasty (before 2000 BC) has wheel patterns, and the oracle bones unearthed in the Shang Dynasty (before 1400 BC) have records of cattle- and horse-drawn cart, which are evident to demonstrate the existence of roads at that time.

Since there are written records, various records about road construction can be found. For example, "Xiaoya" in *the Book of Songs* says, "The national road is as flat as a whetstone and as straight as an arrow", indicating that the road is flat, straight and smooth. After the first Emperor of Qin (259 BC – 210 BC) unified the six countries, in order to consolidate the regime and facilitate the merchants, he began to build a magnificent road network that spread across the country. According to "Jia Shan Biography" in *Han Shu*: "The Qin Dynasty built the fast roads that stretch to Yan&Qi in the east, to Wu&Chu in the south. The fast roads reach rivers, lakes, and as far as the sea."

However, before 1949, China experienced a long feudal society. And because of imperialist aggression, it then further became a semi-feudal and semi-colonial society so that the transportation industry, like other industries, witnessed very limited development. Until the eve of national liberation, only about 130,000 kilometers of roads had been built across the country, and only 75,000 kilometers of roads could barely maintain traffic. For urban roads, only a few large coastal cities had some wide roads.

Since the founding of the People's Republic of China (PRC), after more than 70 years of construction, especially after the reform and opening-up, transportation has achieved great development. By the end of 2020, the total mileage of highways has reached 5,198,100 km, including 161,000 km of freeways, ranking first in the world. China has certainly become a major country in road traffic. However, at present, there are still some problems in road development in China, such as insufficient density, unbalanced development in the east and west regions, and relatively lowclass of local roads, which need to be improved in the future. At the same time, it has entered the stage of paying equal attention to road construction, maintenance, management and service. How to fully utilize the comprehensive transportation system to improve the public's sense of acquisition and satisfaction with transportation has become a new challenge.

By the end of 2020, China has fully completed the tasks formulated in the 13th Five-Year Plan for the development of the modern comprehensive transportation system, and is moving forward to the goal of the 14th Five-Year Plan for the development of the modern comprehensive transportation system.

The report of the 19th National Congress of the Communist Party of China (CPC) clearly proposes to build a country with strong transportation network, which means that China will start a new journey in the new era. On September 19th, 2019, the CPC Central Committee and the State Council have issued the "Program of Building National Strength in Transportation". And China will promote the program in two phrases from 2021 to the middle of this century. By 2035, a developed network, composed of high-speed railways, freeways, and civil aviation, will be formed with high service quality and high operating speed; a satisfactory arterial network, composed of general high-speed railways, general national highways, channels, and oil and gas pipelines, will be formed with high operation efficiency and strong service capacity; a wide range of basic networks, composed of general provincial highways, rural highways, feeder railways, feeder waterways and general aviation, will be formed with broad coverage, deep accessibility and wide benefits. The "National 123 Transportation Circle" with 1-hour commute in metropolitan areas, 2-hour accessibility in urban agglomerations and 3-hour coverage in major cities in China will be formed; and the "Global 123 Express Logistics Circle" with domestic delivery in one day, neighboring countries in two days and major cities in the world in three days, will also be formed. That is, with the "three transportation networks and two transportation circles", China's strength in transportation will be basically established.

## 1.3 Classifications and technical standards of roads

### 1.3.1 Classifications of roads

Roads can be classified according to the administration, location, function, purpose, and nature of traffic, as well as construction materials.

**1) Classified by administration**

Generally, roads can be divided into: (1) national highways; (2) provincial highways; (3) county roads; (4) township roads; (5) exclusive roads.

**2) Classified by the nature of location**

Roads can be divided into: (1) urban roads; (2) highways; (3) agricultural roads; (4) park roads.

**3) Classified by the purpose of roads**

Roads can be divided into: (1) public transportation roads; (2) factory and mine roads; (3) cultural and recreational roads; (4) vehicle exclusive roads; (5) forestry roads; (6) agricultural roads; (7) military roads; (8) commercial roads.

**4) Classification by construction materials**

Roads can be divided into: (1) unpaved roads, which refer to roads without surface course; (2) paved roads, which refer to roads with surface course, and can be divided into the sand pavement, gravel pavement, cement concrete pavement, and asphalt pavement.

### 1.3.2 Function systems and classifications of roads

There is a necessity to classify roads into different operational systems, functional classes, or geometric types to facilitate communications among engineering, administrators, and the general public. Rural and urban areas use different classification schemes for different purposes. Classification of highway by technical class based on the major geometric features (the freeway, Class I highway and other highways) is the most helpful one for the design procedure. Classification by route numbering (e. g. G25, S204) is the most helpful one for traffic operations. Administrative classification (e. g. national, provincial) is used to denote the levels of government responsible for highway facilities and the financing methods. Functional classification, the grouping of roads by the character of service they provide, is developed for transportation planning purposes. Comprehensive transportation planning, an integral part of total economic and social development, uses functional classification as an important planning tool. The emergence of functional classification as the predominant method of grouping roads is consistent with the policies contained in this book and emphasized in the technical standards and specifications.

**1) Functional system of highways**

The two major considerations for the functional classification of highway and urban road networks are access and mobility. The conflict between serving through movement and providing access to a dispersed pattern of trip origins and destinations necessitates the differences and gradations in the various functional types. Regulated limitation of access is needed on arterials to enhance their primary function of mobility.

Conversely, the primary function of local highways and urban roads is to provide access (the implementation of access control causes a limitation of mobility). The extent and degree of access control are thus significant factors in defining the functional category of a highway or urban road.

Allied to the idea of traffic categorization is the dual role that the highway and urban road networks play in providing: access to property and travel mobility. Access is a fixed need for every area served by the highway system. Mobility is provided at varying levels of service. Mobility can incorporate several qualitative elements, such as riding comfort and absence of speed changes, but the most basic factor is operating speed or travel time.

Classification of highway functions shall be based on the expected role of the highway in the network, and the function design shall meet the travel needs. According to the "Technical Standard of Highway Engineering" (JTG B01—2014) issued by the Ministry of Transport of the PRC (hereinafter referred to as the **Standard (2014)**), highways are divided into arterials, collectors and locals according to their functions. Arterials can be further divided into principal arterials and minor arterials, while collectors can also be divided into principal collectors and minor collectors.

The hierarchy of travel distance (long, medium and short) can be related logically to functional specialization in meeting the property access and travel mobility needs. Local rural facilities emphasize the land access function. Arterials for main movements or distribution emphasize the high level of mobility for through movement. Collectors offer approximately balanced service for both functions. This scheme is illustrated conceptually in Figure 1 - 1.

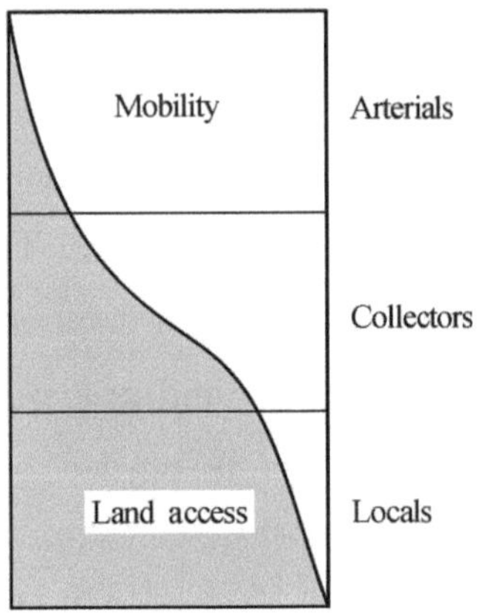


**Figure 1 - 1 Relationship of functional classified systems in serving traffic mobility and land access**

**2) Technical classes of highways**

In the **Standard (2014)**, highways are divided into the following five technical classes according to the traffic volume and the objective and nature of their services:

(1) A **freeway** is a multi-lane highway which is exclusively used for vehicles to drive in different directions and lanes, and with full control of access. The design annual average daily traffic (AADT) of freeways should be more than 15,000 passenger cars.

(2) A **Class Ⅰ highway** is a multi-lane highway for vehicles to drive in different directions and lanes with control of access as needed. The design AADT of Class Ⅰ highways should be more than 15,000 passenger cars.

(3) A **Class Ⅱ highway** is a two-lane highway for vehicles to drive. The design AADT of Class Ⅱ highways should be 5,000 - 15,000 passenger cars.

(4) A **Class Ⅲ highway** is a two-lane highway for mixed traffic of vehicles and non-vehicles. The design AADT of Class Ⅲ highway should be 2,000 - 6,000 passenger cars.

(5) A **Class Ⅳ highway** is a two-lane or single-lane highway for mixed traffic of vehicles and non-vehicles. The design AADT of the two-lane Class Ⅳ highways should be less than 2, 000 passenger cars. The design AADT of the single-lane Class Ⅳ highways should be less than 400 passenger cars.

The relevant technical indexes in the "Design Specification for Highway Alignment" (JTG D20—2017) (hereinafter referred to as the **Specification (2017)**) will also be used in the subsequent chapters.

**3) Functional system of urban roads**

"Code for Design of Urban Road Engineering" (CJJ 37—2012) (hereinafter referred to as the **Code (2012)**) divides urban roads into four classes: expressways, urban arterial roads, urban minor arterial roads (secondary trunk roads), and branch roads according to the status of roads in the system, as well as the traffic mobility function and access service function for vehicles and pedestrians entering and exiting buildings along the route. The four classes of urban roads shall meet the following requirements:

(1) Expressways shall be separated in the middle. All the entrances and exits, as well as the gap and form of entrance and exit, shall be controlled to realize continuous traffic flow. There shall not be less than two lanes in each direction, and the corresponding traffic safety and management facilities shall be provided.

Entrances and exits of public buildings attracting a large number of vehicles and people shall not be provided on both sides of expressways.

(2) Urban arterial roads shall connect the main urban areas, and mainly undertake the traffic mobility function.

Entrances and exits of public buildings attracting a large number of vehicles and people should not be provided on both sides of urban arterial roads.

(3) Urban minor arterial roads shall be combined with arterial roads to form the arterial network, which shall mainly focus on the traffic mobility function, and also consider the access service function.

(4) Branch roads should be connected with urban minor arterial roads and

the internal roads of residential areas, industrial areas and traffic facilities. It shall mainly respond to the traffic demands in local areas and focus on the access service function.

The design indexes of highways of all classes are shown in Table 1 - 1.

**Table 1 - 1 Design indexes of highways**

| Highway class | | | Freeway | | | Class Ⅰ | | | Class Ⅱ | | Class Ⅲ | | Class Ⅳ | |
|---|---|---|---|---|---|---|---|---|---|---|---|---|---|---|
| Design speed/ km • $h^{-1}$ | | | 120 | 100 | 80 | 100 | 80 | 60 | 80 | 60 | 40 | 30 | 30 | 20 |
| Number of lanes | | | ≥4 | | | ≥4 | | | 2 | 2 | 2 | 2 | 2(1)* | |
| Lane width/m | | | 3. 75 | 3. 75 | 3. 75 | 3. 75 | 3. 75 | 3. 5 | 3. 75 | 3. 5 | 3. 5 | 3. 25 | 3. 25 | 3. 0 |
| Stopping sight distance/m | | | 210 | 160 | 110 | 160 | 110 | 75 | 110 | 75 | 40 | 30 | 30 | 20 |
| *R* of circular curve/ m | Max. super-elevation rate | 10% | 570 | 360 | 220 | 360 | 220 | 115 | 220 | 115 | — | — | — | — |
| | | 8% | 650 | 400 | 250 | 400 | 250 | 125 | 250 | 125 | 60 | 30 | 30 | 15 |
| | | 6% | 710 | 440 | 270 | 440 | 270 | 135 | 270 | 135 | 60 | 35 | 35 | 15 |
| Max. grade/% | | | 3 | 4 | 5 | 4 | 5 | 6 | 5 | 6 | 7 | 8 | 8 | 9 |

*Note*: * Two lanes shall be used for the Class Ⅳ highway; when the traffic volume is small or for some difficult section, the single lane may be used.

The design indexes of urban roads of all classes are shown in Table 1 - 2.

**Table 1 - 2 Design indexes of urban roads**

| Class | Indexes | | | |
|---|---|---|---|---|
| | Design speed/ km • $h^{-1}$ | Lane width/m | Separator | Cross-section form |
| Expressway | 100, 80, 60 | 3. 75 (oversize vehicles or mixture of traffic) | Mandatory | Four-road (with side road) |
| | | 3. 5 (passenger cars exclusive, 3. 25 m at a design speed of 60 km/h) | | Two-road (without side road) |
| Urban arterial road | 60, 50, 40 | 3. 5 (oversize vehicles or mixture of traffic) | — | Four-road or three-road |
| | | 3. 25 (passenger cars exclusive) | | |

**Continued Table 1 - 2**

| Class | Indexes | | | |
|---|---|---|---|---|
| | Design speed/ km · $h^{-1}$ | Lane width/m | Separator | Cross-section form |
| Urban minor arterial road | 50, 40, 30 | 3.5 (oversize vehicles or mixture of traffic) | — | Two-road or single-road |
| | | 3.25 (passenger cars exclusive) | | |
| Branch road | 40, 30, 20 | 3.5(oversize vehicles or mixture of traffic) | — | Single-road |
| | | 3.25 (passenger cars exclusive) | | |

## 1.3.3　Control elements of road design

The selection of road classes is directly influenced by the design vehicle, design speed, traffic volume, traffic capacity, level of service, driver characteristics and road boundary, etc.

**1) Design vehicle**

The design vehicle used in the geometric design of highways and intersections shall be selected according to the highway function, vehicle composition, and other factors. The overall dimensions of each representative vehicle type are shown in Figure 1 - 2 and listed in Table 1 - 3. The selection of design vehicles shall meet the following requirements:

(1) The arterial highway and principal collector highway shall accommodate the movement of all design vehicles.

(2) The minor collector highway shall accommodate the movement of passenger cars, trucks, and buses (motor coaches).

(3) The local road shall accommodate the movement of passenger cars and buses.

(4) For highways with special movement requirements, the design vehicles can be determined accordingly.

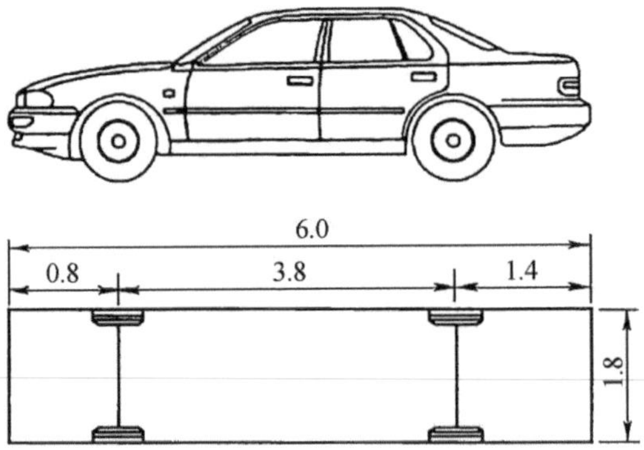

(a) Passenger cars

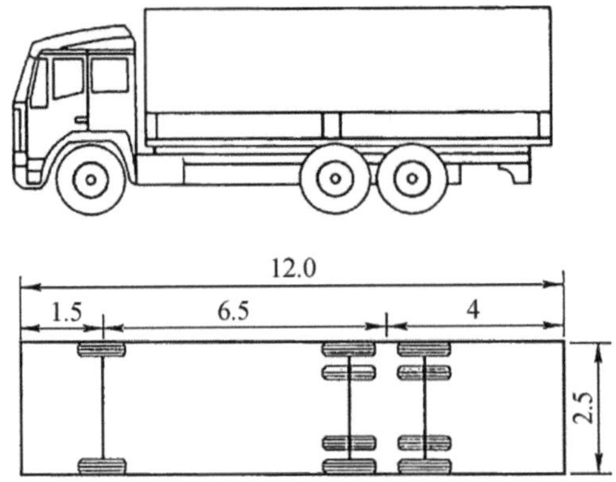

(b) Trucks

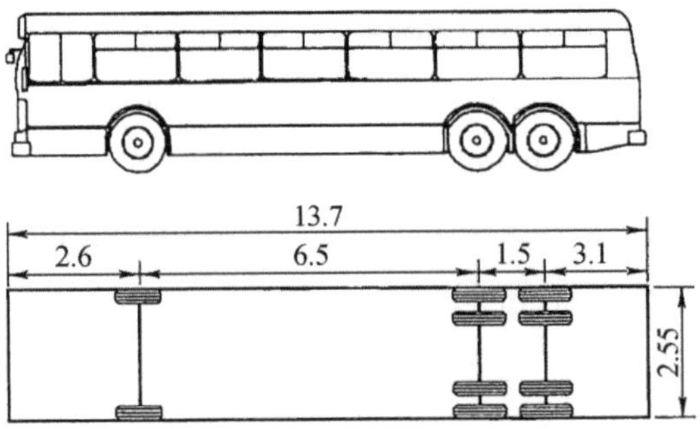

(c) Buses (motor coaches)

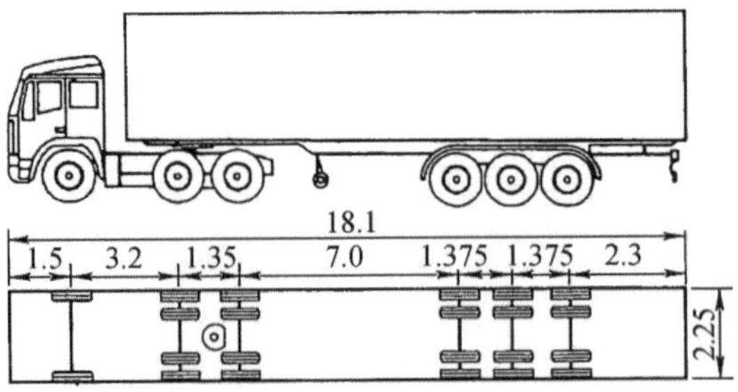

(d) Articulated trucks (trailers)

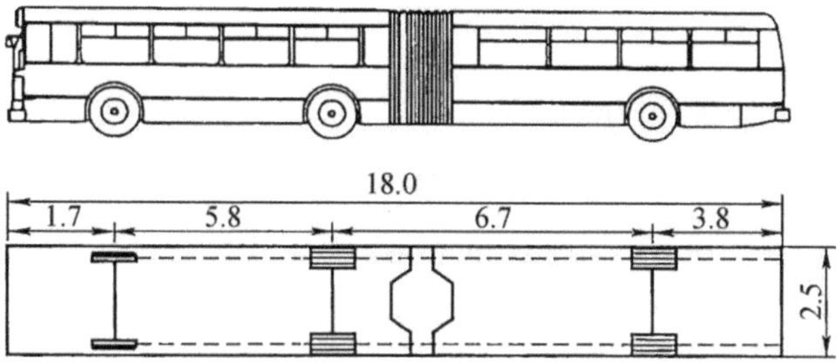


(e) Articulated buses

**Figure 1 - 2 Dimensions of representative vehicles (m)**

*Note*: This figure refers to the **Standard (2014)**.

**Table 1 - 3 Design vehicle dimensions**

| Vehicle type | Length/m | Width/m | Height/m | Front overhang/m | Wheel base/m | Rear overhang/m |
|---|---|---|---|---|---|---|
| Passenger car | 6.0 | 1.8 | 2.0 | 0.8 | 3.8 | 1.4 |
| Bus | 13.7 | 2.55 | 4.0 | 2.6 | 6.5+1.5 | 3.1 |
| Articulated bus | 18.0 | 2.5 | 4.0 | 1.7 | 5.8+6.7 | 3.8 |
| Truck | 12.0 | 2.5 | 4.0 | 1.5 | 6.5 | 4.0 |
| Articulated truck | 18.1 | 2.55 | 4.0 | 1.5 | 3.3+11 | 2.3 |

*Note*: The "3.3 m" of the wheel base of the articulated truck is the distance from the first axle to the articulation joint position, and the "11 m" is the distance from the articulation joint position to the last axle.

*Note*: This table refers to the **Specification (2017)**.

In urban road design, motor vehicles have three types, including passenger cars, buses and articulated vehicles, while non-motor vehicles include bicycles and tricycles.

**Table 1 - 4 Design vehicle dimensions of urban roads**

| Vehicle type | Length/m | Width/m | Height/m | Front overhang/m | Wheel base/m | Rear overhang/m |
|---|---|---|---|---|---|---|
| Passenger car | 6.0 | 1.8 | 2.0 | 0.8 | 3.8 | 1.4 |
| Bus | 12 | 2.5 | 4.0 | 1.5 | 6.5 | 4.0 |
| Articulated vehicle | 18 | 2.5 | 4.0 | 1.7 | 5.8+6.7 | 3.8 |
| Bicycle | 1.93 | 0.60 | 2.25 | — | — | — |
| Tricycle | 3.40 | 1.25 | 2.25 | — | — | — |

*Note*: This table refers to the **Code (2012)**.

**2) Speed**

(1) Speed used in road design

Speed is one of the most important factors considered by a traveler in selecting alternative routes or transportation modes, which is also a very important index for road design. There are three types of speed that are frequently used in road design.

① Design speed

In the **Standard (2014)**, the design speed is defined as the reference speed that determines the geometry of highways, so that they coordinate with each other. In *A Policy on Geometric Design of Highways and Streets* by AASHTO, design speed is a selected speed used to determine the various geometric design features of the roadway. The assumed design speed shall be a logical one with the topography, anticipated operating speed, the adjacent land use, and the functional classification of highways. Once the design speed is selected, all of the pertinent highway features should be related to it to obtain a balanced design. In this way, the design speed serves as the basis for the control design parameters (the minimum radius of horizontal curve, the maximum grade, lane width, sight distance, etc.) of highways at all classes. It is generally deemed as the maximum speed at which a general driver can maintain safe and comfortable driving in the case of good weather and low traffic density.

② Running speed

The running speed refers to the speed at which an individual travels over a highway section, which is calculated by the length of the highway section divided by the running time required for the vehicle to travel through the section. The average running speed of all vehicles is the most appropriate speed measure for evaluating levels of service and road user costs. The average running speed is the sum of the distance traveled by vehicles on a highway section during a specified time period divided by the sum of their running time.

③ Operating speed

The operating speed is the one at which drivers are observed operating their vehicles during free-flow conditions. The 85th percentile of the of observed speed distribution is the most frequently used measurement of the operating speed associated with a particular location or geometric feature.

(2) Selection of design speed

Design speed shall be maintained as consistent as possible for roads of the same class. However, design speed may be changed subject to the topography and traffic conditions. When this situation occurs, it shall be at the place where the traffic volume changes greatly or where the driver can clearly judge that the situation changes and the driving speed needs to be changed. And transition sections shall be provided to facilitate the change. The selection of design speed shall be based on the function and technical class of the highway, combined with the terrain, engineering economy, expected operating speed, land use properties along the route and other factors. The design speed specified in the **Specification (2017)** is listed in Table 1 - 1.

As an important national and provincial arterial highway, or as a national and provincial arterial highway with large traffic volume, or located in plain and hilly areas with good terrain and geology, the design speed of the freeway should be 120 km/h or 100 km/h after technical and economic demonstration. When it is restricted by terrain, geology and other natural conditions, it can be 80 km/h after demonstration.

The design speed of the Class Ⅰ arterial highway should be 100 km/h; when restricted by natural conditions such as terrain and geology, the design speed may be 80 km/h. The design speed of the Class Ⅰ collector highway should be 80 km/h; when restricted by natural conditions such as terrain and geology, the design speed may be 60 km/h.

As a Class Ⅱ arterial highway, the design speed should be 80 km/h; as a collector highway with large mixed traffic volume in the urban-rural fringe area or a section with restricted terrain and other conditions, its design speed should be 60 km/h; when restricted by natural conditions such as terrain and geology, the design speed may be 60 km/h. As a Class Ⅱ highway, the design speed should be 60 km/h; when restricted by natural conditions such as terrain and geology, the design speed may be 40 km/h.

The design speed of the Class Ⅲ highway should be 40 km/h; when restricted by natural conditions such as terrain and geology, the design speed may be 30 km/h.

The design speed of the Class Ⅳ highway should be 30 km/h; when restricted by natural conditions such as terrain and geology, the design speed may be 20 km/h.

Compared with highways, urban roads have the characteristics of diverse functions, complex composition, large pedestrian traffic volume, many vehicles, large speed differences and many intersections. The average running speed is lower than that of highways. The design speed specified in the **Code** (**2012**) is listed in Table 1 - 2, and the larger value should be adopted when conditions permit.

(3) Relationship between design speed and operating speed

It is worth mentioning that the driver often does not drive at the design speed, but selects the speed that adapts to the geometric conditions of the road according to the terrain, traffic and other actual conditions along the way. In other words, the operating speed is not consistent with the design speed. In the sections with low design speed, when the geometric elements of the road exceed the requirements for safe driving and the external conditions (traffic density, terrain, climate, etc.) are good, the operating speed is often close to or exceeds the design speed; the lower the design speed, the higher the probability of the phenomenon. On the contrary, in the sections with high design speed, when the external conditions are poor, the operating speed is generally lower than the design speed; the higher the design speed and the worse the external conditions, the higher the probability of the phenomenon. The above analysis shows that the alignment index obtained by route design with the design speed as the control is likely to be inconsistent with the requirements of operating speed. At present, the commonly used improvement method is to check and evaluate the rationality of the design index by using the difference between the design speed and the operating speed. According to the **Standard** (**2014**), the difference between the operating speeds of adjacent sections shall be less than 20 km/h; the difference between the operating speed and the design speed in the same section should be less than 20 km/h.

**3) Traffic volume**

Traffic volume refers to the traffic flow (number of vehicles) passing through a road section per unit time.

In road survey and design, the most widely used traffic volume is the annual average daily traffic (AADT) volume and the design hourly traffic volume.

(1) Design traffic volume

The design traffic volume refers to the AADT volume that can be achieved by the proposed road to the predicted service life. Its value is predicted based on the traffic observation data over the years. The prediction of design traffic volume shall fully consider the long-term social and economic development planning and the impact of comprehensive transportation system in the corridor. At present, it is usually calculated according to the annual average growth rate, as shown in formula (1 - 1).

$$AADT = ADT \times (1+r)^{n-1} \tag{1-1}$$

where: $AADT$—design traffic volume, vehicle/d;

$ADT$—$AADT$ at the beginning, vehicle/d;

$r$—annual average growth rate, %;

$n$—prediction period, year.

Provisions for the prediction period: the prediction period for design traffic volume of the freeway and Class Ⅰ highway is 20 years; the prediction period of Class Ⅱ and Ⅲ highways is 15 years; the prediction period of the Class Ⅳ highway may be determined by actual conditions. In addition, the starting year of the prediction period is the planned opening year of the project.

(2) Design hourly traffic volume

The design hourly traffic volume of a highway should be the 30th hourly traffic volume of one year or an economic and reasonable one between the 20th and 40th hourly traffic volumes of the year may be adopted according to the characteristics of local highways. The design hourly traffic volume is the basis for determining the width of roads and lanes or evaluating the level of service, which is calculated by formula (1 - 2).

$$N_h = AADT \times D \times k \tag{1-2}$$

where: $N_h$—design hourly traffic volume in the main direction, vehicle/d;

$AADT$—design traffic volume, vehicle/d;

$D$—directional nonuniformity coefficient, 0.5 - 0.6 normally;

*k*—design hourly traffic coefficient, %. If there is no data, it can be selected according to the **Specification (2017)**.

(3) Traffic capacity and level of service

The traffic capacity refers to the maximum number of vehicles that can pass through a section in a certain period of time (usually one hour) under normal highway conditions, traffic conditions, and driving behavior. Traffic capacity reflects the ability of road infrastructure to accommodate traffic flow, which is an important parameter of highway planning, design, and operation management.

The design capacity refers to the maximum number of vehicles that can pass through a certain section per unit time given that the operation state of road traffic is maintained at a certain design level of service. According to the operation state of vehicle flow, the operation state from free flow at small traffic volume to the constrained state is divided into six levels of service, and the traffic volume corresponding to each level of service is called service traffic volume. When the design hourly traffic volume exceeds the design traffic capacity, the road will be congested.

The level of service is a quality index for drivers to feel the running state of highway traffic flow. It is usually characterized by such indicators as average running speed, driving time, driving freedom and traffic delay. Service of highways at all levels shall not be lower than that specified in Table 1 - 5.

**Table 1 - 5 Levels of service of highways**

| Class | Freeway | Class Ⅰ | Class Ⅱ | Class Ⅲ | Class Ⅳ |
|---|---|---|---|---|---|
| Level of service | Three | Three | Four | Four | — |

**4) Driver characteristics**

An appreciation of driver performance is essential to proper highway design and operation. The suitability of a design rests as much on how safely and efficiently drivers can use the highway as on any criterion. When drivers use a highway designed to be compatible with their capacities and limitations, their performance is aided. When a design is incompatible with the capacities of drivers, the chance for driver errors increase, which may lead to crashes or inefficient operation. Driver characteristics include many aspects, such as reaction time, driver expectation, fatigue characteristics, sight distance characteristics, driving age and driver age, which all have a certain impact on the design. The reaction time directly affects the design of sight distance and traffic safety facilities; both alignment design and landscape design need to consider the visual characteristics of drivers.

It takes drivers time to process information. The driver's reaction time increases as a function of decision complexity and the amount of information to be processed increase. Furthermore, the longer the reaction time, the greater the chance for errors. According to a study, when an event is expected, the averages reaction time about 0.6 s, with a few drivers taking as long as 2 s. With unexpected events, reaction times increase by 35%. Thus, for a simple, unexpected decisions and actions, some drivers may take as long as 2.7 s to respond. For this sake, the reaction time takes 3 s in the design to assure safety. And the brake reaction time is recommended to be 2.5 s used in design which exceeds the 90th percentile of reaction time for all drivers.

**5) Road boundary, highway land and boundary line of urban roads**

(1) Road boundary

Road boundary, as known as road clearance, is made up of clearance height and clearance width. It is a space boundary that is not allowed to be invaded by any obstacles within a certain height and width to ensure the normal traffic and safety of vehicles and people on the road. Road boundary is an important basis for the cross-section design. During the design, the interrelationship of road elements and the setting rules of various road facilities shall be fully studied, and reasonable arrangements shall be made in the limited space. Abutments, piers, lighting columns, guardrails, annunciators, signs, street trees, electric poles and other facilities shall not be allowed to intrude into the road boundary.

The clearance height refers to the vertical height that the road must meet to ensure safe passage within the cross section. The clearance height shall be determined by the vehicle loading height, safety height, pavement and other factors. The clearance height of freeways and Class Ⅰ and Ⅱ highways is 5.0 m, and that of Class Ⅲ and Ⅳ highways is 4.5 m. The same clearance height shall be adopted by a highway.

The clearance width refers to the transverse width that must be met by the road to ensure safe passage within the cross section. The clearance width includes the traveled way, shoulder, median, green belt, etc. The shoulder is within the clearance width so that various facilities (guardrails, signs, etc.) on the road shall be set on the protective soil shoulder other than the paved shoulder, and the protruding part must be above the clearance height. The clearance width of bridges, tunnels and viaducts shall generally be the same as that of the road sections.

(2) Highway land

Highway land refers to the land required for highway construction and maintenance and the layout of facilities along the route. Land requisition procedures for highway land must be handled in accordance with relevant national policies. It is not allowed to build non-road buildings within the land used for highways, such as excavating channels, burying pipelines, cables, electric poles or other facilities. When determining the land for highways, it shall not only meet the land scope necessary for highway construction, but also fully consider the precious characteristics of land resources in China to save land. Do not occupy or occupy less high-yield land, and advocate land reclamation through borrowing or waste land.

(3) Boundary line of urban roads

The boundary line of urban roads, also known as the red line, refers to the boundary control line of urban road land. The width between the red lines is the range of road land, which is called the red line width of urban roads or roadway width. The planning of the red line is to determine the road boundary line or the width of the red line. Its purpose is to comprehensively specify the land scope of roads, squares, intersections, etc. at all classes, so as to facilitate the road design, construction and arrangement of buildings on both sides. It is also the main basis for the design, construction and adjustment of various pipeline projects. Once the red line of the road is determined, the land beyond the red line shall be constructed according to the planning, and various pipelines shall be laid according to the red line. Once they are completed, it is difficult to change, so the planning of the red line of the road is very important. The red line of the road is usually determined by the urban planning department according to the road network planning and the functions, properties, trends, locations and other factors of each road. The main contents of the red line planning are as follows:

① Determine the width of the red line: according to the nature and function of the road, considering the appropriate cross-section type, the reasonable width of the traveled lane, non-motor vehicle lane, sidewalk, greenbelt and other components shall be determined, so as to determine the reasonable width of the red line.

② Determine the location of the road red line: on the basis of the overall plan of the city, for the new roads in the new area, the red line shall be drawn according to the proposed width of the red line and the centerline location of the planned road. For the reconstructed roads in the old areas, the location of the red line may be widened to one side or both sides, and it is better to be widened to one side.

③ Determine the intersection form: determine the intersection form, land use scope, specific location and main geometric dimensions according to the short-term and the long-term planning and the specific conditions at the intersection, and draw them on the plane graph in the form of red lines.

④ Determine the coordinates and elevation of control points: the turning point of the planned road centerline and the intersection of each road are the control points. The plane coordinates of control points can be directly measured on the spot, and the control elevation is determined by vertical planning.

### 1.3.4 Understanding and application of technical standards

Technical standards are compilations of design indexes for alignment and various engineering when a certain number of vehicles travel at a certain design speed on roads and are uniformly specified. It is based on the theory and summary of road design and construction experience, which reflects the current technology policies and developments of road construction and shall be abided by in road design.

Due to the vast territory of China, geographical locations and natural conditions vary from place to place. Therefore, on the premise of meeting the basic requirements, the **Standard** (**2014**) and **Specification** (**2017**) shall be applied flexibly according to the specific situation. When they are used, two inclinations shall be avoided: first, low standards are adopted to reduce the project investment without considering the role of the route and the needs of transportation development; second, because of recklessness and greed for perfection, high standards are adopted which not only increase investment, but also occupy more land.

A reasonable application of road technical standards depends on two stages. First, in the preparation of the statement of design tasks, the selection of road technical classes shall be correct and reasonable, because the statement of design tasks is the primary foundation for survey and design. In order to correctly select the road technical class, a feasibility study shall be carried out to provide a scientific basis for the preparation of the statement of design tasks. Second, each specific index shall be determined correctly and reasonably in the process of survey and design according to the selected technical class used in the statement of design tasks. The practice has proved that the correctness of technical classes in the statement of design tasks is not equal to the rationality of a specific road technical index. In addition, according to the importance, prospective scale and topographic characteristics of the road, it is also necessary to further formulate the principle of applying the technical indexes on the premise of the selected technical class so as to prevent the technical indexes from being too high or too low.

A long road (usually a highway) often crosses different zones and connects the distribution pivots of different traffic volumes. When determining the technical class and standards of roads, it is necessary to closely combine the terrain of the zone where the route passes and the traffic volume in between. One technical class may be adopted for the entire length, or different technical classes and standards may be adopted for appropriate sections, but the sections shall not be set too frequently. In general, the freeway and Class Ⅰ highway should not be less than 20 km; the Class Ⅱ highway should not be less than 15 km; the Class Ⅲ highway should not be less than 10 km; and the Class Ⅳ highway should not be less than 5 km. The junction point of technical classes or standard changes shall be selected at the place where the field of vision is wide, the driver can clearly judge the changes and the driving speed is easy to change, and corresponding signs shall be provided. In order to ensure traffic safety, the horizontal, vertical, and cross-sectional technical standards of the route before and after the junction point shall be adopted from high level to low level, or from low level to high level, with a transition section instead of abrupt change. If there is no change in the traffic volume of the route, it will form a bottleneck and cause traffic congestion when a low-standard section is inserted between high-standard sections. Under such circumstances, various engineering measures shall be adopted as far as possible, or the construction workload shall be appropriately increased to make technical standards consistent.

When the traffic volumes of the short term and the long term are greatly different, in order to improve the benefit of investment, phased construction of roads can be an option. When phased construction is adopted, comprehensive consideration and overall arrangement shall be made, so that the short-term project could meet the requirements of recent use, and still be fully utilized in the long term. At the same time, the short-term project shall also be designed to facilitate the construction of the long-term project, so as to prevent traffic interruption due to road expansion. Among the horizontal, vertical and cross-sectional planes, the horizontal and vertical planes are relatively permanent. Once they are built, it is difficult to improve them. The horizontal plane may be partially modified. For example, a small radius is changed to a large one. However, if the vertical section needs to modify grades to improve the technical standard, the original road facilities have to be abandoned. Therefore, higher standards shall be adopted as far as possible to complete the construction at one time.

As for the cross section and other structures, especially the pavement, it is easier to build them in stages. Therefore, when planning the scale of construction by stages, the following principles shall be obeyed according to the requirements of the statement of design tasks: if the project workload is not heavy, the road substructure shall be built at one time according to the plan; if the project is arduous, or in order to save land, it is feasible to temporarily design the overtopping bridge, overflow pavement and use the ferry to accomodate to traffic; when encountering a large deep ditch for the road along the river, it may detour the ditch, lower the technical indexes, and reduce the workload of projects. A bridge directly across the deep ditch could be constructed later when necessary. If the road substructure of the first phase project is not narrowed much, a small bridge and culvert (including the substructure at the culvert) may still be considered to be completed at one time according to the design standard; if the road substructure is made into a single lane first, the protection works can be reserved for the later construction as far as possible, except for the bridge and culvert. If the protection works shall be done in the first phase, it shall be completed at one time according to the design. The pavement may adopt a lower level according to the development of traffic volume and can be gradually completed later. Attention shall be paid to the influence of the alignment, substructure width and pavement structure of the first-phase project on the widening and thickening of future pavement.

When the existing road is unable to meet the needs of traffic, it shall be planned to improve the alignment, rebuild the dangerous bridge, change the ferry to the bridge, reconstruct the pavement, etc., to gradually improve the use quality and traffic capacity. The reconstruction project shall be implemented in accordance with the reconstruction technical class and standard determined in the statement of design tasks, and consider large-scale route change or new route selection only for sections with serious adverse geological and hydrological conditions. It is important to prevent two kinds of decisions: neglecting the use of old roads to pursue high technical standards unilaterally, and excessively accommodating old roads to lower technical standards.

## 1.4 Procedures of road survey and design

The general procedures of road survey and design are illustrated in Figure 1 - 3.

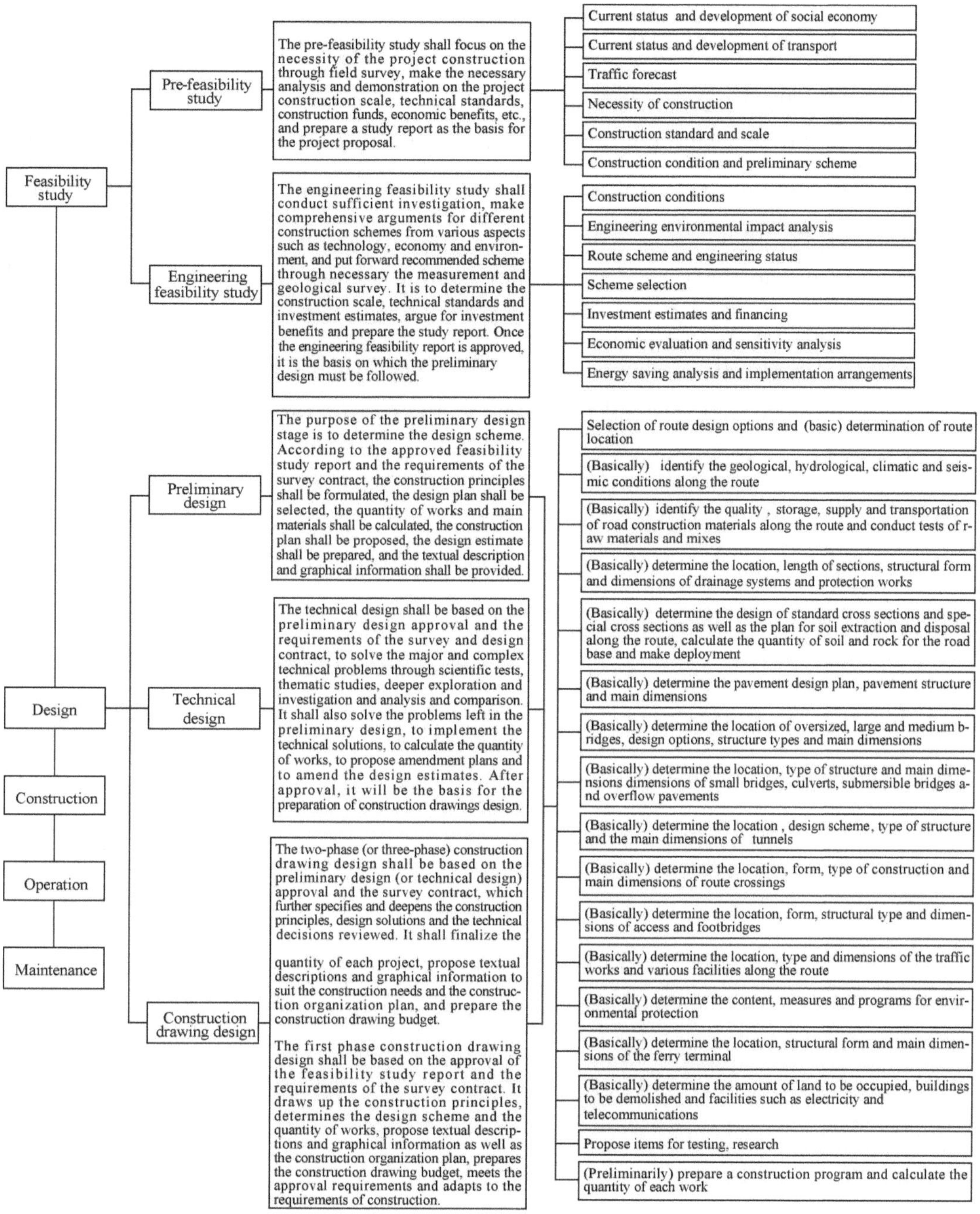


**Figure 1 - 3 Procedures of road survey and design**

## Questions:

1. What are the modes of transportation? What are the characteristics of road transport?
2. What is the basis of road classification? How many classes can highways and urban roads be divided into?

3. As a road design index, what factors shall be considered to determine the design speed? And what is the relationship between the design speed and the operating speed?
4. Why shall road design not only meet the needs of the short-term traffic volume but also the long-term traffic volume? How many years are used for the long-term service life?
5. How to divide the road design stages and what are the main tasks of each stage?
6. Explain the control elements of road design and their functions.
7. Explain the concept of road function and the role of various roads.

# Chapter 2 Horizontal Alignment Design

【**Objectives**: 1. Master the relationship between horizontal alignment elements and vehicle trajectories; 2. Master the characteristics, applications and calculations of the tangent, circular curve and transition curve; 3. Understand the principles of horizontal alignment design, including tangent design method and curve design method; 4. Master typical combinations of horizontal alignment elements and their applications; 5. Understand the deliverables of horizontal alignment, and be able to generate deliverables of horizontal alignment design.】

## 2.1 Introduction

Road alignment refers to the centerline of roads on the ground, which is a spatial line. The projection of the centerline onto the horizontal plane is called the horizontal alignment. It cuts vertically along the centerline and then unfolds to form a vertical profile. The normal section of each point in the centerline of the road is the cross section (see Figure 2 - 1). To meet the requirements of vehicle driving, the horizontal alignment, vertical profile and cross section of the road shall be considered synthetically according to the topographic features.

Horizontal alignment design refers to the design of the road centerline, which is the pioneering work of road design.

The main task of horizontal alignment design is to determine the direction and position of the road centerline. If the stake, coordinate and azimuth angle at any point on the road centerline are determined, then the direction and position of the road centerline are determined accordingly.

© The Author(s), under exclusive license to Springer Nature Singapore Pte Ltd. 2025
B. Yu and J. Cheng, *Road Survey and Design*,
https://doi.org/10.1007/978-981-97-4836-5_2

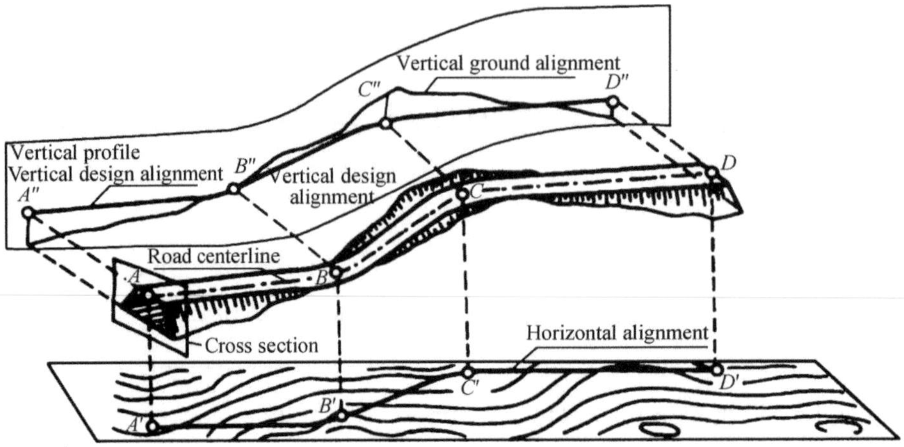


**Figure 2 - 1 Schematic illustration of horizontal alignment, vertical profile and cross section of the road**

### 2.1.1 Elements of horizontal alignment

The ideal alignment design shall be conducted by targeting the three-dimensional road alignment directly to ensure the safety of driving and environmental coordination. However, at present, the alignment design methods widely used in various countries are as follows: first, the horizontal alignment of the road is determined, then the vertical profile and cross section are designed. Due to the demand for coordination among horizontal alignment, vertical profile and cross section, it is possible to regulate the horizontal alignment until the coordination is satisfactory.

The horizontal alignment of the road is closely related to the vehicle trajectory. As shown in Figure 2 - 2, the vehicle trajectory corresponds to the three elements of horizontal alignment: tangent, circular curve and transition curve (normally spiral). The connection of the three elements is that as the vehicle enters the arc (with constant centrifugal force and steering wheel rotation) from the tangent (free of centrifugal force and steering wheel rotation), a transition curve (centrifugal force and steering wheel rotation changes) is required to be set when the arc (circular curve) radius is less than the minimum radius without superelevation. The intuitive explanation is shown in Figure 2 - 3. The tangent and circular curve can be directly connected and there is no need to use the transition curve when the arc radius is large enough so that the centrifugal force is small. Nevertheless, when the arc radius is small and the centrifugal force is large, the transition curve must be added.

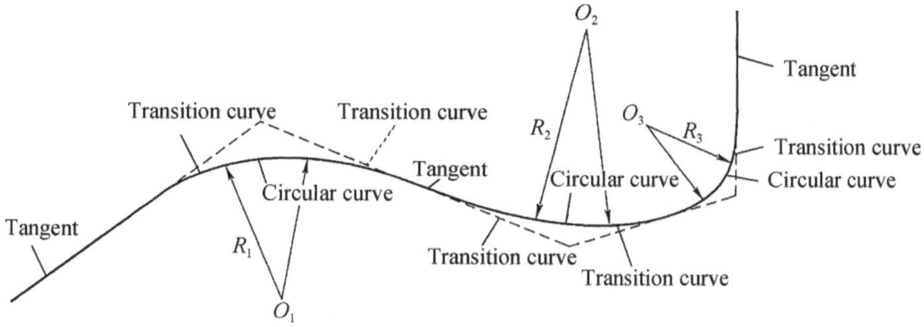


**Figure 2 - 2 Vehicle trajectory**

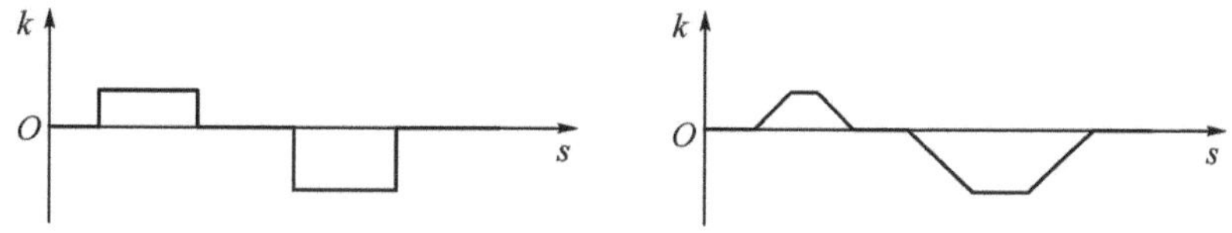


(a) Without transition curve (b) With transition curve

**Figure 2 - 3 Relationship between horizontal alignment elements and curvatures**

The primary task of horizontal alignment design is to determine the direction and position of the road centerline, that is to combine the three horizontal alignment elements organically. The three elements are introduced as follows.

## 2.2 Tangent

A tangent is the simplest element in horizontal alignment while it is not simple to use it well. The tangent section has the advantages of short length, simple surveying and layout, convenient construction and provides favorable passing conditions. It is suitable in levels and unobstructed terrains, such as plain areas, where the tangent is more suitable as the dominating alignment element. However, during road design, whether tangents are too short or too long are unfavorable. The maximum and minimum lengths of tangents shall be restricted.

### 2.2.1 Minimum length of tangents

In the traditional concept of alignment design, a route is considered to be mainly composed of tangents, while curves are only used for smooth connection at the intersection of tangents. There are two types of curves connected by tangents: co-direction curves and reverse curves, as shown in Figure 2 - 4. The **Specification (2017)** stipulates: a tangent should not be too short. When the design speed is greater than or equal to 60km/h, the minimum length of tangents between co-direction curves (in m) should not be less than 6 times the design speed (6 V, V in km/h); and the minimum length of tangents between reverse curves (in m) should not be less than 2 times the design speed (2 V, V in km/h). When the design speed is less than or equal to 40km/h, the minimum length of tangents can be designed referring to the above regulations.

Under constraints when the lengths of 2 V and 6 V are not feasible, the co-direction curves should be replaced by the large radius curve, compound curve, ovate curve or C-shaped curve, and the reverse curves may be replaced by the S-shaped curve.

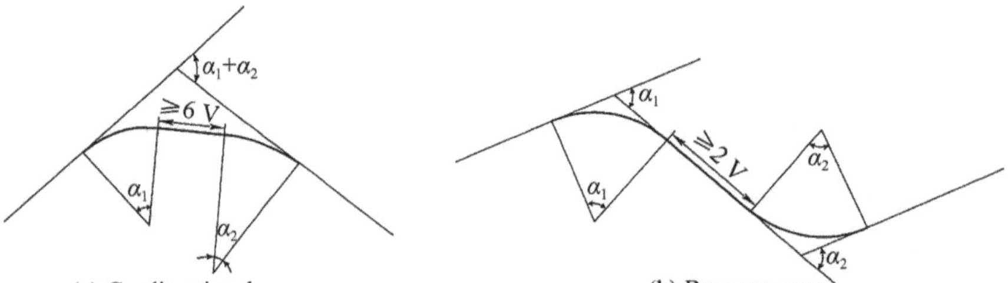


(a) Co-directional curve (b) Reverse curve

**Figure 2-4 Minimum lengths of tangents connecting the co-direction curve and the reverse curve**

If the tangent between co-direction curves is extremely short, it is called the broken-back curve, which is very unfavorable to driving and shall be avoided. When the tangent between reverse curves is significantly short, it is also unfavorable to driving and shall be avoided as well.

## 2.2.2 Maximum length of tangents

A large number of engineering practices indicate that long tangents are unfavorable to the safety and comfort of driving. Table 2-1 shows the provisions of the maximum length of tangents in different countries.

**Table 2-1 Provisions of the maximum length of tangents**

| Country | Specified value | Converted travel time/s | Converted traveled distance/m | Remarks |
|---|---|---|---|---|
| The United States | 180 s traveled distance | 180 | 50 V | Relative value |
| Germany | 20 V (in km/h) traveled distance | 72 | 20 V | Relative value |

**Continued Table 2 - 1**

| Country | Specified value | Converted travel time/s | Converted traveled distance/m | Remarks |
|---|---|---|---|---|
| Japan | 20 V (in km/h) traveled distance | 72 | 20 V | Relative value |
| Soviet Union | 8 km | — | — | Absolute value |
| Spain | 72 s traveled distance | 72 | 20 V | Relative value |

The **Specification (2017)** stipulates the maximum length of tangents shall be restricted. When a long tangent is adopted, corresponding technical measures shall be implemented in combination with the specific conditions along the route in order to make up for the monotonous landscape. As can be seen, the maximum length of tangents is not clearly stipulated, mainly because China is a country with a vast territory, and the situations of diverse regions vary remarkably. Therefore, it is difficult to quantify the maximum length of tangents. At present, on the high-class highway and in monotonous environments, the maximum length of tangents tends to be less than 20 V. In the vicinity of towns and in places with environmental variations, the length more than 20 V is acceptable. While on the roads of the Gobi and prairie, it is extraordinarily tough to limit the length of tangents.

## 2.3 Circular curve

The curve sections on roads are normally composed of circular curves with different radii. According to the **Specification (2017)**, circular curves shall be used for all classes of highways, regardless of the intersection angle.

### 2.3.1 Radius of circular curves

The radius of circular curves is closely related to the driving safety of vehicles on curves, which shall ensure the longitudinal and horizontal stability of vehicles on curves (Figure 2 - 5). The details are shown in the following analysis.

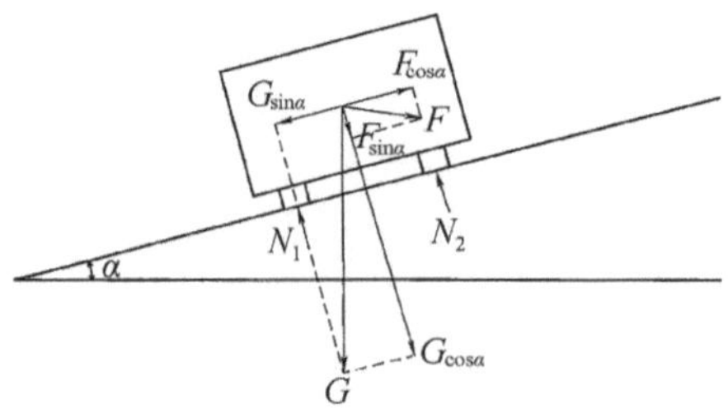


**Figure 2 - 5 Schematic illustration of longitudinal and horizontal forces of vehicles on curves**

$$R=\frac{V^2}{127(\mu\pm i)} \tag{2-1}$$

where, $R$—the radius of circular curves, m;

$V$—design speed, km/h;

$\mu$—sideway force coefficient;

$i$—cross slope of roads, taking "−" on the outer lane and "+" on the inner lane.

According to the above formula, when the vehicle is driving on the outer lane:

$$R_{\text{outer}}=\frac{V^2}{127(\mu-i)} \qquad (2-1)'$$

On the inner lane:

$$R_{\text{inner}}=\frac{V^2}{127(\mu+i)} \qquad (2-1)''$$

Evidently, when the $V$, $i$, and $\mu$ take the same value, the $R_{\text{outer}}$ is greater than $R_{\text{inner}}$. That is, the radius required by the vehicle driving on the outer lane is greater than the one on the inner lane. For this regard, for a circular curve of a two-way cross slope, the radius shall be calculated by formula (2 - 1)′. It is worth mentioning the radius of a circular curve without superelevation in the **Specification (2017)** is calculated by formula (2 - 1)′.

It is a routine practice to raise the outer lane up to the same slope of the inner lane so that a one-way slope is realized. With such a practice, one component force of gravity is opposite to the centrifugal force on both the inner and outer lanes so that the sideway force is reduced. As a result, a smaller radius of a circular curve can be adopted. Such a practice is called the superelevation. Formula (2 - 1)″ is used to calculate the radius of a circular curve with superelevation. In general, the cross slope of a road with superelevation is expressed as $i_h$. Then formula

(2-1)″ can be written as:

$$R_{\min}=\frac{V^2}{127(\mu+i_h)} \tag{2-2}$$

The minimum radius of a circular curve in the **Specification (2017)** is calculated by formula 2-2.

The above formula indicates before calculating the radius of a circular curve, the allowable sideway force coefficient $\mu$ needs to be investigated. The following aspects are vital in determining $\mu$:

**1) Stability of vehicles driving on circular curves**

To ensure the lateral anti-skid stability of vehicles driving on circular curves, the limit value of sideway force $X$ ($X=\mu G$) shall not exceed the lateral friction force $G\varphi_y$ between the tire and pavement. Namely:

$$X=\mu G\leqslant G\varphi_y$$

$$\mu\leqslant\varphi_y$$

where, $\varphi_y$—lateral friction coefficient, the general value being $(0.6-0.7)\varphi$;

$\varphi$—adhesion coefficient.

Hence the stability condition of vehicles on circular curves is:

$$R\geqslant\frac{v^2}{127(\varphi_y+i_h)} \tag{2-3}$$

According to the experience, $\varphi_y$ is normally 0.40-0.80 on dry pavement; it is reduced to 0.25-0.40 when the vehicle runs at a high speed on wet asphalt pavement, and it falls below 0.20 when snow and ice cover pavement.

**2) Psychological effects on drivers and passengers**

When determining the radius of circular curves, in addition to preventing the vehicle from skidding, the comfort of drivers and passengers shall be ensured as well. According to field experiments, the psychological effects of curves on drivers and passengers are as follows:

when $\mu<0.10$, the curve is not felt, and driving is stable;

when $\mu=0.15$, the curve is slightly felt, yet driving is still stable;

when $\mu=0.20$, the curve is felt, and driving is slightly unstable;

when $\mu=0.35$, the curve is clearly felt, and driving is unstable;

when $\mu\geqslant0.40$, driving is very unstable, and there is a risk of overturning.

As a conclusion, sideway force coefficient $\mu$ should not be greater than 0.10, and the maximum $\mu$ shall not exceed 0.20 when conditions are not permissible.

**3) Fuel consumption and tire wear of vehicles**

Fuel consumption and tire wear depend on the $\mu$ suffered. Researches have demonstrated fuel consumption and tire wear on curves are larger than those on tangents, and increase with the increase in the sideway force coefficient.

When $\mu$ is 0.10, fuel consumption increases by 10% and tire wear increases by 1.2 times; when $\mu$ is 0.15, fuel consumption increases by 15% and tire wear increases by 2 times; when $\mu$ is 0.20, fuel consumption increases by 20% and tire wear increases by 2.9 times. Therefore, to enhance vehicle operational economy, $\mu$ should not exceed 0.15.

Synthesizing the above three aspects, it can be concluded $\mu$ should be less than or equal to 0.10 for circular curves without superelevation, and $\mu$ may be raised up to 0.15–0.20 when applying superelevation under complicated terrains.

According to the formula calculation and the consideration of specific situations in China, the radii of circular curves are stipulated in the **Specification (2017)**, as shown in Table 2 - 2.

**Table 2 - 2 Minimum radius of circular curves of highways**

| Design speed/km · $h^{-1}$ | | 120 | 100 | 80 | 60 | 40 | 30 | 20 |
|---|---|---|---|---|---|---|---|---|
| Minimum radius of circular curves (general value)/m | | 1,000 | 700 | 400 | 200 | 100 | 65 | 30 |
| Minimum radius of the circular curve (limit value)/m | $I_{max}=4\%$ | 810 | 500 | 300 | 150 | 65 | 40 | 20 |
| | $I_{max}=6\%$ | 710 | 440 | 270 | 135 | 60 | 35 | 15 |
| | $I_{max}=8\%$ | 650 | 400 | 250 | 125 | 60 | 30 | 15 |
| | $I_{max}=10\%$ | 570 | 360 | 220 | 115 | — | — | — |
| Minimum radius of the circular curve without superelevation/m | Crown slope≤2% | 5,500 | 4,000 | 2,500 | 1,500 | 600 | 350 | 150 |
| | Crown slope≥2% | 7,500 | 5,250 | 3,350 | 1,900 | 800 | 450 | 200 |

When selecting the radii of circular curves for highways of all classes, larger radii shall be adopted in principle to improve the service quality of highways. In general, radii greater than or equal to the radius without superelevation listed in Table 2 - 2 should be adopted. When limited by terrains or other conditions, the general minimum radius listed in the table can be used. When the terrain is restricted or due to special difficulties, and the superelevation value may increase, the limit minimum radius can be used.

To ensure driving safety, the curve radius at the end of the long tangent on level or downhill sections shall be designed according to the possible speed of vehicles after acceleration, instead of using the minimum radius close to or equal to the class of highways.

The minimum radii of circular curves of urban roads are stipulated in the **Code (2012)**, as shown in Table 2 - 3.

**Table 2 - 3 Minimum radius of circular curves of urban roads**

| Design speed/km • h$^{-1}$ | | 100 | 80 | 60 | 50 | 40 | 30 | 20 |
|---|---|---|---|---|---|---|---|---|
| Minimum radius without superelevation/m | | 1,600 | 1,000 | 600 | 400 | 300 | 150 | 70 |
| Minimum radius with superelevation/m | General value | 650 | 400 | 300 | 200 | 150 | 85 | 40 |
| | Limit value | 400 | 250 | 150 | 100 | 70 | 40 | 20 |

## 2.3.2 Calculation of the circular curve

The **Specification (2017)** stipulates when the radius of the circular curve is less than the minimum value without superelevation, a transition curve shall be used. If the radius of the circular curve is greater than the minimum value without superelevation, the circular curve may be connected with the tangent directly. In this sense, the calculation of a circular curve without a transition curve is a special case of the calculation of a circular curve with a transition curve. The calculation of the horizontal curve includes calculations of curve elements, the main points, and coordinates. For the calculation of horizontal curves with a transition curve, it is shown in section 2. 4. 4.

In general, the direct connection between tangents and circular curves is not frequent and the calculation is simple. The primary geometric elements of circular curves without transition curves are determined by the following formulas, as shown in Figure 2 - 6.

Tangent length

$$T=R\tan\frac{\alpha}{2} \tag{2-4}$$

Curve length

$$L=\alpha\frac{\pi}{180}R \tag{2-5}$$

Apex distance

$$E = R\sec\frac{\alpha}{2} - R \tag{2-6}$$

Calibration

$$J = 2T - L \tag{2-7}$$

where, $R$—radius of the circular curve, m;

$\alpha$—deflection angle of the intersection point, °.

In Figure 2 - 6, $JD$ is the intersection point, $ZY$ is the start point of the circular curve, $YZ$ is the end point of the circular curve, $QZ$ is the midpoint of the circular curve, $\alpha$ is the deflection angle of the intersection point, and the arc length from the start point to the end point is the curve length.

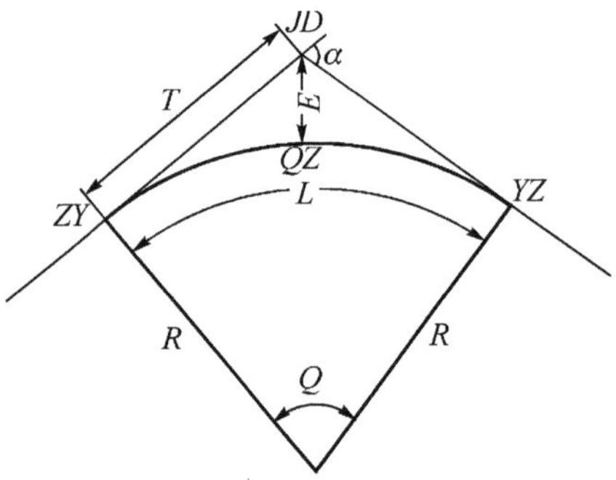


**Figure 2 - 6 Primary geometric elements of circular curves without transition curves**

As can be seen, when $\alpha$ is a constant, the layout of the curve mainly depends on the radius of the circular curve. The circular curve with a large radius may reduce the transportation cost and improve the driving safety and comfort of passengers. However, due to the limitation of topography, ground features, hydrogeology and other conditions, or for the sake of reducing project investment, it is sometimes necessary to adopt a smaller radius, which brings up the first issue to be solved in this section: what is the minimum radius of a circular curve under the premise of maintaining safety, speed, economy and comfort?

As the vehicle drives on a curve section of normal crown, the sideway force of the vehicle driving on the outer lane is larger than that on the inner lane at the same speed, that is, the stability of the vehicle driving on the outer lane is more detrimental than that on the inner lane. Therefore, the curve radius on the normal crown cross section shall be determined by the driving conditions of the outer lane, and its value shall be larger than that determined by the inner lane. To adopt a smaller radius, the outer lane can be raised to the same cross slope of the inner lane so as to obtain a unidirectional cross section, which is called superelevation.

Meanwhile, due to the distinct driving radius of each wheel, the driving radius of the inner rear wheel is the smallest, while that of the outer front wheel is the largest. The width required for a vehicle to travel on curves is greater than that on tangents, so it is required to widen the road.

## 2.4 Transition curve

When the radius of the circular curve is less than the minimum radius of the circular curve without superelevation, it is necessary to set superelevation. Moreover, if the radius is less than or equal to 250 m, the road width shall be widened as well. As a result, the tangent section (normal crown cross section) and the curve section (unidirectional superelevated cross section) shall not be directly connected, while a transition zone shall be supplemented to make the connection. In addition, when a vehicle runs at the design speed on the circular curve, it not only requires to maintain the speed constant, but also needs to keep the front wheels at a constant steering angle $\alpha$. As a result, before entering the circular curve, the vehicle also needs a transition zone to increase the angle of the steering wheel from 0 to $\alpha$.

To sum up, the application of transition curves primarily considers the following three aspects:

(1) A transition zone is needed for the vehicle progressively altering the radius of driving trajectory from a tangent section with a infinite radius to a circular curve with a constant radius. Such a transition zone has to be a curve, which is called a transition curve.

(2) A transition zone that makes the road gradually change from a tangent section with a normal crown to a circular curve with a superelevated cross section, which is called superelevation transition.

(3) A transition zone from a tangent section with a normal width section to a circular curve with a widened section, which is called a curve-widening transition.

Among the above three aspects, the first and second ones play a control role in the desired length of the transition zone. In general, the third one is to make the transition within the specified transition zone, and does not determine the length of the transition zone (except for widening without transition curves and superelevation).

The length of the transition zone uses the longer one in the those required by the above two kinds of transition (i. e. transition curve and superelevation transition), respectively. Once the length of transition zone is determined, various transition shall be performed simultaneously. However, not all types of transition have to be conducted within the transition zone, and the type of transition shall be determined by the actual situation of circular curve section. When the radius of circular curves is sufficiently large (greater than the minimum circular curve radius without superelevation and greater than 250 m), no transition is required. That is, there is no need to supplement a transition zone between the tangent and the curve, and the tangent and the curve can be directly connected.

The application of transition zones is not only needed when tangents are connected with curves, but also when curves of large curvature difference connect. More transition zones are required when the superelevation rates are different, or the pavement widenings are not equal.

### 2.4.1 Form of transition curves

As a vehicle moves from the tangent to the curve, the radial acceleration ($a=v^2/\rho$) changes gradually from zero on the tangent to $a_{\max}$ on the curve (curvature radius of circular curves $\rho=R$, so $a_{\max}=v^2/R$). Clearly, the acceleration $a$ shall be proportional to time if the driving speed remains constant (a hypothetical condition for deriving the equation of transition curves). This means the conditions of transition curves are $v$=constant and $\mathrm{d}a/\mathrm{d}t$=constant. Let $I=\mathrm{d}a/\mathrm{d}t$=constant, it is called the growth rate of acceleration, in $\mathrm{m/s^2}$.

The general equation of transition curves is based on a vehicle running from a tangent (circular curve) to a circular curve (tangent) at a constant speed. The steering angle of the front wheel increases (reduces) gradually from zero (the angle of the horizontal curve) to the angle of the horizontal curve (zero), that is, it is derived from the process that the curvature radius gradually varies from infinity (the radius of the circular curve) to the radius of the circular curve (infinity). It is concluded that the product of arc length and curvature radius is a constant when a vehicle runs at constant speed and angular acceleration, i. e. $rl = C$.

## 2.4.2 Length of the transition curve

According to the function of transition curves, the length ($L_s$) shall be long enough to enable the vehicle to smoothly transit from a tangent to a circular curve, and the growth rate of centrifugal force shall be limited. Additionally, the length of the transition curve shall meet the necessary time for the driver to operate the steering wheel.

**1) $L_s$ based on the growth rate of centrifugal acceleration**

$$L_s = \frac{v^3}{RI} \tag{2-8}$$

where, $L_s$—the length of transition curve;

$v$—design speed, m/s;

$R$—the radius of circular curve;

$I$—the acceleration growth rate.

The acceleration growth rate ($I$) shall not exceed the allowable value considering the smooth driving and the passenger's feeling, and 0.3 - 1.3 m/s$^3$ is recommended in the design of the highway transition curve.

Let $I = 0.6$ m/s$^3$, the length of transition curves can be calculated as follows:

$$L_s = \frac{0.035V^3}{R} \tag{2-9}$$

**2) $L_s$ based on the needed time for drivers to operate steering wheels**

If the length of the transition curve is too short, drivers may not have enough time to adjust and adapt to the change in the front. Hence the shortest travel time of vehicles on the transition curves shall be stipulated.

When driving on transition curves, the most reasonable time for drivers to operate the steering wheels is 3 - 5 s. The length of the transition curve can be calculated as follows:

$$L_s \geqslant vt = \frac{Vt}{3.6} \qquad (2-10)$$

where $t$ is the shortest travel time, 3.0 s is recommended as for the two-lane highway; 4 s for the four-lane highway with no or narrow separators; as short as 2.5 s at intersections or mountain areas.

The stipulated length of transition curves at different design speed in the **Specification** (**2017**) is shown in Table 2 - 4.

**Table 2 - 4 Minimum length of transition curves at different design speed**

| Design speed/km • h$^{-1}$ | 120 | 100 | 80 | 60 | 40 | 30 | 20 |
|---|---|---|---|---|---|---|---|
| Minimum length/m | 100 | 85 | 70 | 50 | 35 | 25 | 20 |

Furthermore, since the length of transition zone adopts the longer value of both the transition curve and superelevation transition, the length of the transition curve shall not be less than that of superelevation transition, which will be presented in Chapter Four.

### 2.4.3 Calculation of the transition curve

The formula and the symbols in the diagram are as follows:

$ZH$—the start point of the first transition curve;

$HY$—the end point of the first transition curve;

$YH$—the start point of the second transition curve;

$HZ$—the end point of the second transition curve;

$l$—the distance from the start point to any point on the transition curve, m;

$L_s$—the length of the transition curve, m;

$r$—the curvature radius at any point on the transition curve, m;

$R$—the curvature radius at the end of the transition curve, m;

$A$—the parameter of the transition curve, m;

$\beta$—the angle between the tangent at any point on the transition curve and the $X$-axis, radian;

$\beta_0$—the angle between the tangent and the $X$-axis at the end of the transition curve (start point of the circular curve), radian;

$\varphi$—the deflection angle of $ZH$ point to any point on the transition curve, radian;

$\varphi_0$—deflection angle of $ZH$ point to $HY$ point, radian.

**1) Calculation formula of the transition curve**

Taking the transition curve from the tangent-spiral point (*ZH*) to the spiral-circular curve point (*HY*) as an example, a coordinate system is established with *ZH* point as the origin, with the tangent direction of *ZH* point as the $x$-axis, and with the normal direction as the $y$-axis, as shown in Figure 2-7:

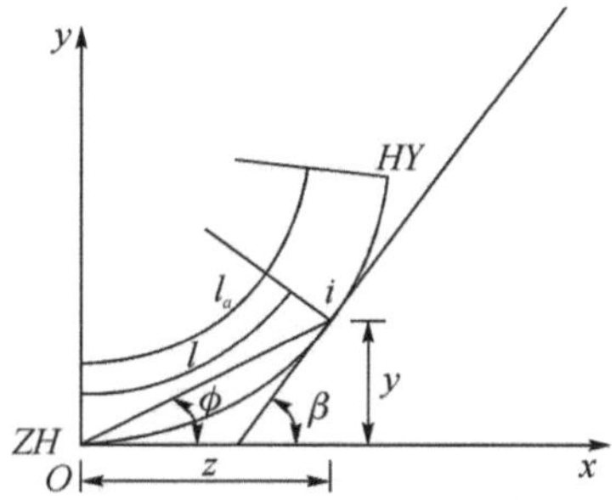


**Figure 2-7 Schematic diagram of the transition curve**

In Figure 2-7, the curve length at any point ($i$) on the transition curve to *ZH* point is $l$, the length of the transition curve is $L_s$, and the curvature radius at the *HY* point is $R$. The **Specification (2017)** stipulates the transition curve adopts a spiral. Curvature radius ($r$) at any point ($i$) on the spiral varies with its length ($l$), with the relationship as:

$$rl=A^2 \tag{2-11}$$

where, $A$—parameter of the spiral, m.

At the end of the spiral,

$$RL_s=A^2 \tag{2-12}$$

In Figure 2-7, differential transformation at any point ($i$) on the spiral is conducted to obtain:

$$\begin{cases} x=\int_0^l \cos\beta \mathrm{d}l \\ y=\int_0^l \sin\beta \mathrm{d}l \end{cases} \tag{2-13}$$

According to the relationship between arc length and radius:

$$\mathrm{d}\beta=\frac{1}{r_i}\mathrm{d}l \tag{2-14}$$

With deduction:

$$\begin{cases} x=l-\dfrac{l^5}{40R^2L_s^{\ 2}}-\cdots \\ y=\dfrac{l^3}{6RL_s}-\dfrac{l^7}{336R^3L_s^{\ 3}}+\cdots \end{cases} \tag{2-15}$$

Formula (2 - 15) is the rectangular coordinate formula of the spiral. In practice, the first two terms are generally taken.

To ensure measurement accuracy in road surveys, the national coordinate system (geodetic coordinate system) is generally used to establish a unified coordinate system through the overall alignment and to calculate unified coordinates of each point according to the geographical position and geometric relationship of the alignment. The tangent rectangular coordinate system ($xOy$) is established from the start point of the spiral (Figure 2 - 7), then coordinates at any point on the spiral in the geodetic coordinate system are as follows:

$$\begin{cases} X = X_{ZH} + x\cos\gamma_{ZH} - \xi y\sin\gamma_{ZH}, \\ Y = Y_{ZH} + x\sin\gamma_{ZH} + \xi y\cos\gamma_{ZH}. \end{cases} \tag{2-16}$$

where, $X$—$x$ coordinates at any point on the spiral in the geodetic coordinate system;

$Y$—$y$ coordinates at any point on the spiral in the geodetic coordinate system;

$X_{ZH}$—$x$ coordinates of $ZH$ point in the geodetic coordinate system;

$Y_{ZH}$—$y$ coordinates of $ZH$ point in the geodetic coordinate system;

$\gamma_{ZH}$—azimuth of the start point ($ZH$) of the spiral, radian;

$x$—$x$ coordinates at any point on the spiral in the tangent rectangular coordinate system ($xOy$);

$y$—$y$ coordinates at any point on the spiral in the tangent rectangular coordinate system ($xOy$);

$\xi$—$\xi=1$, the spiral is on the right side of the tangent of $ZH$ point; $\xi=-1$, the spiral is on the left side of the tangent of $ZH$ point.

**2) Properties of the spiral**

(1) Characteristics of spiral equations

From Equations (2 - 11) and (2 - 12), it can be seen that the parameter $A$ of spiral is constant, which indicates the radius at any point on the spiral is inversely proportional to the distance from the point to the start of the spiral.

(2) Angle of the spiral

The angle of the spiral is equal to the angle between the tangent of $HY$ point and the coordinate axis, and its value is:

$$\beta_0 = \frac{L_s}{2R} \tag{2-17}$$

(3) The deflection angle ($\varphi$) from $ZH$ point to any point ($i$) on the spiral can be calculated by

$$\varphi=\tan\frac{y_i}{x_i} \tag{2-18}$$

Since $\varphi$ is small, it can be approximated as: $\varphi \approx \tan\varphi$ (radian),

$$\varphi=\frac{y_i}{x_i} \tag{2-19}$$

where, $x_i$—$x$ coordinates at any point on the spiral in the tangent rectangular coordinate system ($xOy$);

$y_i$—$y$ coordinates at any point on the spiral in the tangent rectangular coordinate system ($xOy$).

The total deflection angle from $ZH$ point to $HY$ point is:

$$\varphi_0=\frac{y_0}{x_0} \tag{2-20}$$

where, $x_0$—$x$ coordinates of $HY$ point in the tangent rectangular coordinate system ($xOy$);

$y_0$—$y$ coordinates of $HY$ point in the tangent rectangular coordinate system ($xOy$).

(4) Condition for spiral application

About half of the spiral is set outside the circular curve and the other half is set inside the circular curve. Therefore, for a circular curve segment, the spiral can be set when the length of the circular curve (when the spiral is not set) is greater than or equal to that of the spiral. When they are equal, the two spirals will meet at the midpoint of the curve; if the length of the circular curve is less than that of the spiral, the specific spiral could not be applied to this circular curve segment. Under this condition, it is necessary to shorten the length of the spiral or increase the radius of the circular curve.

The condition for setting the spiral can be examined according to the angle of the circular curve ($\alpha$) and the angle of spiral $\beta_0$. After setting the spiral, the angle corresponding to the spiral that extends into the circular curve is $\beta_0$. Therefore, the angle of the circular curve ($\alpha$) will be reduced by $2\beta_0$ as the two ends of the circular curve are shortened by the spiral. Therefore, the spiral could be set under the following condition:

$$\alpha-2\beta_0\geqslant 0 \tag{2-21}$$

(5) Regulation about spiral parameter $A$

Transition curve parameter $A$ determines the curvature change rate of the spiral. The minimum value of $A$ shall be determined according to the requirements for driving comfort on the transition curve, for the minimum driving time and for the allowable superelevation relative gradient. The transition curve parameter shall be consistent with the radius of circular curve. Research shows that as long as the relationship of $R/3 \leqslant A \leqslant R$ holds between transition curve parameter $A$ and circular curve radius $R$, a visually coordinated and comfortable alignment can be obtained.

However, $R$ in the above relationship depends on certain conditions. Experience has proved that when $R$ is about 100 m, $A=R$ is usually taken; if $R<$ 100 m, $A=R$ or $>R$. On the contrary, when the radius of the circular curve is large, $A$ may be about $R/3$; if R exceeds 3,000 m, even if $A$ is less than $R/3$, it is still free of visual concern.

(6) Spiral omission

With the application of the spiral between a tangent and a circular curve, the circular curve generates an inward shift value $P$, which is inversely proportional to the radius of the circular curve when $L_s$ is given; when $R$ is large enough, $P$ value is very small. Even if the tangent is directly connected with the circular curve, vehicles can complete driving with curvature change, because the inward shift value has been included in the spare width of lanes. According to the **Specification (2017)**, the transition curve may be removed in a series of scenarios to form the so-called compound curve, including:

① Between a tangent and a circular curve, when the radius of the circular curve is greater than or equal to the "minimum radius without superelevation".

② Between co-direction circular curves with different radii, when the radius of the small circular curve is greater than or equal to the "minimum radius without superelevation".

③ The radius of the small circular curve is greater than the critical radii of the compound curve listed in Table 2-5, and meets one of the following conditions:

(a) When the small circular curve is provided with a spiral with the minimum transition curve length according to the regulations, the difference between the inward shift values of the large circular curve and the small circular curve does not exceed 0.10 m;

(b) When the design speed is greater than or equal to 80 km/h, the ratio of large circular curve radius to small circular curve radius is less than 1.5;

(c) When the design speed is less than 80 km/h, the ratio of large circular curve radius to small circular curve radius is less than 2.0.

**Table 2 - 5 Critical radius of the circular curve in the compound curve**

| Design speed/km • $h^{-1}$ | 120 | 100 | 80 | 60 | 40 | 30 |
|---|---|---|---|---|---|---|
| Critical radius of the circular curve/m | 2,100 | 1,500 | 900 | 500 | 250 | 130 |

Table 2 - 6 shows the minimum radius of the circular curve without the transition curve specified in the **Code (2012)**.

**Table 2 - 6 Minimum radius of the circular curve without the transition curve of urban roads**

| Design speed/km • $h^{-1}$ | 80 | 60 | 50 | 40 |
|---|---|---|---|---|
| Minimum radius of the circular curve without the transition curve/m | 2,000 | 1,000 | 700 | 500 |

## 2.4.4 Calculation of horizontal curves with the spiral

For horizontal curves, the form of tangent-spiral-circular curve-spiral-tangent is common. The main geometric elements of horizontal curves with the spiral are determined by the following formulas, as shown in Figure 2 - 8.

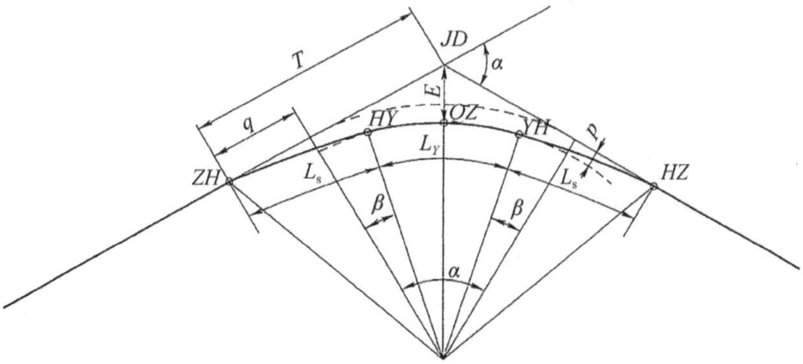


**Figure 2 - 8 Main geometric elements of horizontal curves with the spiral**

### 1) Calculation of curve elements

Tangent increment value can be calculated by

$$q=\frac{L_s}{2}-\frac{L_s^3}{240R^2} \tag{2-22}$$

Inward shift value can be calculated by

$$p=\frac{L_s^2}{24R}-\frac{L_s^4}{2,688R^3} \tag{2-23}$$

Spiral angle can be calculated by

$$\beta_0=\frac{L_S}{2R}\cdot\frac{180°}{\pi} \tag{2-24}$$

Tangent length can be calculated by

$$T=(R+p)\tan\frac{\alpha}{2}+q \tag{2-25}$$

Curve length can be calculated by

$$L=(\alpha-2\beta_0)\frac{\pi}{180}R+2L_s \tag{2-26}$$

Apex distance can be calculated by

$$E=(R+p)\sec\frac{\alpha}{2}-R \tag{2-27}$$

Calibration can be calculated by

$$J=2T-L \tag{2-28}$$

where, $R$—radius of circular curves. m;

$L_s$—length of the spiral, m;

$\alpha$—deflection angle, °.

In Figure 2 - 8, $JD$ is the intersection point, $ZH$ is the start point of the horizontal curve, $HZ$ is the end point of the horizontal curve, $HY$ and $YH$ are the junctions between the spiral and the circular curve, $QZ$ is the midpoint of the horizontal curve, $\alpha$ is the deflection angle of the intersection point, and the arc length from the start point to the end points is the horizontal curve length.

**2) Calculation of stakes of main points**

After completing the calculation of horizontal curve elements, the stakes of main points can be calculated and verified by the following formulas:

$ZH$:

$$ZH=JD-T \tag{2-29}$$

$HY$:

$$HY=ZH+L_s \tag{2-30}$$

$QZ$:

$$QZ = ZH + \frac{L}{2} \tag{2-31}$$

$YH$:

$$YH = HZ - L_s \tag{2-32}$$

$HZ$:

$$HZ = ZH + L \tag{2-33}$$

$JD$:

$$JD = HZ - T + J \tag{2-34}$$

**3) Coordinate calculations**

Due to the introduction of computer softwares, manual coordinate calculation has been gradually abandoned, which will not be presented in this book.

## 2.5 Principle of horizontal alignment design

**1) Meeting requirements of the overall design and pursuing the concept of overall optimization**

Horizontal alignment shall be suitable for terrains and ground features. It should be straight when a tangent is required and curved when a curve is required. A pursuit of the straight or curve blindly is not preferred, which reflects the requirements of aesthetic, economic and environmental protection. Transition is mandatory when the design indexes change, such as the design speed, cross section form.

**2) Maintaining continuous and uniform transition to prevent abrupt and frequent changes**

To make the vehicles operate at a uniform speed on roads, attention shall be paid to maintaining the alignment elements continuous and balanced, so as to prevent the sudden change of technical indexes. The junctions of road sections with different standards shall be selected at the place where the traffic volume changes or where the driver can clearly judge that the speed ahead needs to be changed.

**3) Meeting all types of constraints and considering coordination among horizontal, vertical and cross-section planes**

For horizontal alignment design, the requirement of the vertical profile and cross-section designs shall be taken into account. Especially in the plain and lightly hilly areas, the horizontal curve index is normally high, with a longer curve length. The intersections between tangents with the railways, main roads and rivers normally become the control points of the vertical profile. During the horizontal alignment design, it is necessary to consider the specific requirements of vertical profile design so as to facilitate the combination design of horizontal alignment and the vertical profile.

**4) Enough length of horizontal curves**

(1) Minimum length of horizontal curves required by drivers to operate and passengers to feel comfortable

The horizontal curve is generally composed of three sections, i. e. the front and back spirals and the intermediate circular curve. According to experience, it takes drivers at least 3 s to operate the steering wheel without difficulty on each curve, and therefore 9 s for the whole horizontal curve. If the length of the intermediate circle curve is 0, the minimum length of the horizontal curve is twice the length of the spiral, forming a convex curve. Under such a circumstance, drivers will feel sudden changes in operation and visual discomfort.

In the **Specification (2017)**, the length of horizontal curves should be 5 - 8 times the minimum value, and the minimum length shall not be less than that specified in Table 2 - 7.

**Table 2 - 7 Minimum length of horizontal curves**

| Design speed/km • $h^{-1}$ | 120 | 100 | 80 | 60 | 40 | 30 | 20 |
|---|---|---|---|---|---|---|---|
| General value/m | 600 | 500 | 400 | 300 | 200 | 150 | 200 |
| Limit value/m | 200 | 170 | 140 | 100 | 70 | 50 | 40 |

(2) Length of horizontal curves with the intersection angle $\triangle$ less than 7°

If the intersection angle of horizontal curves is too small, even if the radius is large, the visual length of horizontal curves will be shorter than the actual length, resulting in the illusion of sharp turning. Therefore, when the intersection angle is less than 7°, a longer horizontal curve shall be set, and its length is specified in Table 2 - 8. $\triangle$ in the table is the intersection angle (°); when $\triangle < 2°$, it is calculated as $\triangle = 2°$.

**Table 2 - 8 Length of horizontal curves with the intersection angle △ equal to or less than 7°**

| Design speed/km • $h^{-1}$ | 120 | 100 | 80 | 60 | 40 | 30 | 20 |
|---|---|---|---|---|---|---|---|
| General value/m | 1,400/△ | 1,200/△ | 1,000/△ | 700/△ | 500/△ | 350/△ | 280/△ |
| Limit value/m | 200 | 170 | 140 | 100 | 70 | 50 | 40 |

## 2.6 Method of horizontal alignment design

The task of horizontal alignment design is to reasonably determine the position of the centerline of roads using the existing control points and subject to the constraints, that is, to determine the stake, horizontal coordinates, azimuth, deflection angle, and other information at any point on the centerline. At present, for horizontal alignment design at home and abroad, the tangent design method and the curve design method are commonly used.

### 2.6.1 Tangent design method

The tangent design method, also known as the transverse design method, takes the tangent as the main design element, which determines tangents first and then curves. During the design, tangents (transverses) are used as direction control. Circular curves and spirals act as supplements to tangents and play the role of transition and smoothing. Tangents, circular curves, and spirals are not applied as a unified entirety. Currently, the tangent design method is widely used in practice.

However, in the tangent design method, the route is artificially divided into two components: the tangent section and the curve section. During curve layout, it is artificially divided into several combination types formed by the combination of basic alignment elements. For the ramp design of interchange or the alignment design mainly consisting of curves, it is very difficult to use the tangent design method to lay out and calculate. For instance, when the major alignment of the interchange is curved, the inner ramp often presents an S-shaped alignment. The traverse location of the S-shaped curve is difficult to be determined artificially; hence it is difficult to use the tangent design method to calculate. Another example is the ovate alignment in which there is no traverse.

In addition, because of the limitation of traverses, in sections with strict terrain and features, constraints, the primary consideration in the design is to satisfy the specifications of curve elements, while ignoring the coordination with the terrain and features, conditions. Or in some unconstrained road sections, the values of the design elements are too large, leading to excess fill and cut, destruction of natural environment and an increase in project cost.

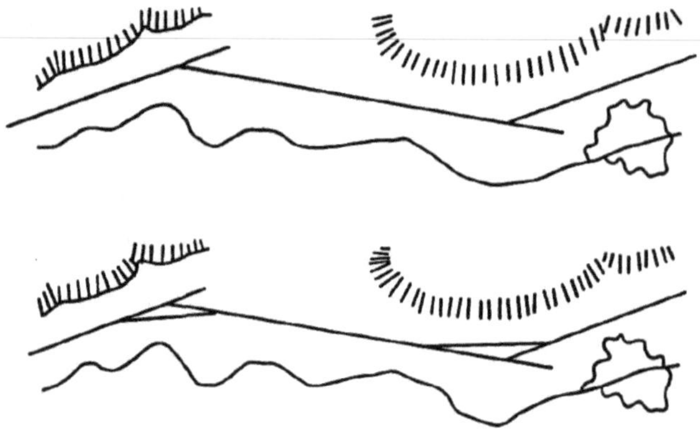

**Figure 2 - 9 Schematic diagram of tangent design method**

**Example** As shown in Figure 2 - 10, the design speed of a highway is 60 km/h; the deflection angle at $JD_1$ is 45°05′20″, and the stake is $K4+895.25$. The deflection angle at $JD_2$ is 38°24′38″. The distance between $JD_1$ and $JD_2$ is 356.00 m. There is an obstacle on the inside of the $JD_2$ curve, of which the distance is 15.00 m away.

If the radius $R_1$ at $JD_1$ is 250 m and the length of the spiral $L_{s1}$ is 70 m. Please calculate the curve elements and the stake of the main points at $JD_1$, and determine the curve radius and spiral length at $JD_2$.

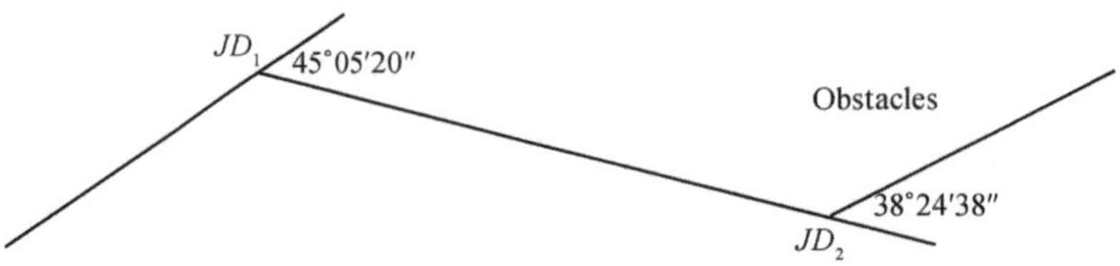


**Fig. 2 - 10 Schematic diagram of the horizontal alignments**

**Solution:**

(1) Curve element calculation:

$$q=\frac{L_s}{2}-\frac{L_s^3}{240R^2}=\frac{70}{2}-\frac{70^3}{240\times 250^2}\approx 34.98(\text{m})$$

$$p=\frac{L_s^2}{24R}-\frac{L_s^4}{2,688R^3}=\frac{70^2}{24\times 250}-\frac{70^4}{2,688\times 250^3}\approx 0.82(\text{m})$$

$$\beta_0=\frac{L_s}{2R}\cdot\frac{180}{\pi}=\frac{70}{2\times 250}\times\frac{180}{3.14}\approx 8.025°=8°1'12''$$

$$T=(R+p)\tan\frac{\alpha}{2}+q=(250+0.82)\tan\frac{45°05'20''}{2}+34.98=139.10(\text{m})$$

$$L=(\alpha-2\beta_0)\frac{\pi}{180}R+2L_s=(45.09-2\times 8.02)\times\frac{\pi}{180}\times 250+2\times 70\approx 266.74(\text{m})$$

$$E=(R+p)\sec\frac{\alpha}{2}-R=(250+0.82)\sec\frac{45°05'20''}{2}-250=21.57(\text{m})$$

$$J=2T-L=2\times 139.10-266.74=11.46(\text{m})$$

(2) Stake of main points:

$$ZH=K4+895.25-139.10=K4+756.15$$

$$HY=K4+756.15+70=K4+826.15$$

$$QZ=K4+756.15+\frac{266.74}{2}=K4+889.52$$

$$YH=K5+022.89-70=K4+952.89$$

$$HZ=K4+756.15+266.74=K5+022.89$$

$$\textit{Check } JD=HZ-T+J=K4+895.25$$

(3) Analysis and calculation of $R_2$ and $L_{s2}$:

The distance between $JD_1$ and $JD_2$ is 356 m, and the tangent length at $JD_1$ is 139.10 m. According to the minimum length requirement of 120 m (2V) between reverse curves in the **Specification (2017)**, if $JD_1$ and $JD_2$ are designed as the S-shaped curve, then the apex distance $E$ of $JD_2$ (distance from the obstacle) is more than 15 m. Therefore, the S-shaped curve is not feasible and $JD_1$ and $JD_2$ shall be designed as reverse curves that meet the requirements of $2V$ tangent length. For the horizontal curve at $JD_2$, the tangent length shall be less than 96.9 m and the apex distance $E$ shall be less than 15 m. Taking $L_{s2}=50$ m, $R_2$ is calculated according to the apex distance $E$ and tangent length and rounded to 200 m. After verification, the results meet the requirements of the **Specification (2017)**.

To sum up, the curve radius of $JD_2$ is 200 m, and the spiral length is 50 m.

### 2.6.2 Curve design method

Compared with the tangent design method, the main difference of the curve design method is that there is no intersection, as shown in Figure 2 - 11. It first lays out arcs according to the terrain, ground object and other constraints, and then connects arcs by spirals. The practice has demonstrated that the horizontal alignment laid out by the curve design method can better accommodate the constraints. At present, the tangent design method is most widely used in engineering practice because of the design habits and the requirements of design document format. However, for the ramp of interchanges (including urban interchanges), due to the limitation of terrain and ground objects, the selected curve parameters are difficult to settle for multiple constraints (including the terrain, ground objects, specification requirements, etc.) when the alignment is controlled by traverses, which results in unnecessary demolition.

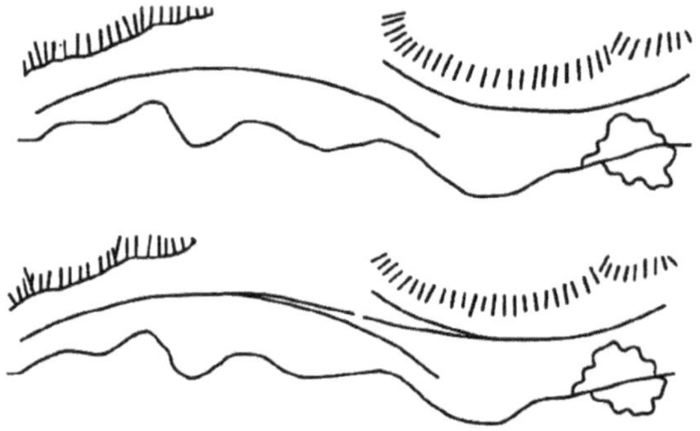

**Figure 2 - 11 Schematic illustration of the curve design method**

Instead, the curve design method is used to construct the alignment framework in combination with the constraints of terrain and ground features, and then selects appropriate spirals for connection, so that the integral alignment can better meet the constraints of all aspects. As a matter of fact, the horizontal alignment dominated by curves can better cooperate with the terrain and ground features, reduce the amount of earthwork, and meanwhile, better coordinate with the scenery. The landscape along the alignment is beautiful; the curve alignment is rich in changes; the driver's attention is concentrated and the traffic accidents are reduced. It is easy to understand that the curve design method is a general method, while the tangent design method is a special case of the former. The similarities and differences between them are shown in the Table 2 - 9.

Table 2 - 9 Similarities and differences between the tangent design method and the curve design method

| Design methods | Intersections | Parameters of alignment elements | Characteristics |
|---|---|---|---|
| Tangent | Yes | Parameters are determined by circular curves and spirals according to experience and constraints; there are multi-solution and multi-parameter combinations. | Tangents are dominating, laying out curves at intersections; coordination with terrain and features is less flexible. |
| Curve | No | Two or more arcs (or tangents) control the route according to constraints; spirals between arcs generate iteratively, normally with only one solution. | Curves are dominating, which coordinate well with terrain and features. |

If the position of the tangent and circle has been determined, namely, the coordinates of the two points on the tangent $(x_1, y_1)$ and $(x_2, y_2)$, as well as the coordinates of the center of circle $(x_c, y_c)$ and its radius $R$ are known, then the shortest distance between the circle and tangent is:

$$\Delta R=\frac{|k(x_c-x_1)-(y_c-y_1)|}{\sqrt{1+k^2}}-R \tag{2-35}$$

where, $k=\frac{y_2-y_1}{x_2-x_1}$.

Based on this, the approximate value of $A$ can be obtained as the initial value of iteration:

$$A=R\cdot\left[56\left(1-\sqrt{1-\frac{6}{7}\cdot\frac{\Delta R}{R}}\right)\right]^{\frac{1}{4}} \tag{2-36}$$

As shown in the Figure 2 - 12, the radius $R$ of the circle and the coordinates of the center of the circle, as well as the position of the tangent, have been determined. Taking the approximate value of $A$ in Formula (2 - 36) to calculate, it obtains:

$$L=\frac{A^2}{R} \tag{2-37}$$

$$\beta=\frac{L}{2\Delta A^2} \tag{2-38}$$

Assuming the coordinates at the tangency point of the spiral and circular curve be $X$ and $Y$, then:

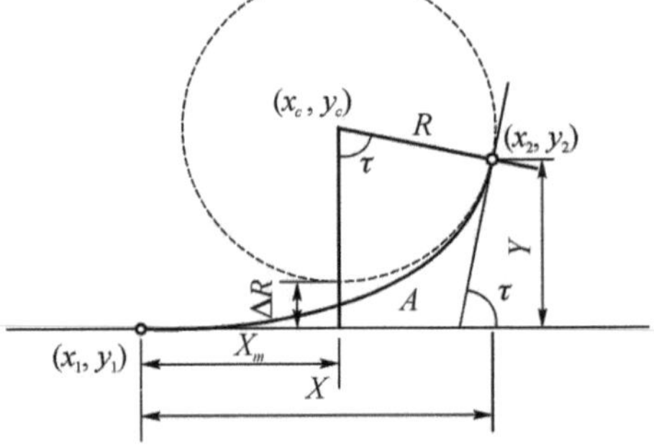


**Figure 2 - 12 Curve design method (tangent-circle type)**

$$\begin{cases} X = \sum_{k=1}^{n} (-1)^{k+1} \cdot \frac{L}{(4k-3)} \cdot \prod_{j=1}^{2k-2} \left(\frac{\tau}{j}\right) \\ Y = \sum_{k=1}^{n} (-1)^{k+1} \cdot \frac{L}{(4k-1)} \cdot \prod_{j=1}^{2k-1} \left(\frac{\tau}{j}\right) \end{cases} \tag{2-39}$$

Then:

$$\Delta R' = Y + R \cdot \cos\beta - R \tag{2-40}$$

The calculated $\Delta R'$ from accurate $A$ value shall be equal to $\Delta R$. Therefore, the approximate value of $A$ can be used as the initial input of iteration. Taking $\varepsilon = |\Delta R - \Delta R'|$ as the objective function, the $A$ value satisfying the precision requirement may be obtained by iteration. It can be found that there is only one transition curve value that meets the requirements. Meanwhile, since the calculation is too cumbersome, it is generally solved by software.

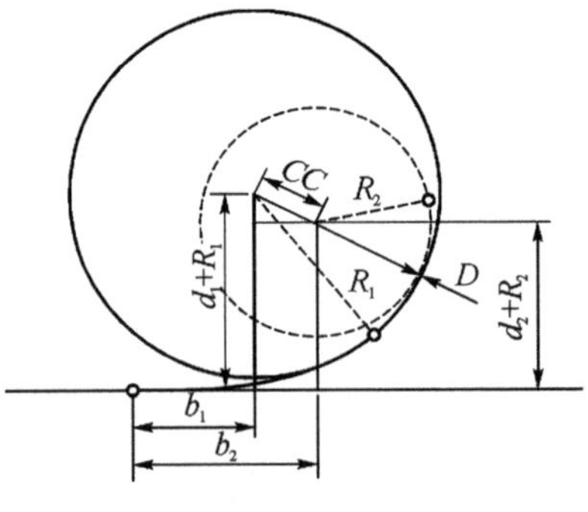


**Figure 2 - 13 Curve design method (ovate type)**

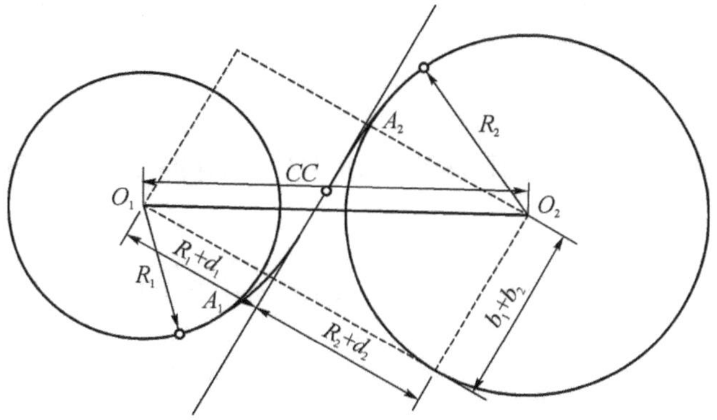


**Figure 2 - 14 Curve design method (S-shaped type)**

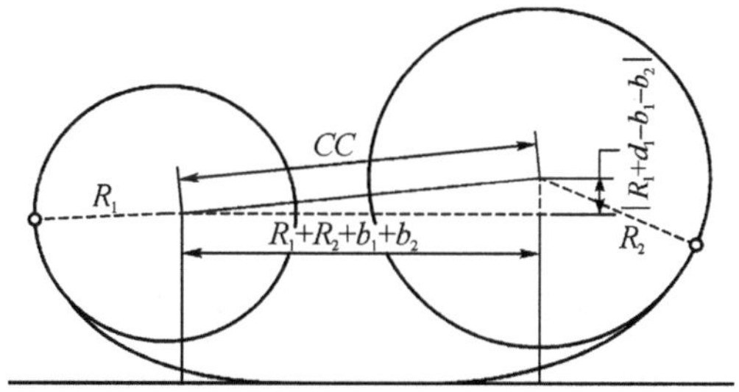


**Figure 2 - 15 Curve design method (C-shaped type)**

Taking the S-shaped curve as an example, the curve design method is to connect the two circles in reverse with two complete spirals. Since the two spiral parameters $A_1$ and $A_2$ may or may not be equal, to search and determine the $A$ values, the ratio of $A_1$ and $A_2$ or one of the $A$ values must be the input first, otherwise, it cannot be solved. At this point, the center position and radii of the two circles are known, thus the distance between the two circles can be obtained as $CC$. The $d_1$, $d_2$, $b_1$, and $b_2$ can be obtained according to the calculation formula of the spiral, and thereby the circular center distance between the two circles ($CC'$) can be obtained.

$$CC'=\sqrt{(R_1+d_1+R_2+d_2)^2+(b_1+b_2)^2} \tag{2-41}$$

The approximate value of $A_1$ can be obtained based on $\Delta R=\frac{CC-R_1-R_2}{2}$ as the initial iterative input. Taking $\varepsilon=|CC-CC'|$ as the objective function, the $A_1$ and $A_2$ values satisfying the precision requirement can be obtained by iteration.

According to the above solving process, it can be found that, given the position of two circles, the radius and the initial value of a certain spiral, through iterative calculation, there is only one set of spiral parameters that meet the requirements. Meanwhile, since the calculation is too cumbersome, $A_1$ and $A_2$ are generally solved through software.

## 2.7 Combination of horizontal alignment elements

### 2.7.1 Combination of the tangent and the curve

Based on the requirements of each horizontal alignment of element, the basic curve, and tangents connecting co-directional curves and reverse curves, this section will further introduce various combinations of horizontal alignment.

Driving comfort on route requires that tangents and curves coordinate with each other and alternate in proportion. The variation of tangents and curves of the route shall be smooth and even. The radius and length of the horizontal curve shall be adaptive to the adjacent tangent length. A long tangent section will make the driver feel tired and is also one of the reasons for accidents, which shall be used cautiously. For instance, the German Highway Design Specification stipulates that the radius $R$ of curves depends on the length ($L$) of the connected tangent: when $L \leqslant 500$ m, $R > L$; when $L > 500$ m, $R \geqslant 500$ m.

Improper alignment combination of tangents and curves needs to be prevented. For example, a small radius curve shall not be placed at the end of a long tangent. A short tangent between co-directional curves could be substituted by a curve with a large radius. There shall be a tangent of appropriate length between reverse curves, which could also be replaced by the spiral.

Therefore, the appropriate combination of tangents and curves may improve the driving quality of alignment. For instance, freeway alignment is dominated by circular curves and spirals with appropriate lengths of tangents, which will give drivers proper stimulation.

### 2.7.2 Combination of curves

Combinations of curves shall make the alignment continuous and uniform without sharp changes. The circular curve is the basic element of curve combinations. Combinations of curves have the following types:

**1) Basic curve**

The basic curve refers to a curve made up of tangent, spiral, circular curve, spiral and tangent in sequence. When the parameters of the two spirals are equal, it is called a symmetric basic curve. Otherwise, it is called an asymmetric basic curve. It is called a simple curve when there are no spirals. The spiral parameters and the minimum length of circular curves shall conform to the relevant provisions in the basic curve. For the coordination of alignment, the ratio of spiral length, circular curve length and spiral length should be 1 : 1 : 1 – 1 : 2 : 1, as shown in Figure 2 – 16.

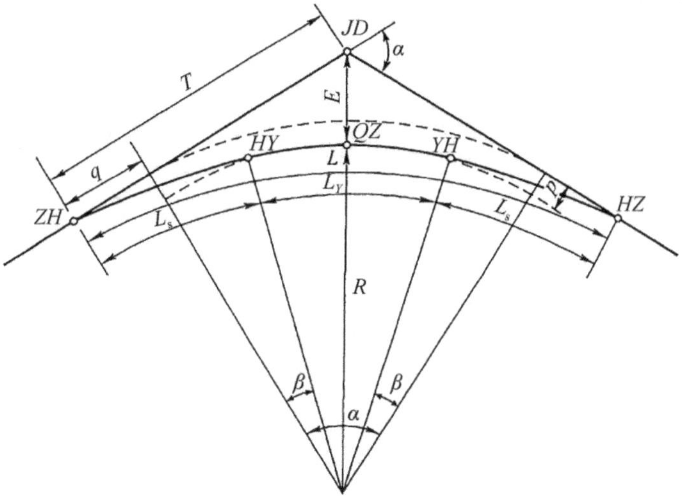


**Figure 2 – 16 Basic curve**

**2) Ovate curve**

The ovate curve refers to a curve made up of two circular curves in co-direction connected by a spiral. The parameter $A$ of the spiral in the ovate curve should be within the range of $R_2/2 \leqslant A \leqslant R_2$ ($R_2$ is the radius of the small circular curve); the ratio of radii of the two circular curves should be $R_2/R_1 = 0.2 - 0.8$; and $D/R_2 = 0.003 - 0.03$ is appropriate ($D$ is the minimum gap between the two circular curves). The spiral of the ovate curve is not a complete spiral from the origin, but rather an incomplete one using the curvature from $1/R_1$ to $1/R_2$, as shown in Figure 2 – 17.

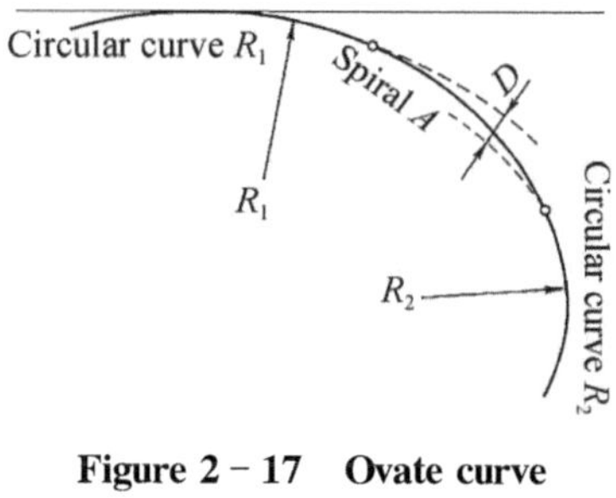


**Figure 2 – 17 Ovate curve**

### 3) Convex curve

The convex curve refers to a curve made up of two co-directional spirals radially connected without a circular curve, as shown in Figure 2 - 18. Although the curvature of convex curve is continuous at junction, the length of the circular curve in the middle is zero, which is unfavorable to vehicle operation. Hence it is only used in places such as spurs where the terrain is limited.

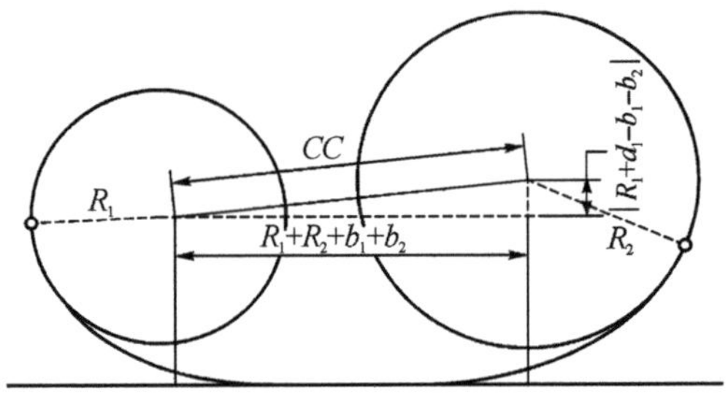


Figure 2 - 18 Convex curve

### 4) S-shaped curve

The S-shaped curve refers to a curve made up of two reverse circular curves connected by two reverse spirals. The parameters of the two adjacent spirals should be equal. The ratio of parameters should be less than 2. 0 when unequal parameters are used, and less than 1. 5 when conditions are allowed. 1/3 - 1 is the appropriate ratio of radii of the two circular curves, as shown in Figure 2 - 19.

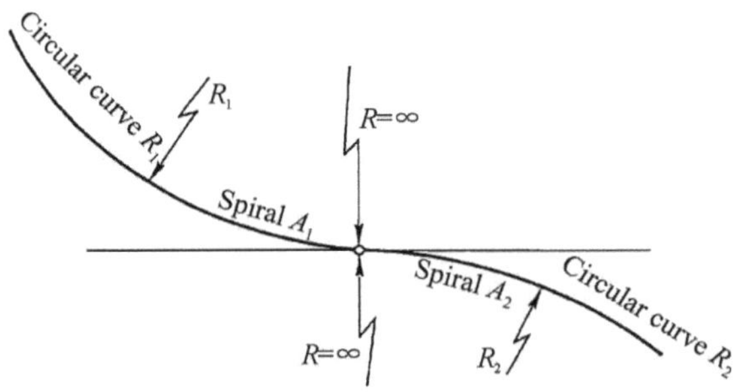


Figure 2 - 19 S-shaped curve

### 5) C-shaped curve

The C-shaped curve refers to a curve made up of two co-directional spirals connected at zero curvature. The curvature of the junction of the C-shaped curve is zero, which means that the tangent length applied in the middle is zero. Thus the C-shaped curve is unfavorable to driving and vision. It is used only in special and tough terrain conditions where the route is strictly restricted, as shown in

Figure 2 - 20.

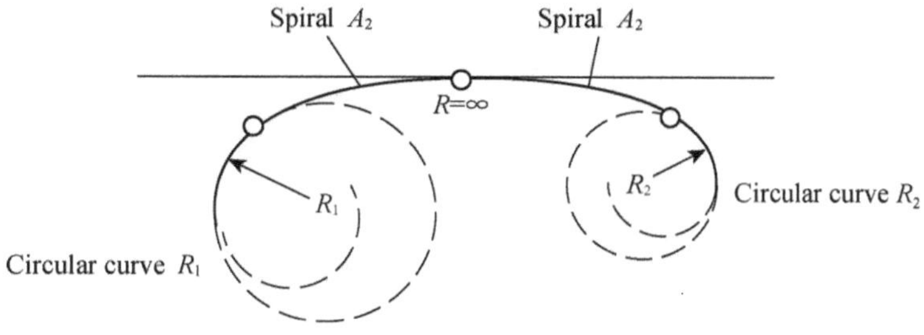


**Figure 2 - 20 C-shaped curve**

### 6) Compound curve

The compound curve refers to a curve made up of the two or more co-directional circular curves connected tangently, as shown in Figure 2 - 21. When the superelevation rates of the two circular curves in the compound curve for Class Ⅳ highways are different, the superelevation transition section shall be inserted from the common tangent point to the larger radius curve according to the superelevated slope difference. Its length is the difference between the two superelevation transition lengths or the superelevation transition length corresponding to the superelevated slope difference.

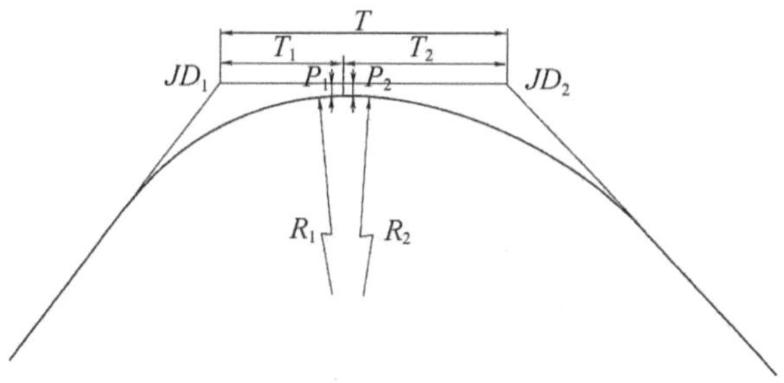


**Figure 2 - 21 Compound curve**

### 7) Composite curve

The composite curve refers to a curve made up of two or more co-directional spirals connected at points of equal curvature, as shown in Figure 2 - 22. The ratio of parameter $A$ of adjacent spirals should be less than 1. 5. The curvature of the spiral in the composite curve changes so that drivers need to change the speed and direction continuously, which is unfavorable to vehicle operation. Except for the interchange ramp, the composite curve is only used when restricted by terrain or other special reasons.

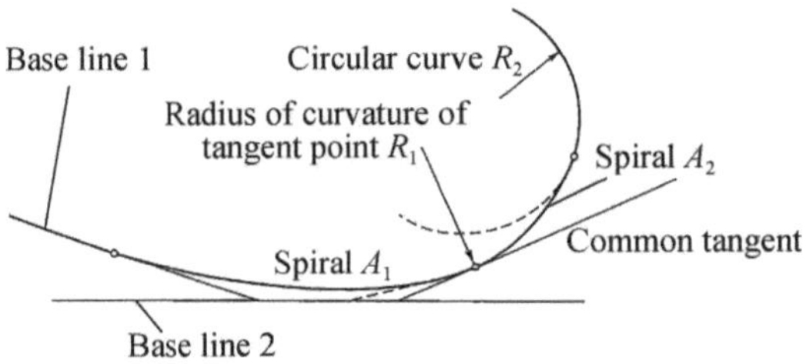


**Figure 2 - 22 Composite curve**

### 8) Switch-back curve

When the terrain and geological conditions of mountains are limited and the natural alignment development is difficult, a pin-shaped curve is used, which is called the switch-back curve, as shown in Figure 2 - 23. The switch-back curve is a curve made up of a principal curve and an auxiliary curve. The angle of the circle is close to or greater than 180°.

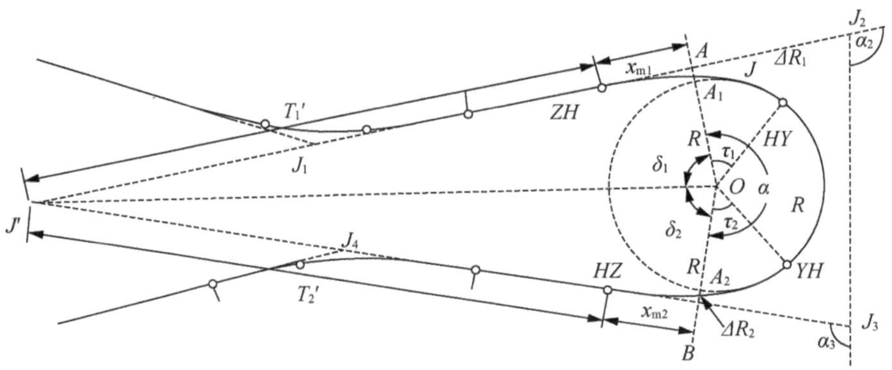


**Figure 2 - 23 Switch-back curve**

The technical indexes of the switch-back curve are shown in Table 2 - 10:

**Table 2 - 10 Design indexes of the switch-back curve**

| Item | Highway class | | | |
|---|---|---|---|---|
| | Ⅲ | | | Ⅳ |
| Design speed/km • $h^{-1}$ | 35 | 30 | 25 | 20 |
| Minimum radius of the principal curve/m | 40 | 30 | 20 | 15 |
| Minimum length of the spiral/m | 35 | 30 | 25 | 20 |
| Superelevated slope/% | 6 | 6 | 6 | 6 |
| Widening of the two-lane road/% | 2. 5 | 2. 5 | 2. 5 | 3 |
| Maximum grade/% | 3. 5 | 3. 5 | 4 | 4. 5 |

A long distance shall be strived for between two adjacent switch-back curves. From the end of a switch-back curve to the start of the next one, the distance shall not be less than 200 m, 150 m, 100 m for Class Ⅱ, Ⅲ and Ⅳ highways, respectively. And the minimum length of the switch-back curve is 120 m, 100 m, 80 m for Class Ⅱ, Ⅲ and Ⅳ highways, respectively.

Continuity of alignment before and after the switch-back curve shall be assured and the transitional curve should be appropriately adopted at both ends. The speed limit signs shall be placed as well, and certain technical measures shall be taken to ensure favorable visibility.

## 2.8 Deliverables of horizontal alignment design

After the completion of horizontal alignment design, several charts and tables are delivered. In this section, the "Table of tangents, curves, and intersection angles", "Table of coordinates by stakes" and "Chart of horizontal alignment" are illustrated.

### 2.8.1 Table of tangents, curves, and intersection angles

This table is one of the main deliverables of horizontal alignment design by the tangent design method, which comprehensively expresses the basic elements of the route and the relationship between tangents and curves, as shown in Table 2-11.

### 2.8.2 Table of coordinates by stakes

This table contains significant data on road laying out and is one of the important deliverables in the alignment design of high-class roads. The table of coordinates by stakes reflects the coordinates of each middle stake, as shown in Table 2-12.

### 2.8.3 Chart of horizontal alignment

The chart of horizontal alignment is one of the main deliverables of horizontal alignment design. It comprehensively reflects the horizontal location, alignment, and size of the route. It also reflects the layout of artificial structures and engineering facilities along the route as well as the relationship between the road and the surrounding environment, terrain, and ground features.

The scale of the design chart of horizontal alignment is 1 : 2,000 for highways, and the scale could be 1 : 5,000 in plain terrain. The design chart shall be marked with the centerline, stake, elements of horizontal curves, benchmark, large and medium bridges, route intersection (indicating the form and structure type), tunnel, location of main facilities along the route, boundary line above the county level, etc. The width of the strip shall be 200 – 250 m on both sides of the centerline.

High-class highways shall also exhibit the grid of coordinates, traverse points, intersection coordinate tables, bridges, culverts, tunnels, route intersections, drainage systems, and layout of main facilities along the route, etc. The position of the route shall be marked with the centerline, medial separator, subgrade edge, toe of slope (or top of slope) line, and the main stake position of the curve. The scale is 1 : 1,000 or 1 : 2,000. The band width is 100 – 200 m on both sides of the centerline.

The scale used in the design chart of urban roads is normally 1 : 500 – 1 : 1,000, and both sides shall be 20 – 50 m beyond the range of the boundary line. The centerline of the road, long-term and short-term plans of the boundary line, traveled way line, sidewalk line, parking lot, green belt, traffic island, crosswalk line, entrance and exit of buildings along the street, direction of various pipelines underground or on the ground, rainwater inlet, cellar well, etc. shall be indicated. Intersections and stakes along the route shall be marked. Curve elements and turning radii of intersection curbs shall be indicated at curves and intersections.

Charts of horizontal alignment design in highways and urban roads are shown in Figure 2 – 24 and Figure 2 – 25.

## Table 2 - 11 Table of tangents, curves, and intersection angles

| Intersection ID | Intersection coordinates | | Stake | Deflection angle | | Geometric elements of horizontal alignment/m | | | | | | | | | Position of alignment | | | | | Length and direction of tangent | | | Note |
|---|---|---|---|---|---|---|---|---|---|---|---|---|---|---|---|---|---|---|---|---|---|---|---|
| | X | Y | | To left /° ′ ″ | To right /° ′ ″ | Radius R | Parameter of the 1st spiral $A_1$ | Length of the 1st spiral $L_{s1}$ | Parameter of the 2nd spiral $A_2$ | Length of the 2nd spiral $L_{s2}$ | Length of the 1st tangent $T_1$ | Length of the 2nd tangent $T_2$ | Length of horizontal alignment L | Apex distance E | BP of the 1st spiral ZH | EP of the 1st spiral HY(ZY) | Middle point of circular curve QZ | BP of the 2nd spiral YH(YZ) | EP of the 2nd spiral (HZ) | Tagent length/m | Intersection distance/m | Azimuth angle /° ′ ″ | |
| BP | 376,038.178 | 331,638.165 | K0+000 | | | | | | | | | | | | | | | | | | | | |
| JD1 | 376,463.539 | 331,196.408 | K0+613.255 | 41°59′55″ | | 400 | 244.949 | 150 | 253.546 | 150 | 229.77 | 224.021 | 438.205 | | K0+383.485 | K0+533.485 | K0+602.588 | K0+671.69 | K0+821.69 | 383.485 | 613.255 | 313°55′0″ | |
| JD2 | 376,478.715 | 330,743.261 | K1+051.07 | 4°22′43″ | | 6,000 | 0 | 0 | 0 | 0 | 229.38 | 229.38 | 458.536 | 4.383 | K0+821.69 | K0+821.69 | K1+050.958 | K1+280.226 | K1+280.226 | 0 | 453.401 | 271°55′5″ | |
| JD3 | 376,387.791 | 328,627.309 | K3+168.751 | 5°32′35″ | | 6,000 | 0 | 0 | 0 | 0 | 290.462 | 290.462 | 580.47 | 7.027 | K2+878.289 | K2+878.289 | K3+168.524 | K3+458.76 | K3+458.76 | 1,598.063 | 2,117.905 | 267°32′22″ | |
| JD4 | 376,308.685 | 328,064.700 | K3+736.441 | | 12°40′34″ | 2,500 | 0 | 0 | 0 | 0 | 277.681 | 277.681 | 553.096 | 15.374 | K3+458.76 | K3+458.76 | K3+735.307 | K4+011.855 | K4+011.855 | 0 | 568.143 | 261°59′47″ | |
| JD5 | 376,354.270 | 327,506.951 | K4+293.782 | 5°22′50″ | | 6,000 | 0 | 0 | 0 | 0 | 281.927 | 281.927 | 563.44 | 6.62 | K4+011.855 | K4+011.855 | K4+293.575 | K4+575.295 | K4+575.295 | 0 | 559.609 | 274°40′21″ | |
| JD6 | 376,333.491 | 325,825.538 | K5+974.91 | | 23°24′46″ | 1,000 | 447.214 | 200 | 447.214 | 200 | 307.518 | 307.518 | 608.63 | 22.943 | K5+667.392 | K5+867.392 | K5+971.707 | K6+076.022 | K6+276.022 | 1,092.097 | 1,681.542 | 269°17′31″ | |
| JD7 | 376,570.531 | 325,259.005 | K6+582.627 | 23°18′46″ | | 1,000 | 447.214 | 200 | 447.214 | 200 | 306.606 | 306.606 | 606.882 | 22.758 | K6+276.022 | K6+476.022 | K6+579.463 | K6+682.904 | K6+882.904 | 0 | 614.124 | 292°42′17″ | |
| JD8 | 376,548.790 | 323,210.015 | K8+625.403 | 45°40′42″ | | 1,000 | 500 | 250 | 500 | 250 | 547.197 | 547.197 | 1047.238 | 87.896 | K8+078.206 | K8+328.206 | K8+601.825 | K8+875.444 | K9+125.444 | 1,195.302 | 2,049.105 | 269°23′31″ | |
| JD9 | 374,932.816 | 321,665.018 | K10+813.954 | | 19°29′10″ | 1,000 | 387.298 | 150 | 387.298 | 150 | 246.853 | 246.853 | 490.095 | 15.585 | K10+567.101 | K10+717.101 | K10+812.149 | K10+907.196 | K11+057.196 | 1,441.657 | 2,235.707 | 223°42′49″ | |
| JD10 | 374,428.435 | 320,666.527 | K11+928.997 | | 6°3′48″ | 4,000 | 0 | 0 | 0 | 0 | 211.851 | 211.851 | 423.307 | 5.606 | K11+717.145 | K11+717.145 | K11+928.799 | K12+140.452 | K12+140.452 | 659.949 | 1,118.653 | 243°11′59″ | |
| JD11 | 374,200.391 | 320,064.197 | K12+572.655 | | 32°27′15″ | 1,250 | 603.023 | 200 | 500 | 200 | 432.203 | 464.991 | 876.792 | | K12+140.452 | K12+340.452 | K12+578.848 | K12+817.244 | K13+017.244 | 0 | 644.054 | 249°15′47″ | |
| JD12 | 374,355.424 | 319,316.712 | K13+315.646 | 13°36′48″ | | 2,500 | 0 | 0 | 0 | 0 | 298.401 | 298.401 | 593.993 | 17.746 | K13+017.244 | K13+017.244 | K13+314.241 | K13+611.237 | K13+611.237 | 0 | 763.393 | 281°43′3″ | |
| JD13 | 374,310.138 | 317,948.631 | K14+681.667 | 23°43′20″ | | 1,250 | 559.017 | 250 | 559.017 | 250 | 387.927 | 387.927 | 767.54 | 29.4 | K14+293.74 | K14+543.74 | K14+677.51 | K14+811.28 | K15+061.28 | 682.503 | 1,368.831 | 268°6′15″ | |
| EP | 374,055.763 | 317,418.141 | K15+261.677 | | | | | | | | | | | | | | | | | 200.397 | 588.324 | 244°22′55″ | |
| Sum | | | | | | | | | | | | | 8,008.224 | | | | | | | 7,253.454 | | | |

**Table 2 - 12 Table of coordinates by stakes**

| Stake | Coordinates | | Azimuth angle | Stake | Coordinates | | Azimuth angle | Stake | Coordinates | | Azimuth angle | Stake | Coordinates | | Azimuth angle |
|---|---|---|---|---|---|---|---|---|---|---|---|---|---|---|---|
| | X | Y | | | X | Y | | | X | Y | | | X | Y | |
| 1 | 2 | 3 | 4 | 1 | 2 | 3 | 4 | 1 | 2 | 3 | 4 | 1 | 2 | 3 | 4 |
| K0+000 | 376,038. 178 | 331,638. 165 | 313°55′0″ | K0+671. 69 | 376,455. 449 | 331,121. 436 | 283°22′38″ | K1+340 | 376,466. 301 | 330,454. 374 | 267°32′22″ | K2+060 | 376,435. 391 | 329,735. 038 | 267°32′22″ |
| K0+020 | 376,052. 050 | 331,623. 758 | 313°55′0″ | K0+680 | 376,457. 289 | 331,113. 333 | 282°13′4″ | K1+360 | 376,465. 443 | 330,434. 393 | 267°32′22″ | K2+080 | 376,434. 532 | 329,715. 056 | 267°32′22″ |
| K0+040 | 376,065. 922 | 331,609. 351 | 313°55′0″ | K0+700 | 376,461. 077 | 331,093. 696 | 279°40′46″ | K1+380 | 376,464. 584 | 330,414. 411 | 267°32′22″ | K2+100 | 376,433. 674 | 329,695. 075 | 267°32′22″ |
| K0+060 | 376,079. 795 | 331,594. 944 | 313°55′0″ | K0+720 | 376,464. 053 | 331,073. 920 | 277°29′51″ | K1+400 | 376,463. 725 | 330,394. 430 | 267°32′22″ | K2+120 | 376,432. 815 | 329,675. 093 | 267°32′22″ |
| K0+080 | 376,093. 667 | 331,580. 537 | 313°55′0″ | K0+740 | 376,466. 336 | 331,054. 052 | 275°40′20″ | K1+420 | 376,462. 867 | 330,374. 448 | 267°32′22″ | K2+140 | 376,431. 956 | 329,655. 112 | 267°32′22″ |
| K0+100 | 376,107. 539 | 331,566. 130 | 313°55′0″ | K0+760 | 376,468. 047 | 331,034. 126 | 274°12′12″ | K1+440 | 376,462. 008 | 330,354. 466 | 267°32′22″ | K2+160 | 376,431. 098 | 329,635. 130 | 267°32′22″ |
| K0+120 | 376,121. 411 | 331,551. 723 | 313°55′0″ | K0+780 | 376,469. 309 | 331,014. 166 | 273°5′27″ | K1+460 | 376,461. 150 | 330,334. 485 | 267°32′22″ | K2+180 | 376,430. 239 | 329,615. 149 | 267°32′22″ |
| K0+140 | 376,135. 284 | 331,537. 316 | 313°55′0″ | K0+800 | 376,470. 245 | 330,994. 188 | 272°20′6″ | K1+480 | 376,460. 291 | 330,314. 503 | 267°32′22″ | K2+200 | 376,429. 381 | 329,595. 167 | 267°32′22″ |
| K0+160 | 376,149. 156 | 331,522. 909 | 313°55′0″ | K0+820 | 376,470. 980 | 330,974. 201 | 271°56′8″ | K1+500 | 376,459. 432 | 330,294. 522 | 267°32′22″ | K2+220 | 376,428. 522 | 329,575. 186 | 267°32′22″ |
| K0+180 | 376,163. 028 | 331,508. 502 | 313°55′0″ | K0+821. 69 | 376,471. 037 | 330,972. 512 | 271°55′5″ | K1+520 | 376,458. 574 | 330,274. 540 | 267°32′22″ | K2+240 | 376,427. 663 | 329,555. 204 | 267°32′22″ |
| K0+200 | 376,176. 900 | 331,494. 095 | 313°55′0″ | K0+840 | 376,471. 622 | 330,954. 212 | 271°44′36″ | K1+540 | 376,457. 715 | 330,254. 559 | 267°32′22″ | K2+260 | 376,426. 805 | 329,535. 222 | 267°32′22″ |
| K0+220 | 376,190. 773 | 331,479. 688 | 313°55′0″ | K0+860 | 376,472. 197 | 330,934. 220 | 271°33′8″ | K1+560 | 376,456. 856 | 330,234. 577 | 267°32′22″ | K2+280 | 376425. 946 | 329,515. 241 | 267°32′22″ |
| K0+240 | 376,204. 645 | 331,465. 281 | 313°55′0″ | K0+880 | 376,472. 706 | 330,914. 226 | 271°21′41″ | K1+580 | 376,455. 998 | 330,214. 595 | 267°32′22″ | K2+300 | 376,425. 087 | 329,495. 259 | 267°32′22″ |
| K0+260 | 376,218. 517 | 331,450. 875 | 313°55′0″ | K0+900 | 376,473. 147 | 330,894. 231 | 271°10′13″ | K1+600 | 376,455. 139 | 330,194. 614 | 267°32′22″ | K2+320 | 376,424. 229 | 329,475. 278 | 267°32′22″ |
| K0+280 | 376,232. 389 | 331,436. 468 | 313°55′0″ | K0+920 | 376,473. 523 | 330,874. 235 | 270°58′46″ | K1+620 | 376,454. 281 | 330,174. 632 | 267°32′22″ | K2+340 | 376,423. 370 | 329,455. 296 | 267°32′22″ |
| K0+300 | 376,246. 262 | 331,422. 061 | 313°55′0″ | K0+940 | 376,473. 831 | 330,854. 237 | 270°47′18″ | K1+640 | 376,453. 422 | 330,154. 651 | 267°32′22″ | K2+360 | 376,422. 512 | 329,435. 315 | 267°32′22″ |
| K0+320 | 376,260. 134 | 331,407. 654 | 313°55′0″ | K0+960 | 376,474. 073 | 330,834. 239 | 270°35′51″ | K1+660 | 376,452. 563 | 330,134. 669 | 267°32′22″ | K2+380 | 376,421. 653 | 329,415. 333 | 267°32′22″ |
| K0+340 | 376,274. 006 | 331,393. 247 | 313°55′0″ | K0+980 | 376,474. 248 | 330,814. 240 | 270°24′23″ | K1+680 | 376,451. 705 | 330,114. 688 | 267°32′22″ | K2+400 | 376,420. 794 | 329,395. 351 | 267°32′22″ |
| K0+360 | 376,287. 878 | 331,378. 840 | 313°55′0″ | K1+000 | 376,474. 357 | 330,794. 240 | 270°12′56″ | K1+700 | 376,450. 846 | 330,094. 706 | 267°32′22″ | K2+420 | 376,419. 936 | 329,375. 370 | 267°32′22″ |
| K0+380 | 376,301. 751 | 331,364. 433 | 313°55′0″ | K1+020 | 376,474. 399 | 330,774. 240 | 270°1′28″ | K1+720 | 376,449. 987 | 330,074. 725 | 267°32′22″ | K2+440 | 376,419. 077 | 329,355. 388 | 267°32′22″ |
| K0+383. 485 | 376,304. 168 | 331,361. 922 | 313°55′0″ | K1+040 | 376,474. 374 | 330,754. 240 | 269°50′0″ | K1+740 | 376,449. 129 | 330,054. 743 | 267°32′22″ | K2+460 | 376,418. 218 | 329,335. 407 | 267°32′22″ |
| K0+400 | 376,315. 614 | 331,350. 017 | 313°47′11″ | K1+060 | 376,474. 282 | 330,734. 240 | 269°38′33″ | K1+760 | 376,448. 270 | 330,034. 761 | 267°32′22″ | K2+480 | 376,417. 360 | 329,315. 425 | 267°32′22″ |

**Continued Table 2 - 12**

| Stake | Coordinates | | Azimuth angle | Stake | Coordinates | | Azimuth angle | Stake | Coordinates | | Azimuth angle | Stake | Coordinates | | Azimuth angle |
|---|---|---|---|---|---|---|---|---|---|---|---|---|---|---|---|
| | $X$ | $Y$ | | | $X$ | $Y$ | | | $X$ | $Y$ | | | $X$ | $Y$ | |
| 1 | 2 | 3 | 4 | 1 | 2 | 3 | 4 | 1 | 2 | 3 | 4 | 1 | 2 | 3 | 4 |
| K0+420 | 376,329.397 | 331,335.525 | 313°16′48″ | K1+080 | 376,474.124 | 330,714.241 | 269°27′5″ | K1+780 | 376,447.412 | 330,014.780 | 267°32′22″ | K2+500 | 376,416.501 | 329,295.444 | 267°32′22″ |
| K0+440 | 376,343.003 | 331,320.867 | 312°23′30″ | K1+100 | 376,473.899 | 330,694.242 | 269°15′38″ | K1+800 | 376,446.553 | 329,994.798 | 267°32′22″ | K2+520 | 376,415.643 | 329,275.462 | 267°32′22″ |
| K0+460 | 376,356.331 | 331,305.955 | 311°7′17″ | K1+120 | 376,473.608 | 330,674.244 | 269°4′10″ | K1+820 | 376,445.694 | 329,974.817 | 267°32′22″ | K2+540 | 376,414.784 | 329,255.481 | 267°32′22″ |
| K0+480 | 376,369.273 | 331,290.708 | 309°28′9″ | K1+140 | 376,473.250 | 330,654.247 | 268°52′43″ | K1+840 | 376,444.836 | 329,954.835 | 267°32′22″ | K2+560 | 376,413.925 | 329,235.499 | 267°32′22″ |
| K0+500 | 376,381.718 | 331,275.053 | 307°26′5″ | K1+160 | 376,472.825 | 330,634.252 | 268°41′15″ | K1+860 | 376,443.977 | 329,934.854 | 267°32′22″ | K2+580 | 376,413.067 | 329,215.517 | 267°32′22″ |
| K0+520 | 376,393.546 | 331,258.927 | 305°1′7″ | K1+180 | 376,472.334 | 330,614.258 | 268°29′48″ | K1+880 | 376,443.118 | 329,914.872 | 267°32′22″ | K2+600 | 376,412.208 | 329,195.536 | 267°32′22″ |
| K0+533.485 | 376,401.108 | 331,247.763 | 303°10′25″ | K1+200 | 376,471.776 | 330,594.266 | 268°18′20″ | K1+900 | 376,442.260 | 329,894.891 | 267°32′22″ | K2+620 | 376,411.350 | 329,175.554 | 267°32′22″ |
| K0+540 | 376,404.628 | 331,242.281 | 302°14′26″ | K1+220 | 376,471.151 | 330,574.276 | 268°6′53″ | K1+920 | 376,441.401 | 329,874.909 | 267°32′22″ | K2+640 | 376,410.491 | 329,155.573 | 267°32′22″ |
| K0+560 | 376,414.871 | 331,225.105 | 299°22′33″ | K1+240 | 376,470.460 | 330,554.288 | 267°55′25″ | K1+940 | 376,440.543 | 329,854.927 | 267°32′22″ | K2+660 | 376,409.632 | 329,135.591 | 267°32′22″ |
| K0+580 | 376,424.242 | 331,207.439 | 296°30′40″ | K1+260 | 376,469.702 | 330,534.302 | 267°43′57″ | K1+960 | 376,439.684 | 329,834.946 | 267°32′22″ | K2+680 | 376,408.774 | 329,115.610 | 267°32′22″ |
| K0+600 | 376,432.718 | 331,189.326 | 293°38′46″ | K1+280 | 376,468.877 | 330,514.319 | 267°32′30″ | K1+980 | 376,438.825 | 329,814.964 | 267°32′22″ | K2+700 | 376,407.915 | 329,095.628 | 267°32′22″ |
| K0+620 | 376,440.278 | 331,170.812 | 290°46′53″ | K1+280.226 | 376,468.867 | 330,514.093 | 267°32′22″ | K2+000 | 376,437.967 | 329,794.983 | 267°32′22″ | K2+720 | 376,407.056 | 329,075.647 | 267°32′22″ |
| K0+640 | 376,446.904 | 331,151.944 | 287°54′60″ | K1+300 | 376,468.018 | 330,494.337 | 267°32′22″ | K2+020 | 376,437.108 | 329,775.001 | 267°32′22″ | K2+740 | 376,406.198 | 329,055.665 | 267°32′22″ |
| K0+660 | 376,452.579 | 331,132.768 | 285°3′7″ | K1+320 | 376,467.160 | 330,474.356 | 267°32′22″ | K2+040 | 376,436.250 | 329,755.020 | 267°32′22″ | K2+760 | 376,405.339 | 329,035.683 | 267°32′22″ |

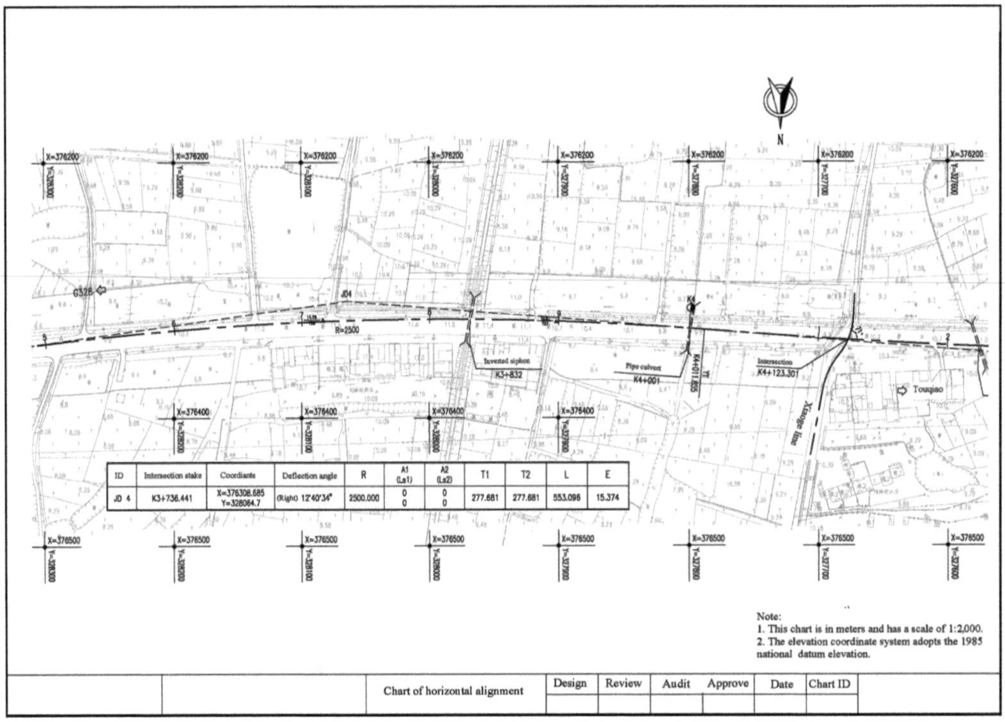


**Figure 2 - 24　Chart of horizontal alignment of a highway**

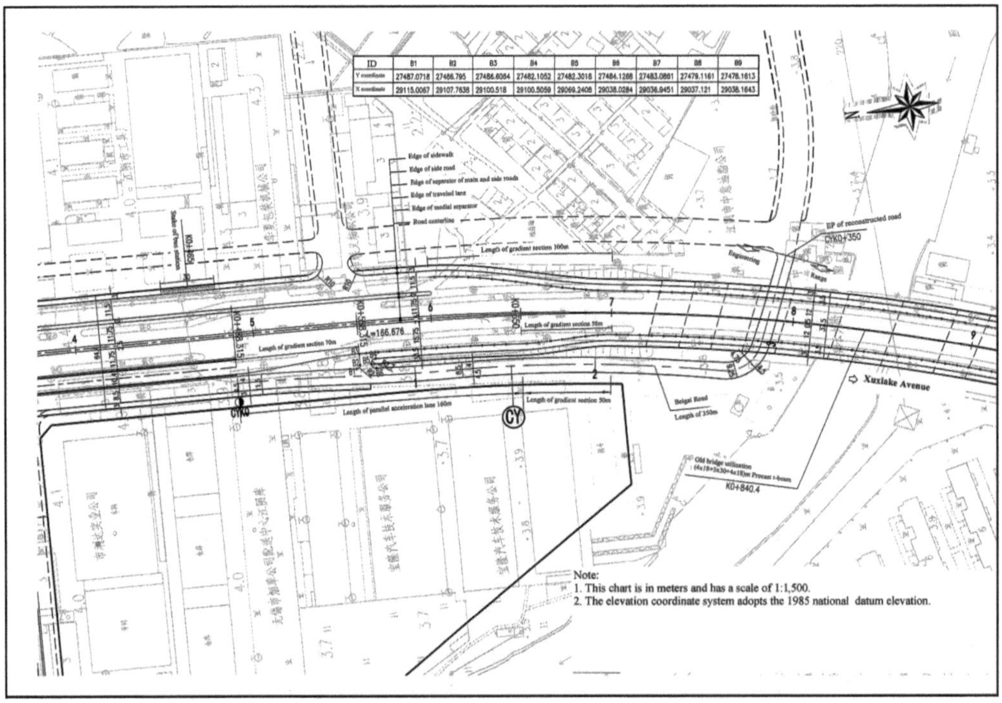


**Figure 2 - 25　Chart of horizontal alignment of an urban road**

**Questions:**

1. What factors shall be considered in the determination of the radii of the circular curve?
2. What is the sideway force coefficient? What is its maximum value? How to choose the sideway force coefficient in the design?
3. Why do vehicles display lateral instability when driving at high speed on curves with small radius? Why are vehicles more stable on the inner side of curves than on the outer side?
4. Why set up a spiral? How to determine its length?
5. What is the difference in the calculation formulas between the limited minimum radius and the minimum radius without superelevation? Why?
6. What are the geometric elements of horizontal alignment? What are the combinations?
7. What is a compound curve and under what circumstances will it be used? What is the basis for the application of the compound curve?
8. Shall the superelevation be set for the co-directional circular curves in a compound curve? If so, how to calculate the superelevation transition length? And how to set it?
9. For an intersection on the road centerline (point $JD$), the deflection angle $\alpha =$ 55°. There is a building close to point $JD$ with a size of 6 m×10 m, as shown in Figure 1. The distance from the outer edge of the building to point $JD$ is 25 m. To retain the building, what are the possible minimum radius and maximum radius of the curve? It is given that the design speed of the road is 40 km/h; the width is 24 m; the cross slope is 2%; and the sideway force coefficient $\mu = 0.1$.

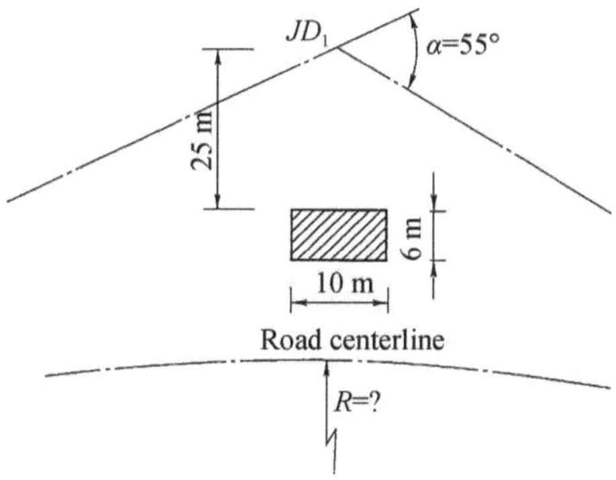


**Figure 1 Diagram of route design**

10. There is a Class Ⅱ highway with a design speed of 80 km/h. There are two adjacent intersection points $JD_3$ and $JD_4$ on the highway, with a distance of 700.00 m. The right deflection angle $\alpha_3$ is $15°3'24''$, the right deflection angle $\alpha_4$ is $16°10'12''$, and the stake of $JD_3$ is K5+281.07. If the circular curve radius at $JD_3$ is 2000 m, try to complete the horizontal curve design of $JD_3$ and $JD_4$, and calculate the stakes of the main points of $JD_4$.

# Chapter 3 Vertical Alignment Design

【**Objectives**: 1. Understand the relationship between ground elevation, design elevation, and heights of fill and cut; 2. Master the technical requirements of control grades for design; 3. Master the characteristics, applicable conditions, and parameter calculations of the vertical curve; 4. Master the principles, forms and requirements of combinations of horizontal alignment and vertical alignment; 5. Understand the relationship between vertical alignment design and the terrain, fill and cut, driving safety, and traffic capacity.】

## 3.1 Introduction

After the completion of the horizontal alignment design, the horizontal position of the road centerline is determined, which reflects the variation of tangents and curves of the road centerline, but does not reflect the ups and downs of the road centerline, so it is necessary to carry out vertical alignment design. Road vertical alignment refers to the profile cutting through the road centerline. Only by combining the design outputs of horizontal alignment and vertical profile can the spatial position of the road centerline be reflected.

The profile diagram reflects the shape, position, and size of vertical alignment of a route, as shown in Figure 3.1. The profile diagram is the final result of vertical alignment design and one important deliverable of route design documents.

In the profile diagram, the elevation of the original ground points on the road centerline is called the ground elevation. The line connecting the ground elevations at all stakes is called the ground line, which is an irregular broken line and reflects the ups and downs of the original ground of the road centerline. Naturally, a ground line shall not be directly used as a route, so the vertical alignment design is necessary.

For the profile diagram, the elevation is expressed by the design elevation.

© The Author(s), under exclusive license to Springer Nature Singapore Pte Ltd. 2025
B. Yu and J. Cheng, *Road Survey and Design*,
https://doi.org/10.1007/978-981-97-4836-5_3

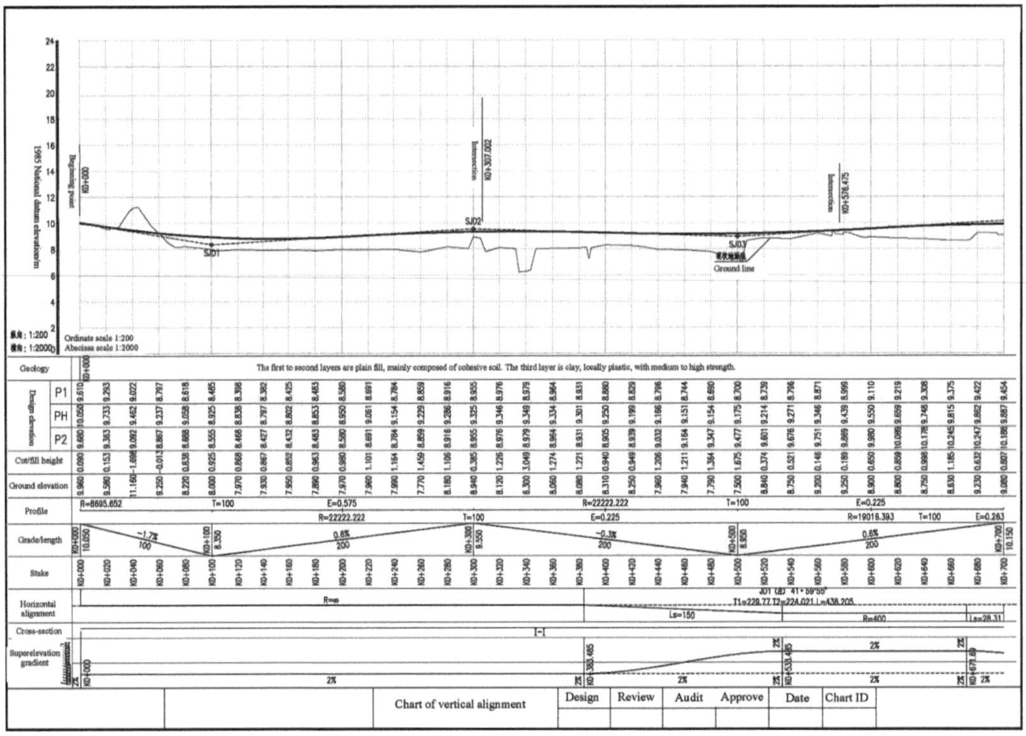


**Figure 3 - 1 Profile diagram of a route**

In the **Specification (2017)**, the design elevation is defined as follows:

(1) Subgrade (roadway: a portion of highway, including shoulders, for vehicular use) design elevation of new highways: the elevation of the outer edge of medial separator should be adopted for the freeway and Class Ⅰ highway; the elevation of the roadway edge should be adopted for Class Ⅱ, Ⅲ, and Ⅳ highways; and the elevation of the roadway edge before superelevation and widening is adopted for the superelevated and widened sections.

(2) Subgrade design elevation of reconstructed highways: it should be implemented according to the provisions of the new highway, or the elevation of the centerline of medial separator or of the centerline of the traveled way (a portion of the roadway for the movements of vehicles, exclusive of shoulders) can be adopted according to the specific situations.

(3) Subgrade design elevation under other special circumstances.

① For sections along the river and sections that may be flooded, the calculated elevation of the subgrade edge on the lowest side shall be higher than the calculated water level according to the flood frequency specified in Table 3 - 1, plus backwater height, wave invasion height and a safe height of 0.50 m.

**Table 3-1 Flood frequency of subgrade design**

| Class | Freeway | Class Ⅰ highway | Class Ⅱ highway | Class Ⅲ highway | Class Ⅳ Highway |
|---|---|---|---|---|---|
| Design flood frequency | 1/100 | 1/100 | 1/50 | 1/25 | Depending on the circumstances |

*Note*: This table refers to the **Specification (2017)**.

② For sections along the upstream bank of the reservoir, the elevation of the subgrade edge on the lowest side calculated according to the design elevation shall consider the groundwater level rising after the reservoir water level rises, and the influence of backwater curve elevation and wave height after reservoir sedimentation; in cold regions, the impact of ice blockage on water level rise shall also be considered.

③ For the approach road of large and medium bridges (within the scope of flood), the elevation of the subgrade edge on the lowest side calculated according to the design elevation shall be at least 0.50 m higher than the design flood level of bridges (including backwater and wave height); and the elevation of the subgrade edge on the lowest side near the small bridge and culvert shall be at least 0.50 m higher than the backwater level in front of the bridge (culvert) (excluding wave height).

④ The design flood frequency of the highway subgrade in the surrounding area of a city shall be determined synthetically according to the urban flood control standard, considering the disaster relief passage, flood drainage, and flood discharge demand.

The difference between the design elevation and the ground elevation of each stake is called the fill and cut height of the stake. When the design line is above the ground line, the subgrade is filled; on the contrary, the subgrade is cut.

The connection of design elevation of each stake is called the design line, which is determined by the designer after considering the technical and economic factors synthetically. A profile is composed of tangent grades and the vertical curve connecting the two adjacent tangent grades. The task of vertical alignment design is to determine the profile, including the grade and vertical curve, which meets the requirements of driving and is economical and aesthetic.

## 3.2 Grade design

The ratio of vertical elevation difference to horizontal length between any two adjacent vertical points of intersection (VPI) is called the tangent grade, or grade in short. The rate and length of grade not only affect vehicle operation, but also relate to economy and alignment elegancy. Grade design is realized by repetitive comparison and optimization.

### 3.2.1 Grade design standards

The grade design standards refer to the technical requirements that shall be complied with during grade design. These indexes include the rate and length of grade, which are the controls of grade design.

**1) Maximum grade**

The maximum grade is the maximum value allowed to be used in the grade design, which is an important index of vertical alignment design. The maximum grade directly affects driving safety, route mileage, use quality, operation cost and engineering investment. Because of the sound ability of passenger cars to overcome the grade, the determination of the maximum grade mainly considers heavy trucks. When trucks are on upgrade, they need to increase the tractive force in order to overcome the grade resistance, which decelerates, hinders the following fast vehicles, and increases the passing demand. It not only reduces traffic capacity, but also jeopardizes driving safety. The determination of the maximum grade considers the climbing capacity, highway traffic capacity, downgrade braking safety, traffic accident rate, fuel consumption, environmental protection, engineering investment, the terrain, traffic composition, economic development and so on. Provisions in the **Specification (2017)** for the maximum grade of highways of all classes are shown in Table 3 - 2. The maximum grade of the traveled way of urban roads shall conform to the provisions in Table 3 - 3.

**Table 3 - 2 Maximum grade of highways**

| Design speed/km · h$^{-1}$ | 120 | 100 | 80 | 60 | 40 | 30 | 20 |
|---|---|---|---|---|---|---|---|
| Maximum grade/% | 3 | 4 | 5 | 6 | 7 | 8 | 9 |

*Notes*: (1) When the design speeds are 120 km/h, 100 km/h, and 80 km/h, the maximum grade may increase by 1% after technical and economic demonstrations.

(2) When the reconstructed highways, with the design speeds of 40 km/h, 30 km/h, and 20 km/h, use the original pavement structures, the maximum grade may increase by 1% through technical and economic demonstrations.

(3) For Class Ⅳ highways with an altitude above 2,000 m or in snowy or frozen areas, the maximum grade shall not be greater than 8%.

(4) This table refers to the **Specification (2017)**.

**Table 3 - 3 Maximum grade of the traveled way of urban roads**

| Design speed/km · h$^{-1}$ | | 100 | 80 | 60 | 50 | 40 | 30 | 20 |
|---|---|---|---|---|---|---|---|---|
| Maximum grade/% | General value | 3 | 4 | 5 | 5.5 | 6 | 7 | 8 |
| | Limit value | 4 | 5 | 6 | 6 | 7 | 8 | 8 |

*Notes*: (1) The general value of the maximum grade shall be used for the new urban roads; the limit value of the maximum grade may be used for the reconstructed roads or when constrained by terrain or other special circumstances.

(2) For urban roads of other classes except for expressways that are constrained by terrain or other special circumstances, the limit value of the maximum grade may increase by 1% after technical and economic demonstrations.

(3) The maximum grade of expressways in snowy or frozen areas shall not be greater than 3.5%, and the maximum grade of urban roads of other classes shall not be greater than 6%.

(4) The general value of the maximum grade of urban roads in the plateau area above 3000 m may reduce by 1%. When the maximum grade is less than 4% after reduction, 4% is used.

(5) This table refers to the **Code (2012)**.

### 2) Grade reduction in plateaus

With the increase in altitudes, the atmospheric pressure, air temperature, and density in the plateau area decrease gradually. The decrease in air density affects the function of engines and the dynamic performance of vehicles. In addition, the air becomes thinner, the heat capacity reduces, and the engine is easy to overheat. Vehicles have to adopt low gear for a long time, which is easy to overheat the engine, boil the water in the vehicle water tank, and damage the cooling system. When highways with a design speed less than or equal to 80 km/h in the plateau area above 3,000 m, the maximum grade shall be reduced according to the provisions in Table 3 - 4. When the maximum grade is less than 4% after reduction, 4% is used.

**Table 3 - 4 Grade reduction in plateau**

| Altitude/m | 3,000-4,000 | 4,000-5,000 | >5,000 |
|---|---|---|---|
| Reduction/% | 1 | 2 | 3 |

*Note*: This table refers to the **Specification (2017)**.

**3) Minimum grade**

In order to ensure the drainage of the cut section, the low-fill section with side ditches, and the section with inadequate horizontal drainage, the highway grade should not be less than 0.3% to prevent water from infiltrating into the base structure and affecting its stability. When the cross slope is 0 or the grade is less than 0.3%, side ditches shall be designed for longitudinal drainage.

**4) Grade length**

The grade length refers to the horizontal distance between two adjacent vertical points of intersections on vertical alignment. The control of grade length is two-fold: the minimum one and the maximum one.

(1) Minimum grade length

In terms of driving, when the length of grades is short and undulating, drivers need to shift gears frequently, which is easy to cause driving fatigue. Gear shifts also cause loss of energy, oil and time, and accelerate the wear of gears, clutches, and tires. Meanwhile, when vehicles run on the sag and crest vertical curves, they are affected by hypergravity and hypogravity, which makes passengers uncomfortable, especially at high speeds. Therefore, it is necessary to limit the minimum grade length and prevent frequent grade changes to ensure the safety and comfort of driving.

For alignment geometry, a profile is generally composed of "vertical curve-tangent grade-vertical curve", and the minimum length of grades shall at least meet the need of setting vertical curves. In order to prevent the profile design from over-emphasizing the engineering economy, it is necessary to limit the minimum grade length.

In the vertical alignment design of highways, the minimum grade length should be equal to the distance traveled in 9 - 12 s at the design speed; and the minimum grade length of highways shall conform to the provisions in Table 3 - 5. The minimum grade length of the traveled way of urban roads is shown in Table 3 - 6.

**Table 3 - 5 Minimum grade length of highways**

| Design speed/km • $h^{-1}$ | 120 | 100 | 80 | 60 | 40 | 30 | 20 |
|---|---|---|---|---|---|---|---|
| Minimum grade length/m | 300 | 250 | 200 | 150 | 120 | 100 | 60 |

*Note*: The table refers to the **Specification (2017)**.

**Table 3 - 6 Minimum grade length of the traveled way of urban roads**

| Design speed/km • $h^{-1}$ | 100 | 80 | 60 | 50 | 40 | 30 | 20 |
|---|---|---|---|---|---|---|---|
| Minimum grade length/m | 250 | 200 | 150 | 130 | 110 | 85 | 60 |

*Note*: This table refers to the **Code (2012)**.

(2) Maximum grade length

A long and steep grade is detrimental to vehicle operation. Continuous upgrades lead to the deceleration of vehicles, especially trucks, which lowers the capacity and level of service, and causes congestion. Therefore, continous upgrades increase the operating speed difference between large and small vehicles, leading to traffic accidents. Therefore, the length of steep grades shall be controlled. The maximum grade length of highways of all classes shall conform to the provisions in Table 3 - 7.

**Table 3 - 7 Maximum grade length of highways**

| Design speed/km • $h^{-1}$ | | 120 | 100 | 80 | 60 | 40 | 30 | 20 |
|---|---|---|---|---|---|---|---|---|
| Grade/% | 3 | 900 | 1,000 | 1,100 | 1,200 | — | — | — |
| | 4 | 700 | 800 | 900 | 1,000 | 1,100 | 1,100 | 1,200 |
| | 5 | — | 600 | 700 | 800 | 900 | 900 | 1,000 |
| | 6 | — | — | 500 | 600 | 700 | 700 | 800 |
| | 7 | — | — | — | — | 500 | 500 | 600 |
| | 8 | — | — | — | — | 300 | 300 | 400 |
| | 9 | — | — | — | — | — | 200 | 300 |
| | 10 | — | — | — | — | — | — | 200 |

*Note*: This table refers to the **Specification (2017)**.

The maximum grade length of the traveled way and non-motorized vehicle lane of urban roads are shown in Table 3 - 8 and Table 3 - 9.

**Table 3 - 8 Maximum grade length of the traveled way of urban roads**

| Design speed/ km · h$^{-1}$ | 100 | 80 | 60 | | | 50 | | | 40 | | |
|---|---|---|---|---|---|---|---|---|---|---|---|
| Grade/% | 4 | 5 | 6 | 6.5 | 7 | 6 | 6.5 | 7 | 6.5 | 7 | 8 |
| Maximum grade length/m | 700 | 600 | 400 | 350 | 300 | 350 | 300 | 250 | 300 | 250 | 200 |

*Note*: This table refers to the **Code** (**2012**).

**Table 3 - 9 Maximum grade length for the non-motorized vehicle lane of urban roads**

| Grade/% | | 3.5 | 3.0 | 2.5 |
|---|---|---|---|---|
| Maximum grade length/m | Bicycle | 150 | 200 | 300 |
| | Tricycle | — | 100 | 150 |

*Note*: This table refers to the **Code** (**2012**).

**5) Transition grade**

When trucks are on an upgrade, the speed decreases gradually. The larger and the longer the upgrade, the more the speed decreases. The purpose of setting the transition grade is to provide a profile environment that can accelerate trucks so that their running speeds may be restored above the allowable minimum speed. And then trucks may continue to operate at a speed not lower than the allowable minimum speed.

For continous upgrades on highways of all classes, the transition grade shall be set for grade lengths not greater than those specified in Table 3 - 7, according to the speed reduction of trucks. Application of the transition grade shall meet the following requirements:

① When the design speed is less than or equal to 80 km/h, the rate of the transition grade shall not be greater than 3%; when the design speed is greater than 80km/h, the rate of transition grade shall not be greater than 2.5%.

② The length of transition grade shall be greater than that stipulated in Table 3 - 5.

**6) Average grade**

An average grade refers to the ratio of elevation difference to horizontal distance of a route within a certain length.

$$i_{\text{Average}}=\frac{H}{L} \tag{3-1}$$

where, $i_{Average}$—average grade;

$H$—elevation difference;

$L$—road length.

Vertical alignment design cannot warrant the quality of use even if it fully conforms to the provisions of the maximum grade, grade length limitation and transition grade. For mountainous road design, because of the large elevation difference, sometimes a stepped profile composed of a steep grade followed by a gentle transition grade is used. Although each single steep grade is not very long, and there is even a gentle transition grade, the average grade is still large. As a result, low gear is used for a long time on upgrade, resulting in the boil of the water tank; and the brake is overheating on downgrade, resulting in traffic accidents. Therefore, it is necessary to limit the average grade, which can not only ensure that the average grade of the route is not too steep, but also avoid large average grades used in certain local segments.

In the **Specification (2017)**, it stipulates: for a ridge crossing route, when the relative elevation difference of continuous upgrade or downgrade is 200 - 500 m, the average grade shall not be greater than 5.5%; when the relative elevation difference is more than 500 m, the average grade shall not be greater than 5%; and the average grade of any 3 km continuous section should not be greater than 5.5%.

Taking account of the weak performance of some trucks and the low continuous braking capacity on downgrade, the average grade and length of continuous long and steep downgrades of freeways and Class Ⅰ highways should not exceed the provisions in Table 3 - 10. Otherwise, it is necessary to perform a traffic safety evaluation, to put forward a speed control and traffic management scheme, to improve traffic engineering and safety facilities, and to implement the truck compulsory parking area at suitable locations.

**Table 3 - 10 Average grade and length of continuous long and steep downgrades**

| Average grade/% | <2.5 | 2.5 | 3.0 | 3.5 | 4.0 | 4.5 | 5.0 | 5.5 | 6.0 |
|---|---|---|---|---|---|---|---|---|---|
| Continuous grade length/km | Unlimited | 20.0 | 14.8 | 9.3 | 6.8 | 5.4 | 4.4 | 3.8 | 3.4 |
| Relative elevation difference/m | Unlimited | 500 | 450 | 330 | 270 | 240 | 220 | 210 | 200 |

*Note*: This table refers to the **Specification (2017)**.

### 7) Resultant grade

A resultant grade is obtained by synthesizing the grade in the vertical profile and the superelevated slope in horizontal alignment where there is superelevation. Since the resultant grade is a vector synthesis, the direction conforms to the parallelogram rule, as shown in Figure 3 - 2. It is calculated by the following formula:

$$i_{\text{Resultant}} = \sqrt{i_{\text{Vertical}}^2 + i_h^2} \tag{3-2}$$

where, $i_{\text{Resultant}}$—resultant grade;

$i_{\text{Vertical}}$—grade in vertical profile;

$i_h$—superelevated cross slope.

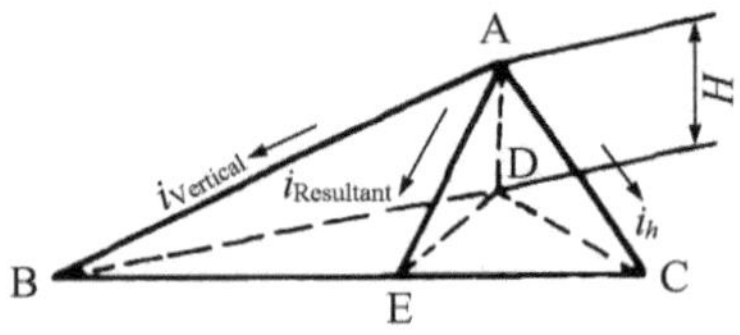


**Figure 3 - 2  Resultant grade**

In order to prevent the adverse effect of the combination of the steep slope and sharp turn on vehicle operation, the resultant grade shall be controlled. The limit of the maximum resultant grade is generally calculated according to the lateral and longitudinal force analysis, and then determined according to the highway class and terrain. The maximum resultant grade specified in the **Specification (2017)** is shown in Table 3 - 11.

**Table 3 - 11  Maximum resultant grade of highways**

| Highway class | Freeways, Class Ⅰ highways | | | | Class Ⅱ, Ⅲ, Ⅳ highways | | | | |
|---|---|---|---|---|---|---|---|---|---|
| Design speed/km • h$^{-1}$ | 120 | 100 | 80 | 60 | 80 | 60 | 40 | 30 | 20 |
| Resultant grade/% | 10.0 | 10.0 | 10.5 | 10.5 | 9.0 | 9.5 | 10.0 | 10.0 | 10.0 |

*Note*: This table refers to the **Specification (2017)**.

When a steep grade coincides with a small radius circular curve, a smaller resultant grade should be adopted. The resultant grade shall be less than 8% under the following conditions:

(1) Areas with snow and ice on roads in winter;

(2) Near mountainous road sections with a steep natural slope;

(3) Road sections with large non-vehicle traffic volume.

The resultant grade is related to pavement drainage. If the resultant grade is too small, drainage is not sound. Water film on the road makes vehicles slip and the splash caused by vehicles in front affects the visibility, increasing the probability of accidents. Therefore, the minimum resultant grade of highways of all classes shall not be less than 0.5%. When the resultant grade is less than 0.5%, comprehensive drainage measures shall be taken to ensure smooth drainage of pavements.

**8) Climbing lane**

(1) Definition of the climbing lane

The climbing lane is an exclusive lane provided for heavy trucks placed on the right side of the traveled way in the steep upgrades.

(2) Justification of the climbing lane

In road sections with large grades, when the proportion of trucks is high, the speed of trucks will decrease and the speed difference between trucks and passenger cars will increase. It increases the passing frequency that is unfavorable to safety. Moreover, mixed traffic with large speed difference also influences the freedom of passenger cars, leading to the reduction of traffic capacity.

In order to eliminate the above adverse effects, it is advisable to add a climbing lane in the steep grade and separate the trucks from traffic flows on the main line, which can improve the speed of passenger cars, ensure driving safety, and increase the traffic capacity of the section.

(3) Condition for addition of the climbing lane

The addition of the climbing lane is a supplementary measure as the length of steep grades is limited, that is, when the speed and capacity of steep grades are not able to meet the normal requirements, it is necessary to consider setting up the climbing lane.

Vehicle operational freedom on highways is not only restricted by traffic volume, but also by the obstruction caused by the speed reduction of heavy trucks on the long and steep upgrades, which is especially prominent in the two-lane highways. The speed of passenger cars on the upgrade does not change much, but trucks will decelerate due to the lack of climbing ability. As a result, the speed difference between the two kinds of vehicles in the traveled way increases, and so do the passing demands. The chances of dangerous passing actions increase which endanger driving safety. Due to the existence of passing lanes on multi-lane highways, trucks may not produce a great impact on vehicle operation unless the traffic volume and proportion of heavy vehicles reach a certain degree.

For continuous upgrades of the four-lane freeway, four-lane Class Ⅰ highway and Class Ⅱ highway, when the grade has a great impact on the operating speed of trucks, capacity and safety, it is necessary to check the operating speed reduction of trucks and the design capacity. A climbing lane should be provided on the right side of the traveled way in the upgrade when one of the following conditions is met:

① When the operating speed of trucks on the upgrade is reduced below the allowable minimum speed specified in Table 3 - 12, the climbing lane should be set.

② When the grade length exceeds the values stipulated in Table 3 - 7 or the design capacity of the upgrade is less than the design hourly traffic volume, the climbing lane should be set.

③ After the technical and economic comparison and demonstration between the setting climbing lane and improving grades of the main line without the climbing lane, the benefit-cost ratio and safety of setting climbing lane are better.

**Table 3 - 12 Allowable minimum speed on the upgrade**

| Design speed/km · $h^{-1}$ | 120 | 100 | 80 | 60 | 40 |
|---|---|---|---|---|---|
| Allowable minimum speed/km · $h^{-1}$ | 60 | 55 | 50 | 40 | 25 |

*Note*: This table refers to the **Specification (2017)**.

To ensure highway capacity, the climbing lane shall be set when the length of the upgrade exceeds the maximum grade length to eliminate the influence of trucks on traffic flows. However, it is not a cost-effective design if the proportion of heavy vehicles is not considered. Although most drivers do not bear the congestion of the whole journey, they can tolerate queuing in some road sections, especially in the mountainous terrain. Therefore, on highways with relatively small traffic volume, even if the grade length exceeds the maximum one, traffic congestion is occasional and the climbing lane is not mandatory.

(4) Method for the addition of the climbing lane

① Cross-section layout of the climbing lane

A climbing lane is located on the right side of the travel way in the upgrade direction, with a width of 3.5 m, which includes the marginal strip width of 0.5 m, as shown in Figure 3 - 3.

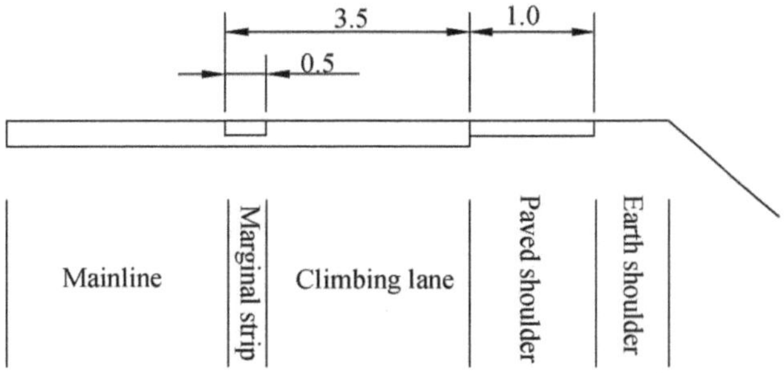


**Figure 3 - 3 Cross section composition of the climbing lane (m)**

Because the running speed of the climbing lane is lower than that of the main line, the superelevation rate of the climbing lane is less than that of the main line, and the superelevated slope shall conform to the provisions in Table 3 - 13. The rotation axis of the superelevated slope shall be the inside edge line of the climbing lane.

**Table 3 - 13 Superelevation rates of the main line and the climbing lane**

| Superelevation rate of the main line/% | 10 | 9 | 8 | 7 | 6 | 5 | 4 | 3 | 2 |
|---|---|---|---|---|---|---|---|---|---|
| Superelevation rate of the climbing lane/% | 5 | | 4 | | | | | 3 | 2 |

*Note*: This table refers to the **Specification (2017)**.

The curve widening value of the climbing lane shall refer to the provisions of curve widening of one-lane.

If the length of the climbing lane of the freeway and Class I highway is more than 500 m, an emergency parking belt shall be set on the right side according to the provisions.

② Horizontal layout of the climbing lane

The horizontal layout of the climbing lane is shown in Figure 3 - 4. The total length of the climbing lane consists of a tapered section for traffic divergence, the climbing lane, and a tapered section for traffic convergence.

The start of the climbing lane shall be located at the place where the operating speed of trucks on the steep grade is reduced to the "allowable minimum speed". The end of the climbing lane shall be located at the place where

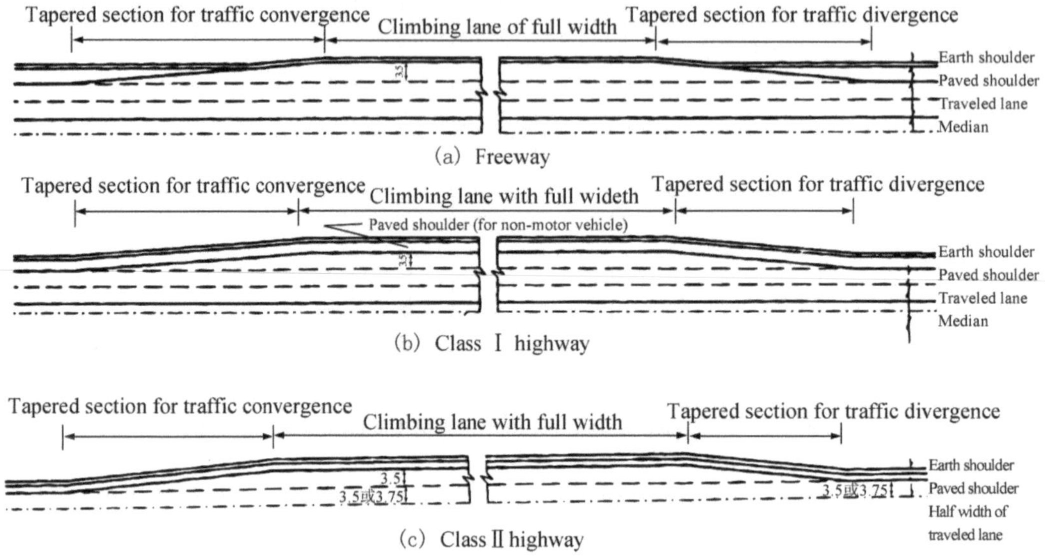


**Figure 3 - 4 Horizontal layout of the climbing lane (m)**

the operating speed of trucks restores to the "allowable minimum speed" after climbing the steep grade, or at the end of the additional length extended after the steep grade, as shown in Figure 3 - 5. The additional length extended after the steep grade is specified in Table 3 - 14.

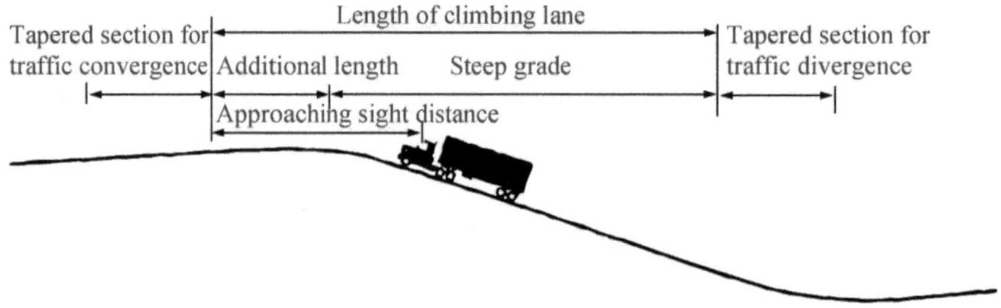


**Figure 3 - 5 Typical climbing lane**

**Table 3 - 14 Additional length extended after the steep grade**

| Grade of additional length/% | Downhill | Flat | Uphill | | | |
|---|---|---|---|---|---|---|
| | | | 0.5 | 1.0 | 1.5 | 2.0 |
| Additional length/m | 100 | 150 | 200 | 250 | 300 | 350 |

*Note*: This table refers to the **Specification (2017)**.

The relationship between the climbing lane and alignment design shall be comprehensively considered when designing the climbing lane. The start and end points of the climbing lane shall be set at spots with good visibility, clear identification, and smooth transition. When two adjacent climbing lanes are close, they should be connected directly. Tapered sections for traffic divergence and convergence shall be set at the start and end points of the climbing lane, and the lengths are specified in Table 3 - 15.

**Table 3 - 15 Length of tapered sections for traffic divergence and convergence**

| Highway class | Length of the tapered section for traffic divergence/m | Length of the tapered section for traffic convergence/m |
|---|---|---|
| Freeway, Class Ⅰ highway | 100 | 150～200 |
| Class Ⅱ highway | 50 | 90 |

*Note*: This table refers to the **Specification** (**2017**).

**9) Emergency escape ramp**

For continuous long and steep downgrades, the emergency escape ramp shall be set in combination with traffic safety evaluation and demonstration. The emergency escape ramp shall be located on the right side of long and steep downgrades with a good visual sight, the width of which shall not be less than 4.5 m. If possible, the refuge lane shall be set parallel to the right side of the emergency escape ramp. Figure 3 - 6 is the schematic diagram of the emergency escape ramp.

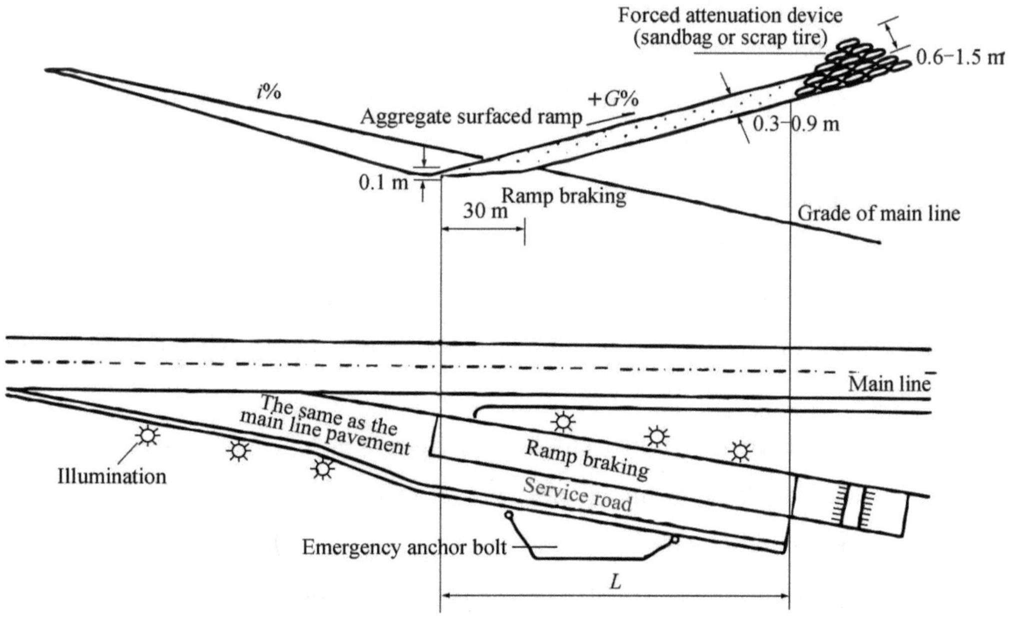


**Figure 3 - 6 Schematic diagram of the emergency escape ramp**

The emergency escape ramp is a kind of fault-tolerant engineering measure, and its function shall be understood correctly. Although the addition of an emergency escape ramp may mitigate and reduce the severity and loss of accidents to a certain extent when the vehicles are out of control, it could not abate the probability of accidents caused by brake failure and out of control of vehicles in essence. Therefore, for continuous long and steep downgrades, traffic safety evaluation is necessary. While improving the traffic engineering and road-side safety facilities, it is important to focus on the implementation of effective traffic organization, speed control, and other measures to solve the traffic safety issues of akin road sections.

**10) Other provisions of grades**

In addition to the above provisions of grades, for some special areas, grade design has special provisions.

The grade of the bridge and its approach shall comply with the following provisions:

(1) The grade of the small bridge shall be designed to comply with the grade of routes.

(2) The grade of large and medium bridges should not be greater than 4%, the grade of the bridge approach should not be greater than 5%, and the alignment of the approach near the bridgehead shall match that of the bridge.

(3) The grade of the bridge that is prone to ice and snow should be appropriately reduced.

(4) The grade of the bridge and its approach near the bridgehead shall not be greater than 3% when there are heavily mixed urban traffic.

The grades of the tunnel and its opening shall comply with the following provisions:

(1) The grade in the tunnel shall be greater than 0.3% and less than 3%, but tunnels shorter than 100 m are not subject to this limit.

(2) For the middle and short tunnels of the freeway and Class Ⅰ highway, when the conditions are constrained, the maximum grade may increase appropriately after the technical and economic demonstrations, but it should not be greater than 4%.

(3) A one-way grade may be used in the tunnel, and a herringbone grade

may be used for tunnels with developed groundwater or long and extra-long tunnels.

For sections with a large proportion of non-vehicle traffic near cities and towns, the grade may reduce appropriately according to specific conditions.

### 3.2.2 Requirement of grade design

Application of the rate of grades shall meet the following requirements:

(1) During vertical alignment design, the terrain and other conditions along the route shall be fully considered, and a gentle grade should be adopted. The minimum grade should not be less than 0.3%. For sections with the flat cross slope or grade less than 0.3%, special drainage design shall be carried out.

(2) The maximum grade and the maximum length corresponding to different grades should not be adopted for highways of all classes. They can be used only when it is necessary to use favorable terrain or detour arduous sections.

Grade design shall meet the following requirements:

(1) The grade of the level terrain shall be uniform and gentle.

(2) The grade of the rolling terrain shall not undulate too much due to excessive accommodation to the terrain.

(3) The grade of the ridge crossing route shall be as uniform as possible. The grade with or close to the maximum rate should not be adopted, and it is not suitable to continuously adopt profiles composed of steep slopes with the maximum grade and transitional grade with short length.

(4) Gentle grades shall be adopted under possible conditions for the routes on the ridge and hillside. Larger grades can only be used to take advantage of favorable terrain.

## 3.3 Vertical alignment design

### 3.3.1 Concepts related to vertical alignment design

A point where the grade changes on the profile is called the vertical points of intersection (grade change point in Chinese), as shown in Figure 3 - 7 (points $A$ and $B$). The algebraic difference between the two adjacent grades on the profile is expressed by $\omega$, namely the algebraic difference in grades (grade change angle in Chinese). $i_1$ and $i_2$ are the two grades, with the positive indicating upgrade and the negative indicating downgrade; thus $\omega_1 = i_2 - i_1$, $\omega_2 = i_3 - i_2$. When a vehicle passes the grade change point, a vertical curve shall be inserted in order to mitigate the impact caused by the change of kinetic energy of the vehicle and to ensure the visual sight. The form of the vertical curve may be a circular curve or parabola. If $\omega$ is "$-$", the grade change point is above the vertical curve, which is a crest vertical curve. When $\omega$ is "$+$", the grade change point is below the vertical curve, which is a sag vertical curve.

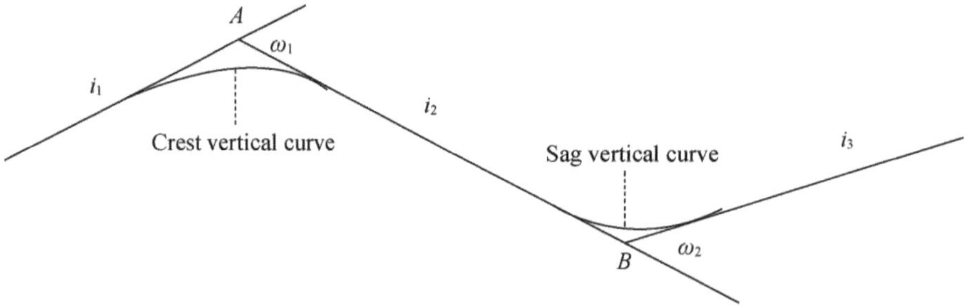


**Figure 3-7 Schematic diagram of the vertical point of the intersection, grade change angle and vertical curve**

### 3.3.2 Design standard of the vertical curve

The design standard of the vertical curve includes the radius and length. Vertical curves are divided into the crest one and the sag one, which have different functions. Therefore, the design standards are different.

**1) Limit minimum radius of the crest vertical curve**

The limit minimum radius of the crest vertical curve shall mainly meet the requirements of three aspects: first, to mitigate vehicle impact; second, to ensure enough illumination distance of headlights for the sake of driving safety at night; third, to have enough sight distance under the overpass bridge. The minimum radii required to ensure these three requirements are derived as follows:

(1) Limit minimum radius required to mitigate vehicle impact

When a vehicle travels on a vertical curve, it is affected by centrifugal force which deviates from the center of circles. Such a force makes the vehicle under hypergravity on the sag vertical curve. When hypergravity reaches a certain degree, it makes passengers feel uncomfortable and poses adverse effects on the suspension system of the vehicle. Therefore, the centrifugal acceleration becomes a control factor in determining the radius of the vertical curve. According to experience, it is appropriate to limit the centrifugal acceleration below 0.28 m/s$^2$.

$$R_{min}=\frac{v^2}{a}=\frac{(V/3.6)^2}{a}=\frac{V^2}{12.96a}=\frac{V^2}{12.96\times 0.28}\approx\frac{V^2}{3.60}(\mathrm{m}) \tag{3-3}$$

where: $v$—design speed (m/s);

$V$—design speed (km/h).

(2) Limit minimum radius required to ensure enough illumination distance of headlights

In order to ensure driving safety at night, the illumination distance of headlights shall not be less than the sight distance. The limit minimum radius can then be deduced, assuming the height of headlights is $h$ and the illumination angle of the headlights is $\beta$, as shown in Figure 3-8.

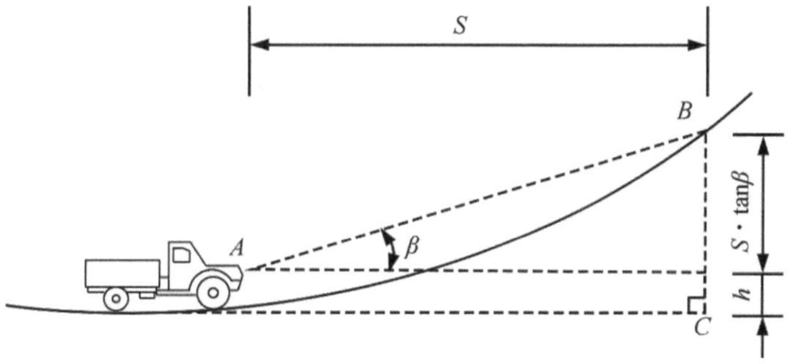


**Figure 3-8 Limit minimum radius required to ensure illumination distance of headlights**

According to the calculation formula of the vertical curve:

$$BC=\frac{S^2}{2R} \tag{3-4}$$

According to the figure:

$$BC=h+S\cdot\tan\beta \tag{3-5}$$

Then:

$$R=\frac{S^2}{2(h+S\cdot\tan\beta)} \tag{3-6}$$

where, $S$—illumination distance of headlights, m, generally taking stopping sight distance;

$h$—height of headlights, m, $h=0.75$ m;

$\beta$—illumination angle of headlights, $\beta=1.5°$.

Inputting $h$ and $\beta$ to formula (3-6):

$$R_{\min}=\frac{S^2}{1.5+0.0524S}(\mathrm{m}) \tag{3-7}$$

(3) Limit minimum radius required to ensure enough sight distance under the overpass bridge

As shown in Figure 3-9, if the radius of the vertical curve under the overpass bridge is too short, it is not able to meet the requirement of sight distance.

When $L<S$, it is obtained:

$$R_{\min}=\frac{2S}{\omega}-\frac{19.27}{\omega^2}\text{(derivation omitted)} \tag{3-8}$$

When $L\geqslant S$, it is obtained:

$$R_{\min}=\frac{S^2}{19.27}\text{(derivation omitted)} \tag{3-9}$$

$R_{\min}=\frac{S^2}{19.27}$ is used as the control.

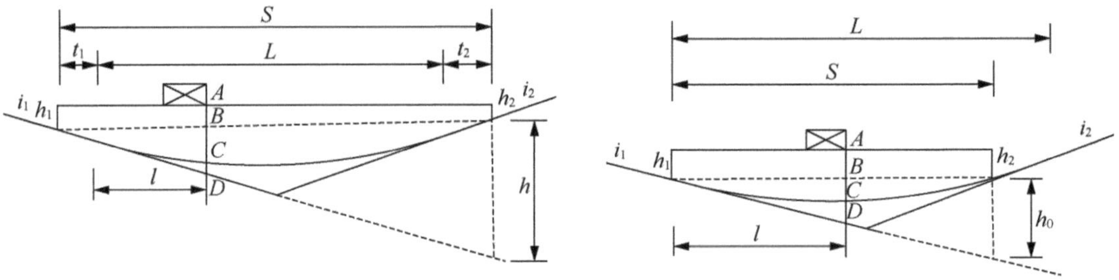


**Figure 3-9 Limit minimum radius required to ensure sight distance under the overpass bridge**

Synthesizing the previous derivations of (1), (2) and (3), the calculation of the limit minimum radius of the crest vertical curve is shown in Table 3-16.

**Table 3-16 Calculation of the limit minimum radius of the crest vertical curve**

| Design speed/ km·h$^{-1}$ | Stopping sight distance/ m | Radius required to mitigate impact/m $\frac{V^2}{3.63}$ | Radius required to ensure illumination distance of headlights at night $R_{\min}=\frac{S^2}{1.5+0.0524S}$ | Radius required to ensure sight distance under the overpass bridge $R_{\min}=\frac{S^2}{19.27}$ | Limit minimum radius used in **Specification (2017)**/m |
|---|---|---|---|---|---|
| 120 | 210 | 4,000 | 3,527 | 2,286 | 4,000 |
| 100 | 160 | 2,778 | 2,590 | 1,328 | 3,000 |
| 80 | 110 | 1,778 | 1,666 | 628 | 2,000 |

**Continued Table 3 - 16**

| Design speed/ km·h$^{-1}$ | Stopping sight distance/ m | Radius required to mitigate impact/m $\frac{V^2}{3.63}$ | Radius required to ensure illumination distance of headlights at night $R_{min}=\frac{S^2}{1.5+0.052,4S}$ | Radius required to ensure sight distance under the overpass bridge $R_{min}=\frac{S^2}{19.27}$ | Limit minimum radius used in **Specification** (**2017**)/m |
|---|---|---|---|---|---|
| 60 | 75 | 1,000 | 1,036 | 292 | 1,000 |
| 40 | 40 | 444 | 445 | 59 | 450 |
| 30 | 30 | 250 | 293 | 47 | 250 |
| 20 | 20 | 111 | 157 | 21 | 100 |

**2) Limit minimum radius of the sag vertical curve**

Determination of the limit minimum radius of the sag vertical curve is mainly to mitigate the vehicle impact and to ensure the vertical horizon. The derivation process is as follows:

(1) Limit minimum radius required to mitigate vehicle impact

When a vehicle travels on the sag vertical curve, centrifugal force produces hypogravity, which also makes passengers feel uncomfortable. According to the previous derivation, the limit minimum radius is:

$$R_{min}=\frac{V^2}{3.60} \tag{3-10}$$

(2) Limit minimum radius to meet sight distance requirements

When the radius of the sag vertical curve is small and the road surface is convex, it will directly affect the sight distance. The formula for calculating the limit minimum radius of the sag vertical curve can be deduced according to the prescribed sight distance. As shown in Figure 3 - 10, there are two scenarios:

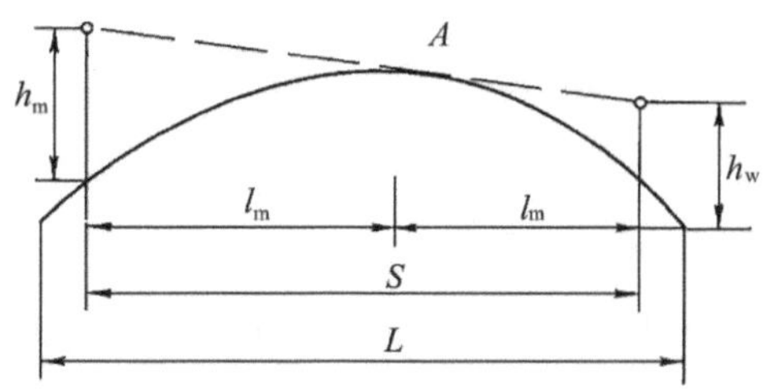


**Figure 3 - 10 Limit minimum radius of the sag vertical curve to meet sight distance requirements**

① $S \leqslant L$

$$h_w = \frac{l_w^2}{2R} \tag{3-11}$$

$$h_m = \frac{l_m^2}{2R} \tag{3-12}$$

With geometric relationship:

$$S = l_w + l_m \tag{3-13}$$

Substituting the above two formulas, we can obtain:

$$S = \sqrt{2R}(\sqrt{h_w} + \sqrt{h_m}) \tag{3-14}$$

where, $h_w$—object height, m, $h_w = 1.0$ m;

$h_m$—eye height, m, $h_m = 1.20$ m;

$l_w$—distance between the vertex of the vertical curve and the object point, m;

$l_m$—distance between the vertex of the vertical curve and the eye point. m;

$S$—required sight distance, m, taking stopping sight distance;

$L$—length of the vertical curve, m.

By substituting $h_w$ and $h_m$, it is obtained:

$$R_{min} = \frac{S^2}{3.98} \tag{3-15}$$

② $S > L$, according to the derivation (omitted), it is obtained:

$$R_{min} = \frac{2S}{\omega} - \frac{3.98}{\omega^2} \tag{3-16}$$

where, $S$—required sight distance, m, taking stopping sight distance;

$\omega$—grade change angle, °.

According to the derivation process, $R_{min} = \frac{S^2}{3.98}$ is taken as the control, and the calculation of the limit minimum radius of the sag vertical curve is shown in Table 3-17.

**3) General minimum radius of the vertical curve**

The limit minimum radius of the vertical curve is the bottom line necessary to ensure safety and comfort of driving, which can be used only under constraints. To increase safety and comfort, 1.5-2.0 times or greater of the limit minimum radius shall be used as the general value in the actual design.

Table 3 - 17 **Calculation of the limit minimum radius of the sag vertical curve**

| Design speed/ km • $h^{-1}$ | Radius required to mitigate vehicle impact/m $R_{min}=\frac{V^2}{3.63}$ | Radius required to meet sight distance requirements/m $R_{min}=\frac{S^2}{3.98}$ | Limit minimum radius used in the **Specification (2017)** |
|---|---|---|---|
| 120 | 4,000 | 11,080 | 11,000 |
| 100 | 2,778 | 6,432 | 6,500 |
| 80 | 1,778 | 3,040 | 3,000 |
| 60 | 1,000 | 1,413 | 1,400 |
| 40 | 444 | 402 | 450 |
| 30 | 250 | 226 | 250 |
| 20 | 111 | 101 | 100 |

### 4) Minimum length of the vertical curve

When the grade change angle is small, even with a large radius, the length of the vertical curve is short. Drivers may have the feeling that the grade changes quickly, and also, a short vertical curve will have an impact on driving. Generally, the minimum length of the vertical curve is determined as the distance traveled in 3 s at the design speed.

The radius and length of the vertical curve stipulated in the **Specification (2017)** are shown in Table 3 - 18.

Table 3 - 18 **Minimum radius and length of the vertical curve**

| Design speed/km • $h^{-1}$ | | 120 | 100 | 80 | 60 | 40 | 30 | 20 |
|---|---|---|---|---|---|---|---|---|
| Radius of the sag vertical curve/m | General value | 17,000 | 10,000 | 4,500 | 2,000 | 700 | 400 | 200 |
| | Limit value | 11,000 | 6,500 | 3,000 | 1,400 | 450 | 250 | 100 |
| Radius of the crest vertical curve/m | General value | 6,000 | 4,500 | 3,000 | 1,500 | 700 | 400 | 200 |
| | Limit value | 4,000 | 3,000 | 2,000 | 1,000 | 450 | 250 | 100 |
| Length of the vertical curve/m | General value | 250 | 210 | 170 | 120 | 90 | 60 | 50 |
| | Limit value | 100 | 85 | 70 | 50 | 35 | 25 | 20 |

*Notes*: (1) The table refers to the **Specification (2017)**.

(2) the "general value" in the table is the one used in normal conditions, and the "limit value" is the one used after technical and economic demonstrations when the condition is limited.

### 3.3.3 General requirements for vertical alignment design

Whether the vertical alignment is smooth or not becomes the main factor that affects the appearance of vertical alignment visually. Poor vertical alignment often blames for too many vertical curves and small radii or short lengths of vertical curves. Vertical curve design shall meet the following general requirements:

(1) When the design speed is greater than or equal to 60 km/h, the vertical alignment design should adopt the combination of long vertical curves and long tangent grades. If possible, the radius of the vertical curve greater than or equal to the visual requirements listed in Table 3 - 19 should be adopted.

**Table 3 - 19 Minimum radius of the vertical curve for visualization**

| Design speed/km · $h^{-1}$ | Radius of the vertical curve/m | |
|---|---|---|
| | Crest | Sag |
| 120 | 20,000 | 12,000 |
| 100 | 16,000 | 10,000 |
| 80 | 12,000 | 8,000 |
| 60 | 9,000 | 6,000 |

*Note*: The table refers the **Specification (2017)**.

(2) For vertical alignment design, an appropriate radius shall be determined first. Under the condition of not excessively increasing the workload, a larger radius of the vertical curve should be selected, especially when the grade change angle is small. When the terrain is constrained, the radius larger than or close to the "general value" of the vertical curve shall be adopted, and the "limit value" can only be adopted when the terrain conditions are special and very difficult.

(3) Between co-directional vertical curves, especially between co-direction sag vertical curves when the length of tangent grades approximates to or reaches the minimum value, the vertical curves shall be merged and replaced by a single vertical curve or compound curve so as to avoid the broken-back curve.

(4) Between adjacent reverse vertical curves, it is better to insert a tangent grade section in the middle to facilitate the transition between hypergravity and hypogravity. If the radii of the two vertical curves approximate to the limit value, the length of the tangent grade section shall be at least 3s traveled distance. When the radius is large, they can also be connected directly.

(5) The demand of passing sight distance shall be taken into account for the two-lane highway in sections with passing demand. A larger radius of the crest vertical curve may be adopted or necessary signs, markings, and other facilities may be provided.

### 3.3.4 Design and calculation of the vertical alignment

The design and calculation of the vertical curve include the selection of radius, calculation of geometric elements, calculation of design elevation of each stake, and dotted plot vertical curves.

**1) Selection of radius**

(1) The selected radius shall meet the requirements in the **Standard (2014)** and **Specification (2017)**;

(2) In order to make driving comfortable, a larger radius shall be adopted as far as possible without excessively increasing the quantity of earthwork.

(3) Sometimes it is necessary to integrate the requirements of elevation control to determine the appropriate external distance, and then to calculate the radius according to the determined external distance.

(4) Sometimes it is necessary to consider the connection of adjacent vertical curves and the radius shall be calculated according to the tangent length.

(5) A very large radius of the vertical curve may bring adverse effects on drainage, and attention shall be paid to the selection of radius.

(6) For sections with large traffic volume at night, considering that the illumination distance of headlights is limited, the radius shall increase appropriately to offer longer illumination distance.

**2) Calculation of geometric elements (Figure 3 - 11)**

Grade change angle

$$\omega = i_1 - i_2 \tag{3-17}$$

Curve length

$$L = R\omega \tag{3-18}$$

Tangent length

$$T = \frac{L}{2} \tag{3-19}$$

External distance

$$E=\frac{T^2}{2R}=\frac{R\omega^2}{8}=\frac{T\omega}{4} \tag{3-20}$$

Start stake of the vertical curve = Stake of the grade change point − $T$ (3 - 21)

End stake of the vertical curve = Stake of the grade change point + $T$ (3 - 22)

Elevation of the start stake = Elevation of the grade change point $\pm T \cdot i_1$ ("+" for the sag vertical curve and "−" for the crest vertical curve) (3 - 23)

Elevation of the end stake = Elevation of the grade change point $\pm T \cdot i_2$ ("+" for the sag vertical curve and "−" for the crest vertical curve) (3 - 24)

Correction value of elevation at any point

$$y=\frac{x^2}{2R} \tag{3-25}$$

$x$ at any point = stake at any point − start stake of the vertical curve (3 - 26)

where, $\omega$—grade change angle;

$L$—length of the vertical curve;

$R$—curvature radius of the parabola vertex;

$T$—tangent length;

$E$—external distance of the vertical curve;

$y$—elevation correction value at any point;

$x$—distance from any point to the start point of the vertical curve.

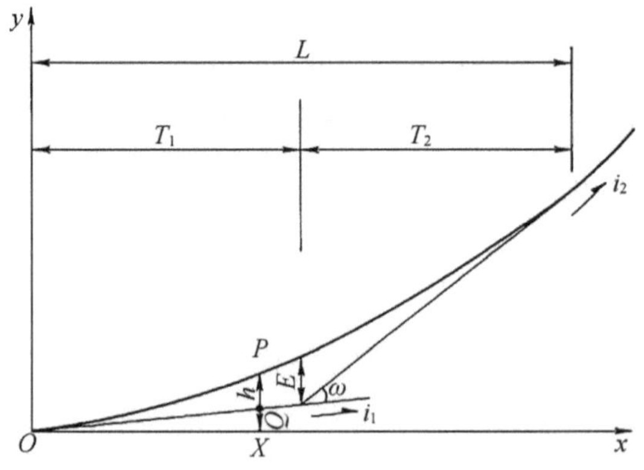


**Figure 3 - 11 Calculation of Geometric Elements of Vertical Curve**

**3) Design elevation calculation at any point**

Tangent grade elevation = elevation of the grade change point $\pm(T-x)i$ ("+" for the sag vertical curve and "−" for the crest vertical curve) (3 - 27)

Design elevation = tangent grade elevation $\pm y$ ("+" for the sag vertical curve and "−" for the crest vertical curve) (3 - 28)

**4) Dotted plot of vertical diagram**

The next step is to plot the calculated design elevation onto the profile diagram, and connect all the design elevation points to obtain the design vertical curve.

**Example 1:** Rates of two adjacent grades of a Class Ⅱ highway are $i_1=+3\%$ and $i_2=-2\%$, as shown in Figure 3 - 12. The stake at the grade change point (or VPI) is K4+480 with an elevation of 213.45 m. If the radius of the vertical curve is 2000 m, please calculate the design elevation of the vertical curve.

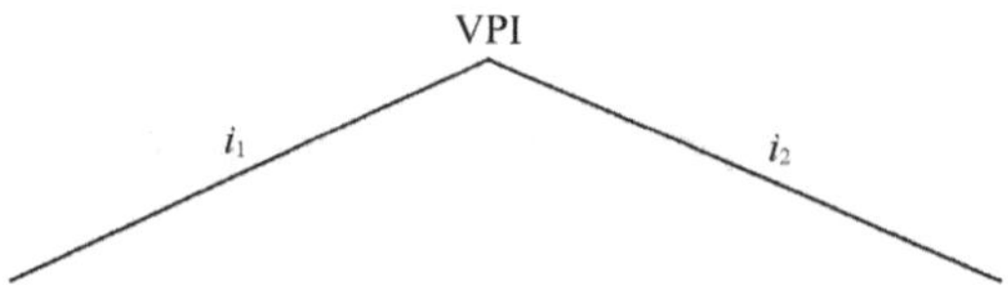


**Figure 3 - 12 Example of vertical curve calculation**

**Solution:**

Grade change angle $\omega=i_1-i_2=3\%-(-2\%)=5\%$

Curve length $L=R\omega=2{,}000\times0.05=100(\text{m})$

Tangent length $T=L/2=100/2=50(\text{m})$

External distance $E=\frac{T^2}{2R}=\frac{50^2}{2\times2{,}000}=0.625(\text{m})$

Start stake of the vertical curve = K4+480−50 = K4+430

End stake of the vertical curve = K4+480+50 = K4+530

Elevation of start stake $=213.45-50\times3\%=211.95(\text{m})$

Elevation of end stake $=213.45-50\times2\%=212.45(\text{m})$

The design elevation of each stake is calculated at an interval of 10 m, and the change of elevation on both sides of the vertical curve is symmetrical. The results are shown in Table 3 - 20.

**Table 3 - 20 Calculation results of vertical curve**

| Stake | Tangent grade elevation/m | Offset distance $x$/m | Elevation correction/m | Design elevation/m |
|---|---|---|---|---|
| K4+430 | 211.950 | 0 | 0 | 211.95 |
| K4+440 | 212.250 | 10 | 0.025 | 212.225 |
| K4+450 | 212.550 | 20 | 0.100 | 212.450 |
| K4+460 | 212.850 | 30 | 0.225 | 212.625 |
| K4+470 | 213.150 | 40 | 0.400 | 212.75 |
| K4+480 | 213.450 | 50 | 0.625 | 212.825 |
| K4+490 | 213.250 | 40 | 0.400 | 212.850 |
| K4+500 | 213.050 | 30 | 0.225 | 212.825 |
| K4+510 | 212.850 | 20 | 0.100 | 212.750 |
| K4+520 | 212.650 | 10 | 0.025 | 212.625 |
| K4+530 | 212.450 | 0 | 0 | 212.450 |

## 3.4 Combination of horizontal and vertical alignments

After route selection and alignment determination, road alignment design takes place in the horizontal, vertical, and cross-section planes in sequence. However, the final impact on driving behavior not only originates from the individual influence of the horizontal alignment, vertical profile and cross section, but also from the spatial alignment and surrounding environment. Even if the design of horizontal and vertical alignments is carried out following the **Standard** (**2014**) and **Specification** (**2017**), an ideal spatial alignment needed for vehicle operation may not be obtained if the combination of horizontal and vertical alignments is not good. Therefore, the combination of horizontal and vertical alignments is of great significance and shall be paid attention to. For horizontal alignment design, it shall consider the vertical alignment; similarly, for vertical alignment design, it shall coordinate with the horizontal alignment. Especially when the design speed is greater than or equal to 60 km/h, attention shall be paid to the reasonable combination of horizontal and vertical alignments, as well as the visual and psychological requirements of drivers.

### 3.4.1 Principles of the combination of horizontal and vertical alignments

(1) For the alignment combination design, each technical index shall not only comply with the specified value of horizontal and vertical alignments, but also consider the influence of the cross section on the combination and driving safety. Combinations of limit values of the horizontal alignment, vertical alignment and cross section shall be avoided.

(2) When determining the respective technical indexes of horizontal and vertical alignments, in addition to their own relative balance and continuity, the balance and continuity of the technical indexes of adjacent sections shall also be considered.

(3) In addition to maintaining the internal relative balance and the coordination of change rhythm among various elements, attention shall also be paid to the coordination with the natural landscape and geological conditions along the highway.

(4) Alignment shall naturally induce the driver's line of sight and maintain its continuity.

### 3.4.2 Types of combination of horizontal and vertical alignments

The horizontal alignment is composed of the tangent and horizontal curve; the vertical alignment is composed of the tangent grade and vertical curve, which could be divided into the crest vertical curve and sag vertical curve. Therefore, horizontal and vertical alignments have the following combinations, as shown in Table 3 - 21.

(1) Combination of the tangent and tangent grade;

(2) Combination of the tangent and sag vertical curve;

(3) Combination of the tangent and crest vertical curve;

(4) Combination of the horizontal curve and tangent grade;

(5) Combination of the horizontal curve and sag vertical curve;

(6) Combination of the horizontal curve and crest vertical curve.

**Table 3 - 21 Types of spatial combinations**

| No. | Horizontal alignment element | Vertical alignment element | Stereoscopic alignment |
|---|---|---|---|
| 1 | Tangent | Tangent grade | Tangent with constant grade |
| 2 | Tangent | Vertical curve | Tangent with sag vertical curve |
| 3 | Tangent | Vertical curve | Tangent with crest vertical curve |
| 4 | Horizontal curve | Tangent grade | Horizontal curve with constant grade |
| 5 | Horizontal curve | Vertical curve | Horizontal curve with sag vertical curve |
| 6 | Horizontal curve | Vertical curve | Horizontal curve with crest vertical curve |

For combination (1), it is a tangent on the plane and a tangent grade on the profile, which is the simplest combination. However, such a combination is monotonous. If the surrounding landscape lacks change, drivers will feel tired or pass frequently, resulting in driving accidents. In the design, the monotonous vision can be improved through roadside facilities or greening.

For combination (2), due to the poor sight distance, it shall be avoided as far as possible. A large radius of the vertical curve shall be used when it is unavoidable.

For combination (3), it not only offers a good sight distance condition, but also has a sag vertical curve so that the spatial alignment is no longer monotonous. The monotonous alignment of combination (1) can be improved by such combination (2).

For combination (4), as long as the radius of the horizontal curve is proper and the grade is not too steep, a sound visual and psychological environment can be obtained. The combination shall check whether the resultant grade exceeds the limit.

Combinations (5) and (6) are more common and complex forms of horizontal and vertical alignment combinations. For such combinations, if the position is appropriate and the index is balanced, a graceful spatial alignment can be obtained. On the contrary, there may be some adverse consequences, which shall be alerted in the design.

### 3.4.3 Requirements for the combination

(1) Horizontal and vertical alignments should correspond to each other, and the horizontal curve should be longer than the vertical curve. When the radii of horizontal and vertical curves are small, the degree of correspondence shall be stricter; with the increase in the radii of horizontal and vertical curves simultaneously, the degree of correspondence may be relaxed appropriately; when the radii of horizontal and vertical curves are large, they are not required to correspond to each other strictly.

As shown in Figure 3 - 13 and Figure 3 - 14, an ideal combination of horizontal and vertical alignments is that the vertices of horizontal and vertical curves coincide, the phases match, and the horizontal curve is slightly longer than the vertical curve. Because the obtained stereoscopic alignment has the same three-dimensional curvature, the alignment is continuous, not distorted, in space. If the problem of mutual coordination is not well considered in the design of horizontal and vertical alignments, and the route is not treated as a three-dimensional alignment, it is difficult to realize the one-to-one correspondence of horizontal and vertical curves. Naturally, sometimes horizontal and vertical

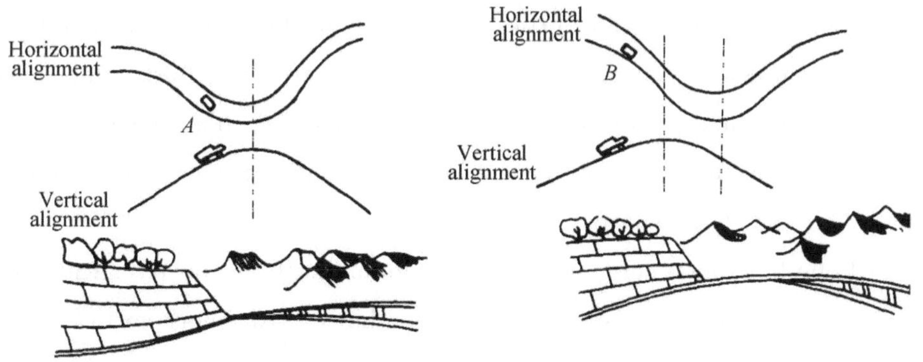


(a) Coincidence of horizontal and vertical curves (b) Offset of horizontal and vertical curves

**Figure 3 - 13 Coincidence of vertices of horizontal and vertical curves**

alignments do not follow the "one-to-one correspondence" principle, nor does it damage the quality of stereoscopic alignments. Perspective inspection and practice show that when the grade difference is small or the radius of the vertical curve is very large, the combination of horizontal and vertical curves is insensitive to stereoscopic alignments.

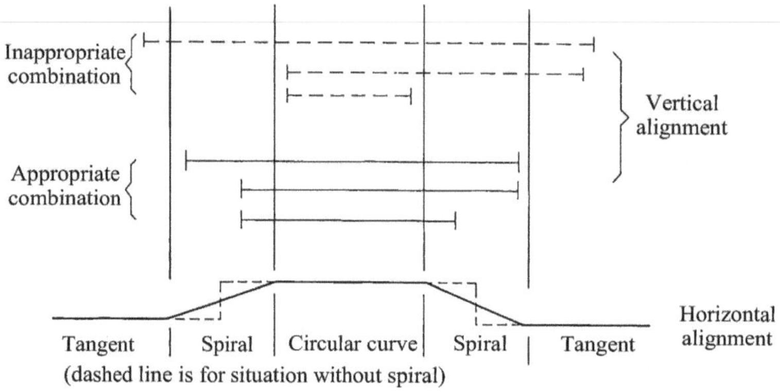


**Figure 3 - 14 Position of horizontal and vertical curves**

Balance between the radii of horizontal and vertical curves is the basic requirement to ensure the coordination, smoothness, and continuity of stereoscopic alignments. When the profile fluctuates repeatedly, the grade length is not long, and the radius of the vertical curve is not large, only by using the horizontal curve with a large radius does not warrant a coordinated, continuous, and smooth alignment, which may lead to the increase of engineering cost. According to experience, if the radius of the horizontal curve is not greater than 1000 m and the radius of the vertical curve is 10 - 20 times the former, alignment balance may be harvested.

(1) When horizontal and vertical curves are difficult to correspond, they should be separated by an appropriate distance so that the horizontal curve is located on a tangent grade or the vertical curve is located on a tangent.

(2) Long tangent should not be combined with steep grades or vertical curves of small radii and short lengths.

(3) Long horizontal curves should not contain multiple short vertical curves, and short horizontal curves should not be combined with short vertical curves. Figure 3 - 15 shows the improper combination of horizontal curves and multiple vertical curves.

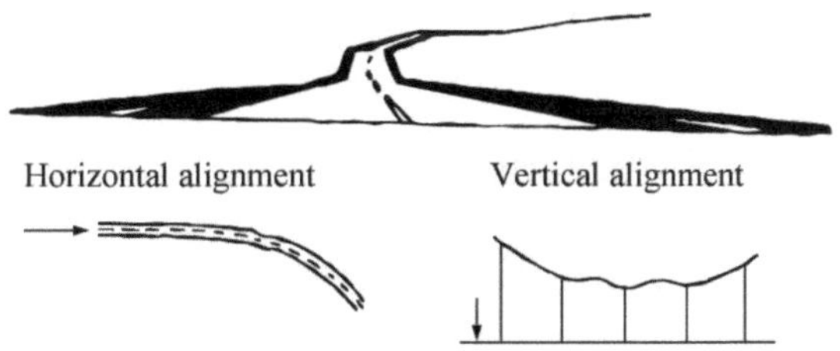


**Figure 3 - 15 Improper combination of horizontal curves and multiple vertical curves**

(4) A circular curve with a small radius should not be close to or located at the top of the sag vertical curve or at the bottom of the crest vertical curve. The former combination exhibits a discontinuous alignment and loses the guidance of the line of sight; in addition to the visual distortion, the latter combination also produces an unsafe alignment of sharp turns at the end of the downgrade.

(5) Horizontal curves with a small radius should not be set in long vertical curves.

(6) The top of the sag vertical curve or the bottom of the crest vertical curve should not coincide with the inflection point of the reverse horizontal curve.

Such combinations have different degrees of distorted appearance. The former combination may not correctly induce the line of sight, which will make the driver misoperate; the latter combination suffers from poor drainage, which affects driving safety.

(7) When the left-turning circular curve in the compound curve and S-shaped curve is not superelevated, it shall be checked by operating speed.

(8) The combinations of small radius circular curves at the end of long downgrade, long tangent, or large radius circular curve shall be avoided.

## 3.5 Design method of the vertical alignment

The main content of vertical alignment design is to determine the appropriate elevation of the route, the rate and length of the tangent grade, and the vertical curve according to the road class, natural conditions along the route and the control elevation of structures. The basic requirements are that the grade is uniform and smooth, the fluctuation is gentle, the length of the grade and vertical curve is proper, the combination of horizontal and vertical alignments is coordinated, and the fill and cut is economical and balanced. Although these requirements are considered in the route selection and alignment location, they shall be realized in vertical alignment design.

## 3.5.1 Requirements of the vertical alignment design

(1) Vertical alignment shall be smooth and visually continuous, adapt to the terrain and coordinate with the surrounding environment.

(2) Grade design shall consider the balance of fill and cut, and deploy the cut volume to the nearest fill section to reduce the impact on the cross slope of natural ground and environment.

(3) A large radius of vertical curve shall be used when the algebraic difference of adjacent grades is small.

(4) For sections with continuous long and steep grades, the upgrade shall meet the requirements of traffic capacity, and the downgrade shall consider driving safety. Combined with the technical indicators of the front and rear sections, traffic capacity and driving safety of the long and steep grade shall be evaluated by operating speed.

(5) The route grades before and after the intersection shall be gentle.

(6) For roads located in snowy or frozen areas, steep grades shall be avoided.

## 3.5.2 Methods and Steps of Vertical Alignment Design

The most important deliverable of vertical alignment design is the profile diagram. The method and step of vertical alignment design are the method and step of completing the profile diagram, which mainly involves tangent grades and vertical curves.

**1) Preparation**

Prior to vertical alignment design, the first step is to collect and be familiar with relevant data and understand the design intent and requirements; the next is to draw the dotted plot of the ground line and fill in relevant contents.

**2) Mark of control points**

Control points refer to the elevation control points that affect the grade design, which are divided into two categories. The first category is the compulsory control points of which grades of the route shall pass through, above, or below it. Such control points mainly include the start and end points of the route, the pass of the ridge crossing line, the minimum fill height of important bridges and culverts, geological adverse sections, the maximum cut depth, the design flood level along the stream, the entrance and exit of tunnels, at-grade intersections, grade separation, railway crossing, urban planning elevation and other factors restricting the elevation control points that the route shall pass. The second category is the reference control points, which are the economic points. In addition to the compulsory control points, the economic points are searched to obtain transverse balance of fill and cut on each cross section in order to reduce the project cost. There are three scenarios of economic points:

(1) When the ground slope is small, the elevation of fill and cut balance can be found above and below the elevation of the center stake. When the grade passes through such an elevation, the earthwork quantity of cut on the cross section is basically equal to that of fill, which is the economic point, as shown in Figure 3 - 16(a).

(2) when the ground slope is steeper, fill is often difficult to stay stable, and sometimes the slope toe extends far. The practice of more cut and less fill or even full cut of the subgrade is more economical than slope protection using stone masonry. Therefore, the elevation of the subgrade with more cut and less fill or full cut is the economic point, as shown in Figure 3 - 16(b).

(3) when the ground slope is very steep and may not be filled, a retaining wall shall be built. Therefore, it is preferred to cut the subgrade completely or deeply. The elevation of the full or deep cut subgrade is its economic point, as shown in Figure 3 - 16(c).

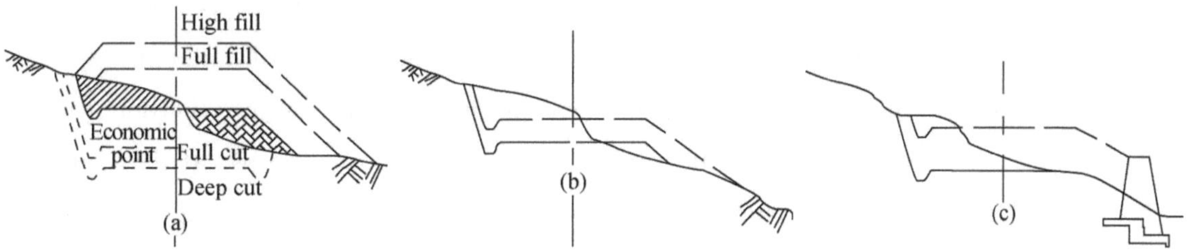


(a) Partial cut and part-fill (b) More cut and less fill (c) Full cut

**Figure 3 - 16 Economic points on cross section**

**3) Grading**

On the profile diagram marked with "control points" and "economic points", grading is performed to determine several tangent grade lines by interspersing between these points. The process shall be conducted according to the technical indicators and route selection intention in combination with the ground fluctuation. The principle of grading is to treat "control points" as control and accommodate most "economic points". Then these candidate grade lines are compared and updated repeatedly. Finally, the design grade lines which meet the requirements of technical standards, control points and earthwork saving are determined. Vertical points of the intersection are preliminarily determined by extending the grade lines.

**4) Adjustment**

Comparing the results of grading with the planned grades of route selection, the two shall be basically consistent. If they are substantially different, it is necessary to analyze them comprehensively, weigh the advantages and disadvantages, and decide the choice. The next step is to check whether the maximum grade, minimum grade, limit of grade length, etc. meet the requirements, whether the combination of horizontal and vertical alignments is appropriate, and whether the grades at the route intersection and approaches of bridges and tunnels are reasonable. If there is any problem, it shall be adjusted. The adjustment method is to raise, lower, extend, shorten or change the rate of the initial grade lines.

**5) Verification**

Key control cross sections, such as high fill and deep cut, the subgrade with steep ground slopes, retaining walls, important bridges and culverts, and other important control points, shall be selected to check whether the fill and cut is too large, whether the slope toe is unsupported or too far away, whether the retaining wall is too large, whether the bridge is too high or too low. If there are any problems, it is necessary to adjust the grade in time. It is more important to check sections with steep cross slopes.

**6) Determination of the grade**

After adjustment and verification, rates of the tangent grade and stakes and elevations of vertical points of intersections are determined by sections.

Generally, vertical points of intersections shall be adjusted to a stake rounded by 10 m, and the stake difference of adjacent vertical points of intersections is the grade length.

**7) Insertion of the vertical curve**

After the grade line and vertical point of intersections are determined, the vertical curve is applied at the vertical point of intersections. The followings are to determine the radius of the vertical curve, to calculate elements and design elevations of the vertical curve and to plot the dotted vertical curve.

## 3.5.3 Profile diagram

The profile diagram is one of the important technical documents of road design and the final deliverable of vertical alignment design. The diagram adopts rectangular coordinates, with the abscissa representing the stake and the ordinate representing the elevation. In order to clearly reflect the undulation of ground along the centerline, the abscissa scale is usually 1 : 2,000 (1 : 500 – 1 : 1,000 for urban roads) and the ordinate scale is 1 : 200 (1 : 50 – 1 : 100 for urban roads).

The profile diagram is composed of the upper and lower parts. The upper part is mainly used to draw the ground line and grade design line, and also to mark the vertical curve and its elements, rate and length of grades (sometimes marked in the lower part); the location, structure, number of holes and apertures of bridges and culverts and the artificial structure along the route; the stake and name of roads that cross other roads and railways; the name, stake, normal water level and maximum flood level of rivers crossed along the route; the position, number and elevation of leveling point, and so on. The lower part, generally from bottom to top, is mainly used to fill in relevant contents, the tangent and horizontal curve, stake, ground elevation, design elevation, height of fill and cut, description of soil and geology, bottom line, slope, distance, elevation, and flow direction (if required) of the design drainage ditch. Figure 3 – 17 shows the profile diagram.

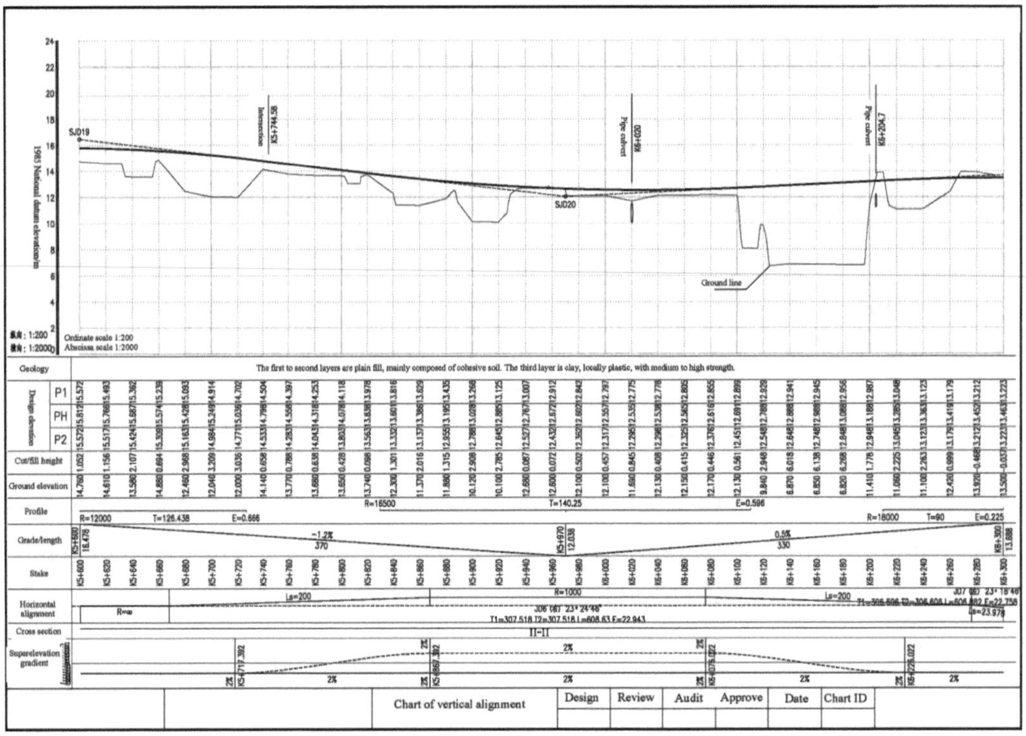

Figure 3 - 17 Profile diagram

**Questions:**

1. What aspects shall be comprehensively considered when designing the profile? How shall these requirements be addressed?
2. In the plateau area, what special difficulties will the vehicle encounter when driving? Why should reduce the maximum allowable grade?
3. Please discuss the general principles and steps of vertical alignment design.
4. What is the broken back curve? What kind of adverse phenomenon will be produced by the broken back curve in the horizontal and vertical alignments?
5. There is a vertical point of intersections (VPI) at the stake of K4+800 with a design elevation of 100 m. Before the VPI is an upgrade with a rate of 3% and after the VPI is a downgrade with a rate of 2%. A sag vertical curve with a radius of 4,000 m is inserted. Please calculate the elements, the start and end stakes of the vertical curve, and the design elevations at an interval of 20 m.
6. The design speed of a highway crossing a railway is 60 km/h, and at least one 20 m level road is required at both ends of the intersection. It is known that

the distance from one end of intersection $A$ to the turning point $O$ of the vertical curve is 30 m, and the grade difference is 0.02 (Figure 1). Please calculate the maximum radius of the vertical curve and check whether the calculated radius meets the requirements of the vertical curve radius.

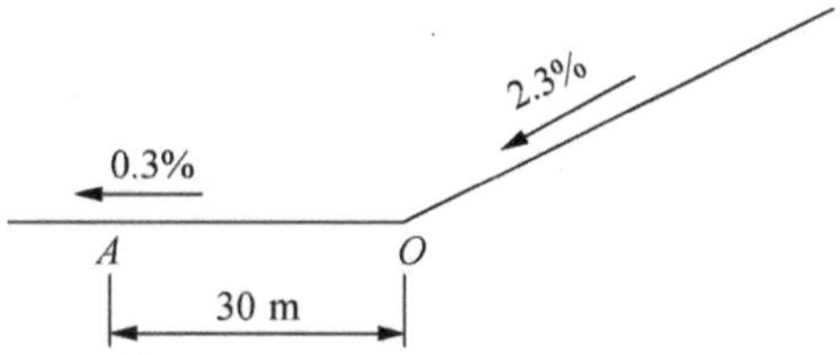


**Figure 1**

# Chapter 4 Cross Section Design

【**Objectives**: 1. Master the layout and form of cross sections of highways and urban roads; 2. Master the main components and corresponding parameters of cross sections; 3. Master the calculation of superelevation and widening; 4. Master the calculation and evaluation of sight distance on the curve; 5. Understand the cross-section diagram, subgrade design table, and master earthwork calculation and allocation.】

The alignment of road is a spatial line, which refers to the centerline of the road. In fact, a road is a strip structure with a certain width. The section cutting along the direction of road width and perpendicular to the road centerline is called the cross section.

In the course "Road Base and Pavement Engineering", the contents of the road base and pavement design will be discussed in detail. Therefore, this chapter mainly discusses the width and layout of the cross section, cross slope and road crown, cross section transition, subgrade design table, calculation and allocation of earthwork, etc.

## 4.1 Cross section layout and form

The normal section of any point on the centerline of roads is the cross section of roads, which consists of the design line and ground line. The design line includes the traveled way, non-motor vehicle lane, sidewalk, shoulder, separator, side ditch, side slope, intercepting ditch, berm and so on. The ground line is the one that represents the undulation of the original ground. The combination of the design line and ground line reflects the fill and cut of cross sections.

© The Author(s), under exclusive license to Springer Nature Singapore Pte Ltd. 2025
B. Yu and J. Cheng, *Road Survey and Design*,
https://doi.org/10.1007/978-981-97-4836-5_4

## 4.1.1 Layout and form of the highway cross section

**1) Layout of the Highway Cross Section**

Layout and dimensions of the highway cross section shall be determined according to the factors such as the highway function, design traffic volume, traffic composition, design speed, terrain. The primary task is to ensure the highway function, highway capacity, and safe and smooth operation. At the same time, it needs to take into account the factors of land saving and less investment, so as to maximize the economic and social benefits of highway engineering.

The freeway and Class Ⅰ highway usually separate the two-way vehicles, and the standard cross section of the subgrade (roadway) is classified into the integral type and divided type. The integral cross section separates the two-way vehicles with a separator, and the divided cross section places the two-way vehicles on different horizontal and vertical sections to form two "independent roads". The standard cross section of the integral subgrade is composed of the traveled way, medians (including medial separator and left marginal strip), shoulders (right paved shoulder and earth shoulder), and so on. The standard cross section of the divided subgrade is composed of the traveled way, shoulders (right paved shoulder, left paved shoulder, and earth shoulder) and so on. The standard cross section of the Class Ⅱ highway is composed of the traveled way, shoulders, and so on. When the proportion of trucks is high, a passing lane may be provided locally according to the need, and a slow-speed lane may also be provided by widening the paved shoulder as needed when there are many low-speed vehicles. The standard cross section of subgrades of Class Ⅲ highway and Class Ⅳ highways is composed of the traveled way, shoulders, and so on. When the Class Ⅳ highway adopts one-lane, a passing bay shall be set. The typical cross section of highways is shown in Figure 4 - 1.

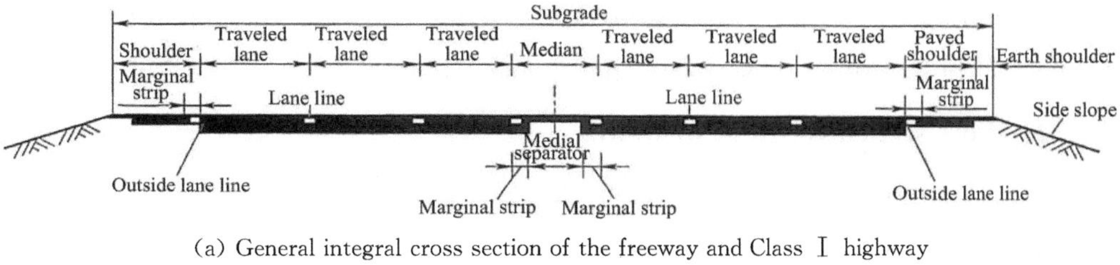


(a) General integral cross section of the freeway and Class Ⅰ highway

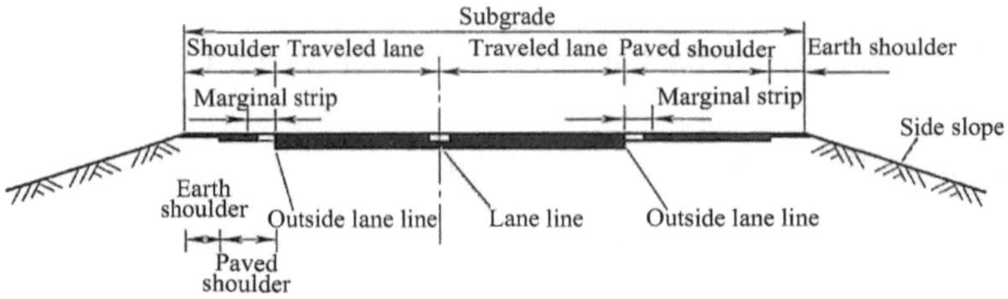


(b) General divided cross section of the freeway and Class Ⅰ highway (right-side section)

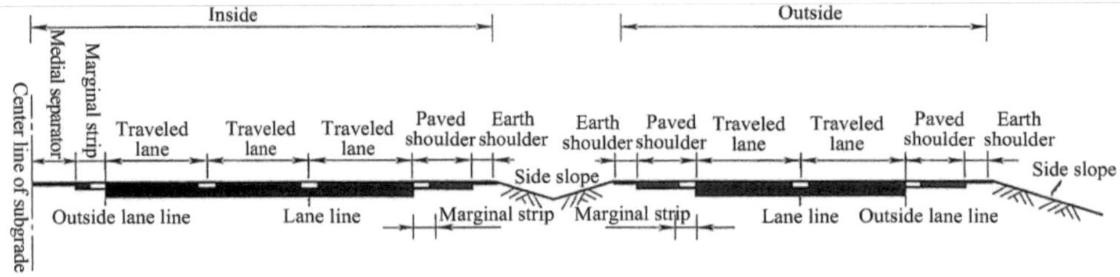


(c) Divided composite cross section of the freeway (right-side section)

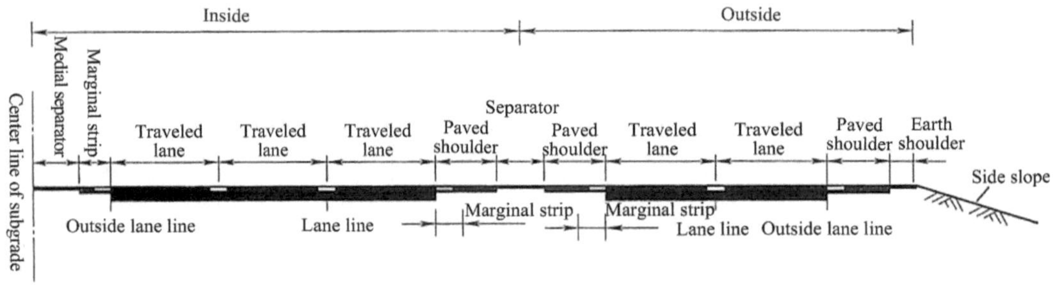


(d) Integral composite cross section of the freeway (right-side section)

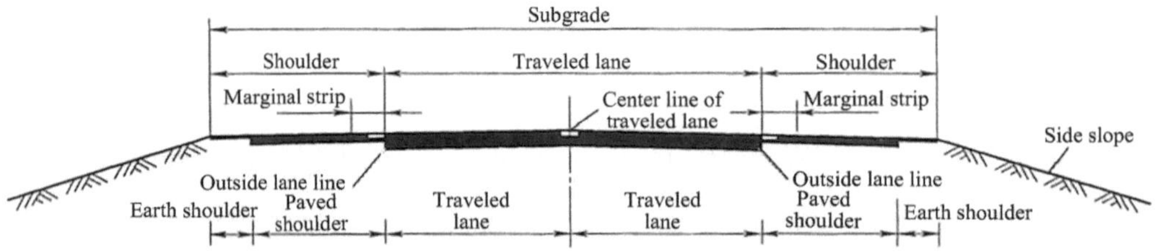


(e) General cross section of Class Ⅱ, Class Ⅲ and Class Ⅳ highways

**Figure 4 - 1 Typical compositions of the highway cross sections**

*Source*: **Specification (2017).**

The subgrade (roadway) width is the sum of width of the traveled way and shoulder. When there are the medial separator, marginal strip, speed-change lane, climbing lane, emergency parking strip, slow-speed lane or roadside facilities, the width of these parts shall be counted.

**2) Cross section forms of highways**

(1) One-roadway two-lane highway

The one-roadway two-lane highway refers to an integral dual-lane highway for two-way traffic, which is used for a large traffic range, up to 15,000 passenger car per day. The range of design speed is 20 – 80 km/h. For road sections where the mixed traffic interferes with each other, the side separator, non-motor vehicle lane and pedestrian lane may be applied so as to isolate non-motor vehicles and pedestrians from vehicles.

(2) Two-roadway multilane highway

The two-roadway multilane highway refers to a highway of four or more lanes that is either divided or with the separator. Some divided roads are designed into two independent one-way highways to take advantage of the terrain or scenic spots. This type of roads adapts to high speeds and large traffic capacity. Each lane can afford more traffic than a two-lane highway, the traffic is smooth, and the accident rate is low. It is suitable for the freeway and Class Ⅰ highway.

**Figure 4 – 2 Two independent one-way highways**

(3) Single – lane highway

The single – lane highway refers to a mountainous highway or local road with small traffic volume, in complex terrain, and of arduous engineering. It adopts a single lane with a passing bay. One lane is suitable for the Class Ⅳ highway with difficult terrain or very small traffic volume. The cost of such a highway is low, but the traffic volume is small and the speed is low. In order to meet the demands of giving way (one vehicle yields the right-of-way to another) and passing in the same direction, it is necessary to supply the passing lane.

## 4.1.2 Layout and form of the urban road cross section

**1) Layout of the urban road cross section**

The cross section of urban roads shall be reasonably designed within the range of the planned boundary line according to the road class, service function, traffic characteristics, and various control factors. At the same time, it shall meet the needs of long-term traffic. The short-term and long-term construction shall be combined to make the short-term project an integral part of the long-term project. Pipeline space shall be reserved, and the road land shall be planned to leave room for long-term construction. Roads in urban built-up areas should not be constructed in stages. The cross section of the reconstructed road shall be arranged by combining engineering measures with road traffic management.

The cross section of urban roads should be composed of the traveled way, non-motor vehicle lane, sidewalk, lane separator, facility belt, green belt, etc. The emergency lane, shoulder, drainage ditch, etc. can be used in special situations.

**2) Cross Section Form of Urban Roads**

Cross sections of urban roads can be divided into the single roadway, two-roadway, three-roadway, four-roadway and special forms of cross sections, as shown in Figure 4 - 3.

The principles of selecting cross sections for different classes of urban roads are as follows:

(1) When there are side roads on both sides of the expressway, a four-roadway cross section shall be adopted; when there are no side roads on both sides, a two-roadway cross section shall be adopted.

(2) The urban arterial road should adopt a three-roadway or four-roadway cross section, the urban minor arterial road should adopt a single roadway or two-roadway cross section, and the branch road should adopt the single roadway cross section.

(3) For the road with a bus lane, the layout of cross sections shall be combined with the location and type of bus lane, and priority shall be given to the bus lane.

(4) The same cross section should be adopted for the same road. When the road cross section changes, a transition section shall be set.

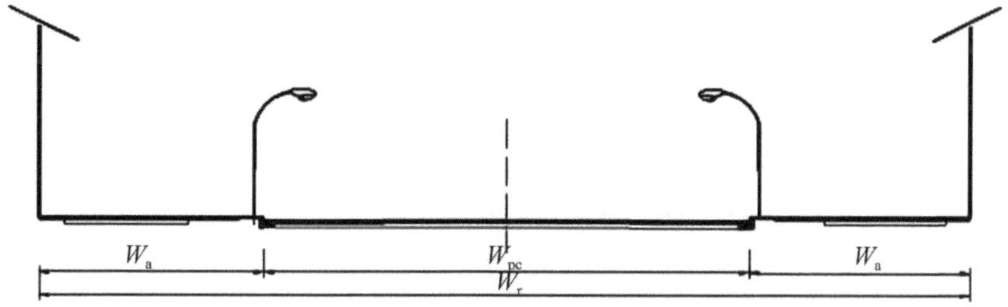


(a) Single roadway

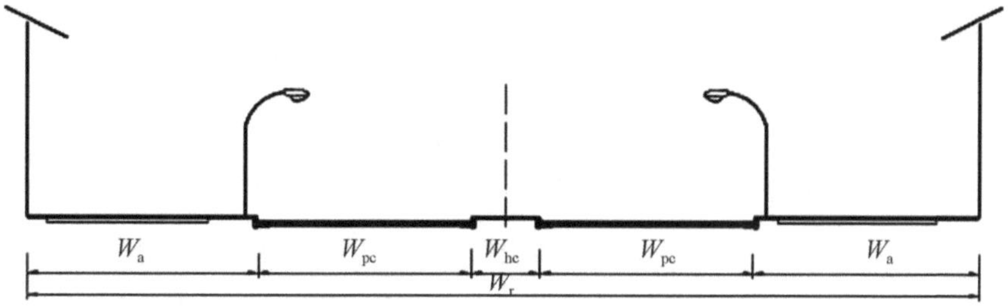


(b) Two-roadway

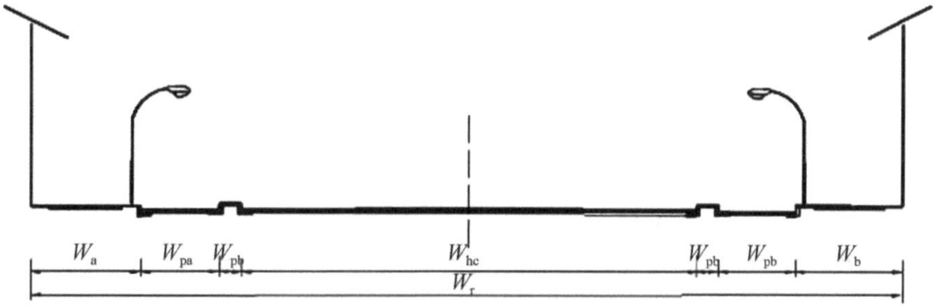


(c) Three-roadway

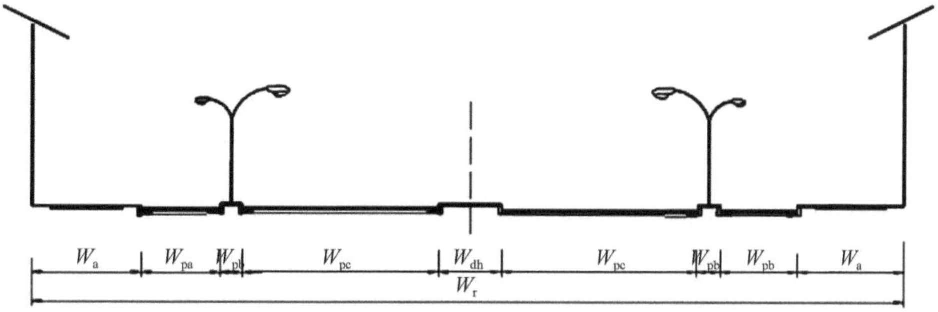


(d) Four-roadway

**Figure 4 - 3 Cross section forms of urban roads**

(5) The cross section and width of the traveled way and marginal strip of bridges and tunnels shall be the same as those of the road cross section.

(6) The separator width of super-large, large and medium bridges can be narrowed appropriately, but it shall meet the requirements of bridge protection facilities.

## 4.2 Width of cross section components

### 4.2.1 Width of the traveled way

#### 1) Lane width

The exclusive strip on the road for safe, smooth, and comfortable operation of the longitudinal vehicle fleet is called a (vehicular) lane; the width necessary for a lane is called the width of the (traveled) lane, viz. lane width.

Lane width mainly depends on the overall dimension and speed of vehicles, that is, when determining the lane width, the width of vehicles and the lateral swing amplitude of vehicles at different speeds shall be considered so that a certain distance between the wheel and road margin, and between two parallel vehicles, can be warranted for safe driving.

The lane widths in the **Specification (2017)** and **Code (2012)** are shown in Table 4-1 and Table 4-2.

**Table 4-1 Lane width of highways**

| Design speed/km · h$^{-1}$ | 120 | 100 | 80 | 60 | 40 | 30 | 20 |
|---|---|---|---|---|---|---|---|
| Lane width/m | 3.75 | 3.75 | 3.75 | 3.50 | 3.50 | 3.25 | 3.00 |

*Notes*: (1) When the inner lanes (1 and 2 lanes on the inner side) of an eight- or above eight-lane highway are limited to passenger cars, the lane width may be 3.50 m.

(2) For the highway mainly accommodating medium and small passenger vehicles with a design speed of 80 km/h and above, the lane width may be 3.50 m after demonstration.

(3) When the single lane is adopted for the Class Ⅳ highway, the lane width shall be 3.50 m.

(4) For the Class Ⅱ highway with a slow-speed lane, the width of the slow-speed lane shall be 3.50 m.

(5) For roads requiring non-motor vehicle lanes and sidewalks, the width of the non-motor vehicle lanes and sidewalks should be determined according to the actual situation.

**Table 4-2 Minimum lane width of urban roads**

| Vehicle type and lane type | Design speed/km · h$^{-1}$ | |
|---|---|---|
| | >60 | ≤60 |
| Oversize vehicle or mixed traffic lane | 3.75 | 3.50 |
| Passenger car exclusive lane | 3.50 | 3.25 |

**2) Width of traveled ways**

The portion of the roadway for the movement of vehicles, exclusive of shoulders, is the traveled way. And there are some consideration in vehicle lane design:

(1) Number of lanes in two directions should not exceed 4 - 6. In some urban arterial roads, there are multilanes in one direction. Vehicles in one lane will affect the traffic capacity of another lane due to passing, detouring, parking and other reasons. Generally speaking, the lane closer to the road centerline has less impact. Therefore, on the traveled way without a separation belt, the traffic capacity of the lane near the road centerline is the largest, the traffic capacity of the lane near the roadside is the smallest, and the impact is expressed by the reduction coefficient $a$. According to the observation, if the reduction coefficient of the first lane from the road centerline is 1. 00, then the reduction coefficients of the second, third, fourth and fifth lanes are 0. 80 - 0. 90, 0. 70 -0. 80, 0. 60 - 0. 70, and 0. 45 - 0. 60, respectively.

When the number of design lanes is more, the reduction coefficient of the lane near the roadside is smaller. For example, the capacity of the fourth and fifth lanes calculated from the road centerline is only half of that of the first lane after reduction. Therefore, the design of too many lanes has little effect on increasing road capacity; on the contrary, it will cause excessive traffic concentration and traffic chaos, and bring difficulties to traffic organization and management. If the requirements of traffic development are not satisfied by the design lane number, it shall be solved by improving the road network, building parallel roads, adjusting traffic organization, improving the urban layout, etc.

(2) The number of lanes is even and equal in two directions generally. Nevertheless, odd lanes may be used sometimes, such as: ① reversible lanes that change the road direction over time; ② tri-lane roads with small and mixed traffic, with one lane used for passing; ③ road diet which refers to a reduction in the lane number available on a road, usually by converting one or more existing lanes into turning lanes, bike lanes, or street parking.

(3) Traffic volume and conditions of each road section are inconsistent so that the width of the travel way shall not be forcibly uniform. However, for the same route, especially on the tangent, too many or abrupt changes are

unfavorable for driving.

### 4.2.2 Width of the non-motor vehicle lane

The non-motor vehicle roadway is a road for bicycles, tricycles, and other vehicles, with the bicycles dominating urban roads in China. Therefore, how to deal with bicycle passage is an important aspect of managing urban traffic, and it is also the fundamental solution to non-motor traffic.

The width of the non-motor vehicle lane shall mainly consider bicycles and cater for other non-motor vehicle traffic.

**1) Width of the bicycle lane**

The width of the bicycle lane is equal to the number of bicycle lanes multiplied by the width of one bicycle lane.

The clearance height for people riding on bicycles is 2.25 m, plus a safety clearance height of 0.25 m. Therefore, the clearance height required for the bicycle lane is 2.50 m. The handlebar width of bicycles is 0.50 m, plus the lateral swing safety clearance of 0.25 m on both sides, so the width of one bicycle lane is 1.0 m. And a safe distance of 0.25 m shall be left on both sides of the bicycle lane. Therefore, the width of a single bicycle lane is 1.5 m, the width of two bicycle lanes is 2.5 m, the width of three bicycle lanes is 3.5 m and so on. The standard width of bicycle lanes is shown in Figure 4-4.

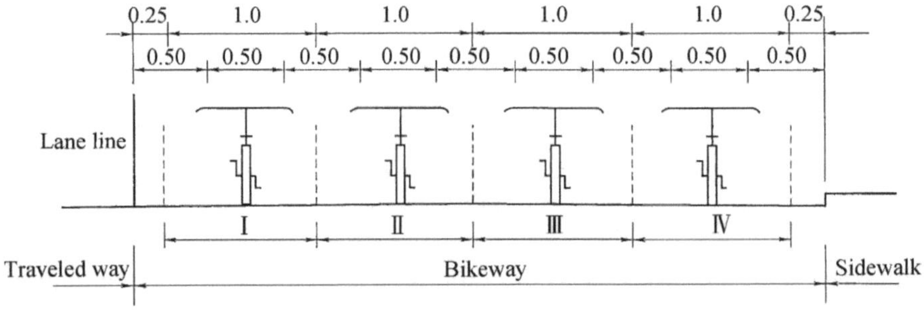


**Figure 4-4 Standard width of bicycle lanes (m)**

**2) Width of non-motor vehicle lanes**

Lane widths of various non-motor vehicles are shown in Table 4-3.

**Table 4 - 3 Width of non-motor vehicle lanes**

| Type of vehicle | Width of one lane/m |
|---|---|
| Bicycle | 1.0 |
| Tricycle | 2.0 |

For the non-motor vehicle lane combined with the traveled lane, the number of lanes in one direction shall not be less than two, and the width shall not be less than 2.50 m; the width of the exclusive non-motor vehicle lane shall include the lane width and the width of the marginal strip on both sides, and should not be less than 3.50 m in one way.

**3) Layout of non-motor vehicle lanes on the cross section**

Generally, the non-motor vehicle lane is symmetrically arranged between the traveled lane and sidewalk along both sides of urban roads. However, when the vehicle speed is high, or the traveled way is wide, or one of the traveled lanes and the non-motor vehicle lane reaches saturation, a separate non-motor vehicle lane may be set in order to avoid mutual interference between vehicles and bicycles. The layout is shown in Figure 4 - 5.

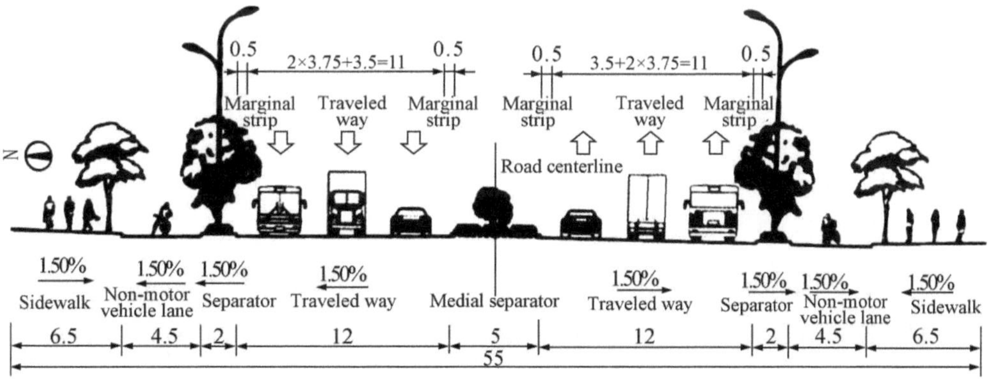


**Figure 4 - 5 Layout of the non-motor vehicles lanes (m)**

## 4.2.4 Width and layout of sidewalk

Urban traffic is mainly composed of vehicle traffic and pedestrian traffic. The sidewalk is an important part of urban roads. Whether the sidewalk design is reasonable or not will affect vehicle traffic on the traveled way. For example, if the sidewalk is not wide enough or the pavement is uneven, flooded, or muddy, it is inevitable that pedestrians will occupy the traveled way to walk, which will affect both driving and walking safety.

The primary function of the sidewalk is for pedestrian passage, followed by planting trees, erecting poles, arranging traffic signs, bulletin boards, etc. The space below it can also be used to bury underground pipes.

Sidewalk design mainly includes the width, cross slope, and layout of the sidewalk on the cross section.

**1) Sidewalk width**

According to the above-mentioned functions of the sidewalk, the total width of the sidewalk consists of the width of pedestrian walkway, the land occupation for planting vegetation, laying poles, setting bulletin boards, and the apron width of building foundation along the street. In addition, the width required to bury pipelines under the sidewalk shall be considered.

(1) Width required for pedestrian passage

Similar to the calculation of the lane width, the width of the sidewalk is equal to the width of a pedestrian strip multiplied by the number of pedestrian strips; the number of pedestrian strips depends on the peak hour passenger flow (person/h) and the capacity of a pedestrian strip (person/h). The calculation formula is as follows:

$$W_p = \frac{N_w}{N_{w1}} \tag{4-1}$$

where, $W_p$—sidewalk width(m);

$N_w$—peak hour passenger flow(person/h);

$N_{w1}$—design pedestrian capacity of 1 m wide sidewalk (person/h).

The required width of a pedestrian strip is related to the way pedestrians carry items, generally ranging from 0.60 - 0.90 m.

On general roads, the average width of a pedestrian strip is 0.75 m; on the roads near railway stations, ports, wharves, large shopping malls, and on the city-wide trunk roads, the average width of a pedestrian strip is 0.90 m considering the large number of pedestrians carrying goods.

(2) Width required for trees, poles and pipelines

For planting green belts, laying electric poles and columns, and setting bulletin boards on the sidewalk, their widths depend on the specific needs; the apron width of building foundation along the street is generally 0.50 m.

The overall sidewalk width is obtained by summarizing the sidewalk width in (1) and the additional widths for planting trees, erecting poles, arranging bulletin boards, and building a foundation apron.

When pipelines are buried under the sidewalk, the sidewalk width shall meet both needs of pedestrian traffic and buried pipelines. For example, the minimum width required for burying the four pipelines of power, telecommunication, cable and water supply is 4.50 m, and the minimum width of trees and lamp posts is 1.50 m. Then the sidewalk width on one side is at least 6.00 m (Figure 4 - 6). In cities with tight land use, it is often necessary to consider burying more pipelines under sidewalks of the same width.

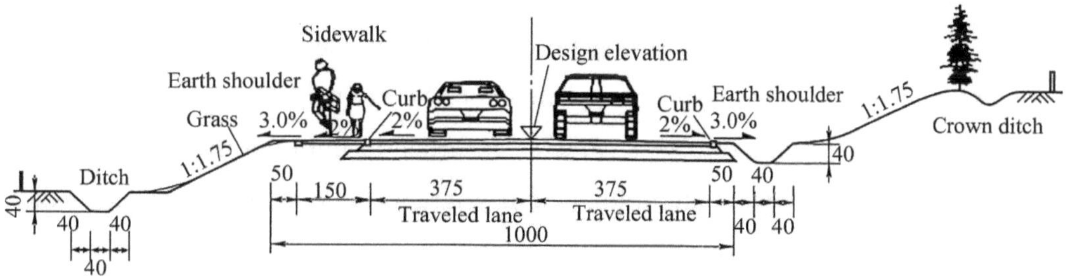


**Figure 4 - 6 Width of the green belt and pipelines on sidewalk (cm)**

(3) Minimum width of the sidewalk

In order to coordinate the width of each part of urban roads, it is generally considered the ratio of the total width of the urban road to the width of the one-side sidewalk is appropriate in the range of 5 : 1 - 7 : 1.

Table 4 - 4 and Table 4 - 5 list the occupied widths of different facilities when they are set independently and the minimum width of the sidewalk, respectively.

**Table 4 - 4 Width of different facilities**

| Item | Width/m |
|---|---|
| Pedestrian guardrail | 0.25 - 0.5 |
| Lamp post | 1.0 - 1.5 |
| Mailbox, garbage can | 0.6 - 1.0 |
| Bench, seat | 1 - 2 |
| Street tree | 1.2 - 1.5 |

Table 4 - 5 Minimum width of the sidewalk

| Object | Minimum width of the sidewalk/m | |
|---|---|---|
| | General value | Minimum value |
| Road of different classes | 3.0 | 2.0 |
| Section in commercial or public areas | 5.0 | 4.0 |
| Section near the railway station and wharf | 5.0 | 4.0 |
| Coach station | 4.0 | 3.0 |

**2) Cross slope of the sidewalk**

In order to ensure that pedestrians and vehicles do not interfere with each other, the sidewalk is generally 0.08 –0.20 m higher than the traveled way, often taking 0.15 m. The cross slope is generally unidirectionally inclined to the right side and depends on the masonry materials and the drainage requirements, which is generally in the range of 0.3%– 3.0%. In order to improve the drainage, the cross slope of the sidewalk should be slightly larger, and on the paved sidewalk, the cross slope is generally 2%.

**3) Layout of the sidewalk**

The sidewalk shall be separated from the traveled way, and the vehicle and its protruding part of the loaded goods shall not collide with pedestrians nearby. Efforts shall also be made to prevent pedestrians from crossing the street at non-specified places.

Most of sidewalks are symmetrically arranged on both sides of the traveled way. Under the restriction, sidewalks may be unequal in width on two sides or arranged only on one side. The one-side sidewalk is used on narrow roads near mountains or rivers.

The crosswise position of the sidewalk varies depending on the nature of buildings along the street and the width of the red line. For example, if the street is residential, the sidewalk should be 3 - 5 m away from the house. If there are shops along the street, it is advisable to set a sidewalk close to shops. Otherwise, two parallel sidewalks may be used. One is close to the shops for shopping and visiting and the other is close to the traveled way for crossing the street, and a green belt is sandwiched in the middle.

(1) Figure 4 - 7(a) shows a single row of roadside trees planted in tree holes

along the curb, which is suitable for narrow roads with shops and public cultural institutions on both sides.

(2) Figure 4 - 7(b) shows that the sidewalk and the traveled way are separated by a green belt, and the green belt is open at the crossing for the entrance and exit of pedestrian, which is beneficial to the separation of people and vehicles. The dust and muddy water caused by vehicles on the traveled way have little impact on pedestrians, which is suitable for sections with high crossing demands of pedestrian and high traffic density.

(3) Figure 4 - 7(c) shows that a green belt is arranged in front of the building. A revetment shall be built along the foot of the house wall to prevent house foundation from being affected by water. It is suitable for roads in residential areas.

(4) Figures 4 - 7(d) and (e) are two sidewalks to avoid interference between pedestrians of different natures. They are suitable for sections with many large malls and public cultural institutions in the city center.

(5) Figure 4 - 7(f) shows that the sidewalk is set under the arcade of high-rise buildings to help pedestrians avoid wind, rain, and sun. It is suitable for the expansion of old roads or in hot and rainy areas in southern China.

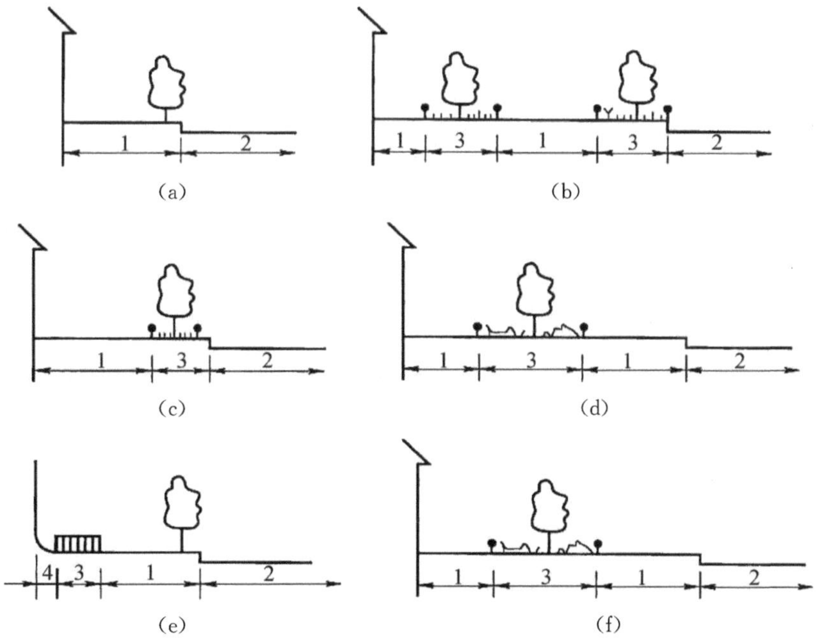

(1—sidewalk; 2—traveled way; 3—green belt; 4—apron; 5—arcade)

**Figure 4 - 7 different forms of the sidewalk layout**

### 4.2.5 Width of the shoulder

A shoulder is the proportion of the roadway contiguous with the traveled way that accommodates stopped vehicles, emergency use, and lateral support of the pavement structure. In suburban roads without sidewalks and high-class highways, shoulders shall be provided. Shoulders include paved shoulders (including marginal strips) and protective shoulders (known as earth shoulders), as shown in Figure 4 - 8. The left shoulder is applicable to the sections with a median between two or four roadways or a drainage ditch between two roadways.

**1) Shoulder function**

(1) Protecting and supporting the pavement structure. The width required to support the pavement structure is 0.50 m.

(2) Temporary parking in an emergency. The effective width of the paved shoulder shall not be less than 3.00 m for highways with heavy transportation burden and high operating speed and arterial highways bearing a large number of large trucks. However, shoulders with a width greater than 3.00 m may be misused as a traveled lane.

(3) Providing lateral width to ensure driving speed and capacity. Researches show that when the shoulder width is greater than 1.8 m, it may provide the sufficient lateral width for driving and ensure operating speed and capacity of the traveled way. In the cut section, the curve sight distance could be increased by shoulders to reduce traffic accidents.

(4) Being used as an emergency rescue channel. When the width of a continuous paved shoulder is greater than or equal to 2.50 m (2.00 m), it can meet the passage of large (small) rescue vehicles.

(5) Providing sites for road maintenance and burying underground pipelines.

(6) For roads without sidewalks, shoulders may be used by pedestrians and bicycles.

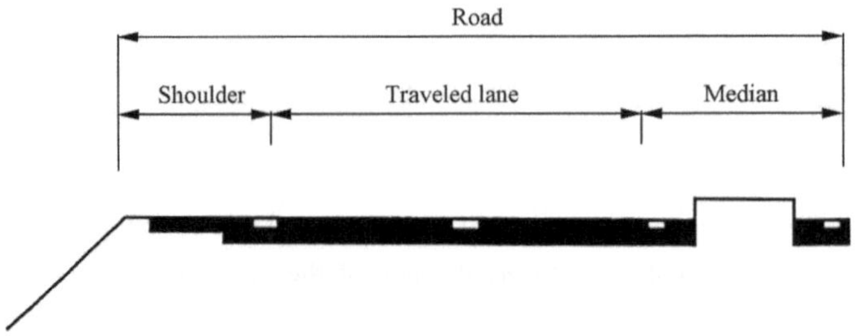


**Figure 4 - 8 Schematic diagram of shoulder**

**2) Shoulder width**

For the suburban ground expressway, the paved shoulder and earth shoulder should be set outside the traveled lane. The width of the paved shoulder shall not be less than 2.50 m, and that of the earth shoulder shall not be less than 0.75 m.

Paved shoulders shall generally have the ability to bear the vehicle load. Under normal weather conditions, it shall have sufficient strength and stability, without pumping, skid, mud, and dust. The structure and thickness of the marginal strip within the paved shoulder shall be the same as those of the pavement, and the other parts may be appropriately thinned.

When the paved shoulder is not used, the width of the earth shoulder shall not be less than 1.25 m.

The width of the protective shoulder shall meet the needs of installing guardrails, poles, traffic signs and planting trees, etc. The minimum width shall not be less than 0.5 m. If there is a bus bay, the protective shoulder shall be designed separately, which is generally the earth shoulder or block masonry.

According to the **Standard (2014)**, the right and left shoulder widths of highways are shown in Table 4-6 and Table 4-7.

**Table 4-6 Right shoulder width of highways**

| Highway class (function) | | Freeway | | | Class Ⅰ (arterial) | | Class Ⅰ (collector) and Class Ⅱ | | Class Ⅲ and Class Ⅳ | | |
|---|---|---|---|---|---|---|---|---|---|---|---|
| Design speed/km • h⁻¹ | | 120 | 100 | 80 | 100 | 80 | 80 | 60 | 40 | 30 | 20 |
| Paved shoulder/m | General value | 3.00 (2.50) | 3.00 (2.50) | 3.00 (2.50) | 3.00 (2.50) | 3.00 (2.50) | 1.50 | 0.75 | — | — | — |
| | Minimum value | 1.50 | 1.50 | 1.50 | 1.50 | 1.50 | 0.75 | 0.25 | | | |
| Earth shoulder/m | General value | 0.75 | 0.75 | 0.75 | 0.75 | 0.75 | 0.75 | 0.75 | 0.75 | 0.50 | 0.25 (two-lane) |
| | Minimum value | 0.75 | 0.75 | 0.75 | 0.75 | 0.75 | 0.50 | 0.50 | | | 0.50 (one-lane) |

*Notes*: (1) Under normal conditions, the "general value" shall be adopted; the "minimum value" can be adopted after demonstration for sections with the climbing lane, speed-change lane, and passing section, sections limited by terrain, ground features, and other conditions, and multilane highway bridges.

(2) When the freeway and Class Ⅰ highway as arterial are dominated by passenger cars, the width of right paved shoulder can adopt the value in brackets.

(3) For some local sections of the freeway with a design speed of 60 km/h, the width of the right paved shoulder shall not be less than 1.50 m.

**Table 4-7 Left shoulder width of divided cross sections of the freeway and Class Ⅰ highway**

| Design speed/km • $h^{-1}$ | 120 | 100 | 80 | 60 |
|---|---|---|---|---|
| Left paved shoulder/m | 1.25 | 1.00 | 0.75 | 0.75 |
| Left earth shoulder/m | 0.75 | 0.75 | 0.75 | 0.50 |

## 4.3 Lane separator

The lane separator is a strip-shaped component set within the road width for organizing and separating traffic that is free of vehicles. The lane separator occupies a certain width in the road cross section, which is different from the road marking, movable vehicle belt and pier. With the application of lane separators, the road is divided into two-, three- or four-roadway roads, while the road without lane separators is called the single roadway.

### 4.3.1 Application of the lane separator

Application of the lane separator mainly depends on traffic conditions. When the traffic volume on each lane is more than 500 pcu/h, the design speed is more than 50 km/h, the peak hour traffic volumes of vehicles and bicycles in either the same or the opposite direction at the same time are close, the road width is more than 35 m, and the two lanes arranged in the same direction may be used for passing, the lane separator could be used. For the situation of low speed, small traffic volume, and narrow road width, it is unnecessary to set the fixed lane separator but to adjust lane widths between the traveled lane and the non-motor vehicle lane. The benefit is to save land and construction investment and centralize street lamps for lighting. Movable vehicle separating piers may be applied to divide lanes and organize traffic temporarily for further reconstruction.

### 4.3.2 Function of the lane separator

The main functions of the lane separator are as follows:

(1) It isolates the scope of traffic flows, channelizes vehicle paths, prevents vehicles from interfering with each other, and hinders vehicles from turning halfway and crossing.

(2) It provides a refuge area for pedestrians crossing the street.

(3) Median separates the opposite traffic flows to prevent collision caused by vehicles entering the wrong way on the multilane road.

(4) Median width can be borrowed for the addition of left-turning lane at intersections, which is conducive to traffic organization.

(5) It provides space for road traffic signs, lighting poles, greening, etc. For the low and leveling median in suburban areas, it allows emergency vehicle intrusion to mitigate or prevent accidents if necessary.

## 4.3.3 Width of the lane separator

The lane separator is composed of separating strips and marginal strips on both sides. According to the functions and positions, the lane separator can be divided into the one in the middle (median) and the one on the side (side separator).

The width of the lane separator is composed of the upper and lower parts, which varies with the vehicle speed. Table 4 - 8 lists the minimum width of the lane separator of urban roads.

**Table 4 - 8 Lane separator width of urban roads**

| Type | | Median | | Side separator | |
|---|---|---|---|---|---|
| Design speed/km • $h^{-1}$ | | ≥60 | <60 | ≥60 | <60 |
| Marginal strip width/m | Traveled lane | 0.50 | 0.25 | 0.50 | 0.25 |
| | Non-vehicle lane | — | — | 0.25 | 0.25 |
| Safety strip width/m | Traveled lane | 0.50 | 0.25 | 0.25 | 0.25 |
| | Non-vehicle lane | — | — | 0.25 | 0.25 |
| Lateral clearance width/m | Traveled lane | 1.00 | 0.50 | 0.75 | 0.50 |
| | Non-vehicle lane | — | — | 0.50 | 0.50 |
| Minimum width of the separator/m | | 2.00 | 1.50 | 1.50 | 1.50 |
| Minimum width of the lane separator/m | | 3.00 | 2.00 | 2.50(2.00) | 2.00 |

*Notes*: (1) The lateral clearance width is the sum of the marginal strip width and safety strip width.

(2) For the side separator width, the values outside the bracket are taken when the traveled lanes are on both sides; the value in the bracket is taken when the traveled lane is on one side and the non-vehicle lane is on the other side.

(3) The minimum width of the separator is determined assuming the width of the facility strip is 1 m, and the specific width shall be determined according to the actual width of the facility strip.

The "Technical Standard of Highway Engineering" (2003 Edition) has made specific provisions on the width of the medial separator of the freeway and Class I highway, including the "general value" to be adopted for general conditions and the "minimum value" to be adopted for sections with limited conditions. When the "Technical Standard of Highway Engineering" was revised (2014 Edition), it was found through national investigation that there were great disputes over the width index of the medial separator, which reflected both the "general value" is wide and the "minimum value" is insufficient. Therefore, the recommended width of the medial separator is no longer specified, but it is emphasized that the width of the medial separator shall be determined from the main functions of opposite-traffic isolation and safety protection, and by comprehensively considering the protection form and protection capacity of the median guardrail.

## 4.4 Rate of the cross slope and superelevation

### 4.4.1 Rate and form of the cross slope

The rate of the sidewalk, traveled way, and green belt rising or falling within a unit transverse length of roads is called the cross slope $i$, as calculated by:

$$i = \mathrm{tg}\alpha = h/d \tag{4-2}$$

Where, $\alpha$—angle of the cross slope, °;

$h$—design elevation difference between the edge line and centerline, m;

$d$—horizontal distance from the edge line to the road centerline, m.

The rate of the cross slope is expressed in percent or a decimal value, such as 2.0% or 0.02.

In order to drain rainwater on the sidewalk, traveled way and green belt smoothly into the ditch or side ditch, they shall be designed with a certain cross slope. The rate of the cross slope is mainly determined by the pavement material and road class. The width of the sidewalk, traveled way, green belt, and the influence of local weather conditions shall also be considered.

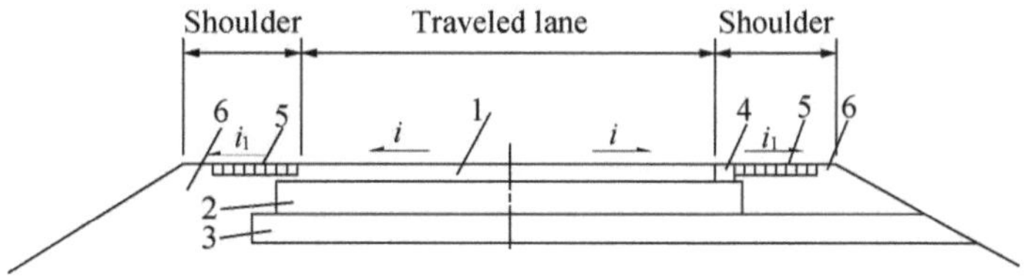


[1—pavement surface(including surface course, bottom course, and binding course); 2—base (including subbase); 3—cushion; 4—curb; 5—paved shoulder; 6—earth shoulde; $i$—cross slope of the traveled way; $i_1$—cross slope of the shoulder.]

**Figure 4 - 9 Diagram of the road cross slope**

If the cross slope of the green belt is too large, the soil at the root of plants will be washed away easily. The general rate is 0.5%–1.0%.

The cross slope of the sidewalk is usually inclined to the side stone in a straight line. In order to facilitate drainage and prevent pedestrians from slipping, the rate of the cross slope varies with pavement material and rainfall intensity, ranging from 1.5% to 3.0%.

Vehicles are prone to slipping on roads with water on surface; also, once the surface water penetrates into the subgrade, it will cause rapid damage to the road and greatly shorten its service life. In order to drain the surface water as soon as possible, the traveled way generally adopts a bidirectional cross slope, which has a crown or high point in the middle and a cross slope downward toward both edges. The crown height is the vertical drop from the center crown line to the edge of the traveled way.

The rougher the road surface and the smaller the grade, the slower the water flows on the road surface, which requires a larger cross slope. However, for the sake of driving safety, the road surface is required to be as flat as possible. Therefore, when designing the cross slope, this contradiction shall be reasonably resolved according to the principle of being beneficial to road drainage and ensuring driving stability.

The rates of the cross slope are shown in Table 4 - 9.

**Table 4 - 9 Rate of the cross slope for different pavements**

| Pavement surface | Cross slope/% | Pavement surface | Cross slope/% |
|---|---|---|---|
| Cement concrete pavement | 1.0 - 2.0 | Semi-even and uneven stone block pavement | 2.0 - 3.0 |
| Asphalt concrete pavement | 1.0 - 2.0 | | |
| Other black pavement | 1.5 - 2.5 | Aggregate pavement such as crushed stone, gravel | 2.5 - 3.5 |
| Stone block pavement | 1.5 - 2.5 | low-class pavement | 3.0 - 4.0 |

The cross slope of the protective shoulder (earth shoulder) on highways and suburban roads shall generally be 1.0%- 2.0% larger than the cross slope of the traveled way.

The forms of road crowns include the parabola, tangent connecting to the parabola, broken line and so on, which are adopted according to the width and type of pavement: the low-class highway adopts the parabolic crown, the high-class highway generally adopts the crown of tangent connecting to the parabola, and the multilane cement concrete pavement adopts the broken-line crown.

## 4.4.2 Road widening design

Offtracking is a characteristic, common to all vehicles, although much more pronounced in larger design vehicles, in which the rear wheels do not precisely follow the same path as the front wheels when the vehicle negotiates a horizontal curve or makes a turn. And the path radius gradually decreases from front wheels to rear wheels, forming the offtracking toward the inside. Large trucks and buses driving on highways and urban roads are very likely to cause traffic accidents if the impact of offtracking is not considered during the turning process. To ensure smooth and safe driving, the road width inside the curve shall be increased.

**1) Widening Calculation**

The geometric relationship of widening of a dual-axle vehicle is shown in Figure 4 - 10:

By geometric relationship

$$b=R-(R_1+B) \tag{4-3}$$

Since

$$R_1+B=\sqrt{R^2-A^2}=R-\frac{A^2}{2R}-\frac{A^4}{8R^3}-\cdots \tag{4-4}$$

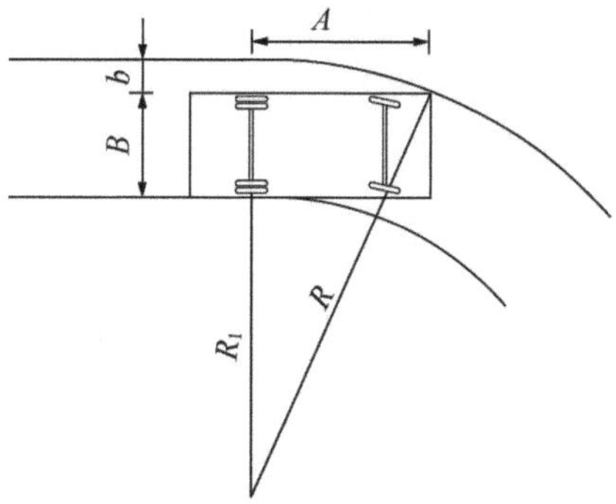


**Figure 4 - 10 Widening of a dual-axle vehicle**

Therefore

$$b=\frac{A^2}{2R}+\frac{A^4}{8R^3}+\cdots \tag{4-5}$$

The second term of the above formula can be ignored, so the widening of one lane is:

$$b=\frac{A^2}{2R} \tag{4-6}$$

where, $A$—distance from the rear axle to the front bumper, m;

$R$—radius of the circle curve, m.

When the number of lanes is $N$:

$$b_N=\frac{NA^2}{2R} \tag{4-7}$$

According to the field measurement, offtracking during turning is also related to vehicle speed. The extra width allowance is an additional radial width of pavements to allow for the difficulty of maneuvering on a curve and the variation in driver operation. This additional width is an empirical value that varies with the speed of traffic and the curve radius. The additional width allowance of one lane is expressed as:

$$b'=\frac{0.05v}{\sqrt{R}} \tag{4-8}$$

Therefore, considering the influence of vehicle speed, the road widening on the circular curve is calculated according to the following formula:

$$b_N=N\left(\frac{A^2}{2R}+\frac{0.05v}{\sqrt{R}}\right) \tag{4-9}$$

When the radius of the circular curve of Class Ⅱ, Class Ⅲ and Class Ⅳ highways is less than or equal to 250 m, these highways shall be widened. The pavement widening of a two-lane highway shall conform to Table 4 - 10, and the widening of the circular curve shall be determined according to the highway function, technical class and traffic composition.

**Table 4 - 10 Widening of a two-lane highway**

| Widening Type | Design vehicle | Radius of the circle curve/m | | | | | | | | |
|---|---|---|---|---|---|---|---|---|---|---|
| | | 200 - 250 | 150 - 200 | 100 - 150 | 70 - 100 | 50 - 70 | 30 - 50 | 25 - 30 | 20 - 25 | 15 - 20 |
| 1 | Passenger car | 0.4 | 0.5 | 0.6 | 0.7 | 0.9 | 1.3 | 1.5 | 1.8 | 2.2 |
| 2 | Truck | 0.6 | 0.7 | 0.9 | 1.2 | 1.5 | 2.0 | — | — | — |
| 3 | Articulated vehicle | 0.8 | 1.0 | 1.5 | 2.0 | 2.7 | — | — | — | — |

The arterial Class Ⅱ highway adopts the widening value of the third type. Collector Class Ⅱ and Class Ⅲ highways adopt the widening value of the third type in consideration of accommodating articulated vehicles; the widening value of the second type could be used when articulated vehicles are not considered. Local Class Ⅲ and Class Ⅳ highways adopt the widening value of the first type. The widening value of special roads accommodating special vehicles shall be determined according to the calculation of special vehicles.

**2) Widening transition**

Widening may happen for the road from a tangent to a circular curve, which requires a widening transition section where the road width changes gradually. The design of widening transition may adopt different methods according to the nature and class of the road.

(1) Proportional transition

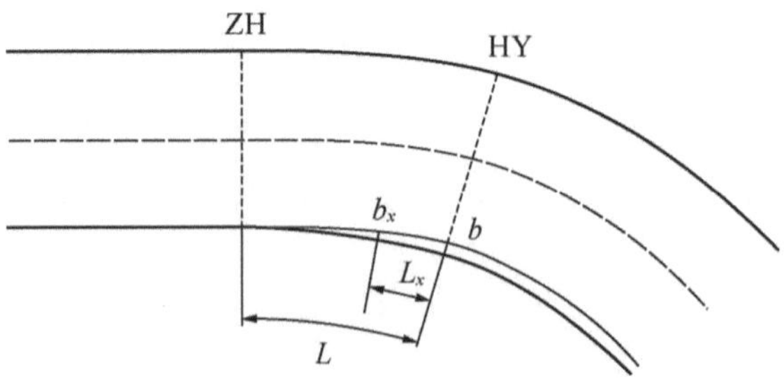


**Figure 4 - 11 Proportional transition**

The widening transition section is gradually widened in proportion to the length, and the widening value at any point is:

$$b_x = \frac{L_x}{L} b$$

where: $L_x$—distance between any point and start of widening transition section, m;

$L$—length of widening transition section, m;

$b$—widening on the circular curve, m.

The proportional transition is simple but the inside edge line of the road does not completely match the driving trajectory using this method. The start and end points of the transition section are not smoothly connected, and the road alignment is not elegant. This method is suitable for Class Ⅱ, Class Ⅲ and Class Ⅳ highways.

(2) High-order parabolic transition

A high-order parabola is used in the widening transition section, and the widening value of any point on the parabola is:

$$b_x = (4k^3 - 3k^4)b \qquad (4-10)$$

where $k = \frac{L_x}{L}$.

The inside edge line of the road is connected smoothly and the road alignment is elegant using this method, which is suitable for the freeway and Class Ⅰ highway.

(3) Spiral transition

A spiral is used in the widening transition section. The inside edge line of the road is consistent with the driving trajectory. The driving is comfortable and the alignment is elegant. It is suitable for the freeway and Class Ⅰ and Class Ⅱ highways of which/that are:

① Road sections located in the suburb of big cities;

② Bridge, viaduct, retaining wall, tunnel, and other structures;

③ Road sections with various safety protection facilities.

**3) Length of the widening transition section**

When the horizontal curve has a spiral or superelevation transition section, the widening transition section adopts the same length as the spiral or superelevation transition section; when the horizontal curve has a superelevation transition section but no spiral, the widening transition section adopts the same length as the superelevation transition section; when the horizontal curve has neither a spiral nor a superelevation transition section, the widening transition section shall be set according to the requirements that the length is not less than 10 m and the longitudinal gradient is 1 : 15.

### 4.4.3 Superelevation design

**1) Function and calculation of superelevation**

In order to resist centrifugal force, it is necessary to set superelevation when a vehicle drives on a curve. When the vehicle runs at a constant speed, the centrifugal force generated on the circular curve is constant, and the superelevated slope shall match the radius of the circular curve. Because of variation nature of curvature and centrifugal force on the spiral, the superelevation on the spiral also needs to change gradually. The section from the normal cross slope on a tangent to the straight cross slope on a circular curve is called the superelevation transition section.

The relationship between superelevation rate $i_h$ and sideway force coefficient $\mu$ is as follows:

$$i_h+\mu=\frac{v^2}{127R} \tag{4-11}$$

To determine the value of $i_h$, it is necessary to make clear the allocation between $i_h$ and $\mu$, and to select different allocation methods for different roads.

**2) Superelevation transition method**

(1) Superelevation transition of roads without a median

When the superelevation rate is equal to the cross slope of the road, it is only necessary to raise the outside lane on the circular curve section about the road centerline until it is equal to the cross slope of the inside-lane, forming a straight cross slope, as shown in Figure 4 - 12.

When the superelevation rate is greater than the cross slope of roads, the following three transition methods can be adopted:

① Revolve about the inside-edge profile

The first method, as shown in Figure 4 - 13(a), let the traveled way revolve about the inside-edge profile. First, the outside traveled way revolves about the

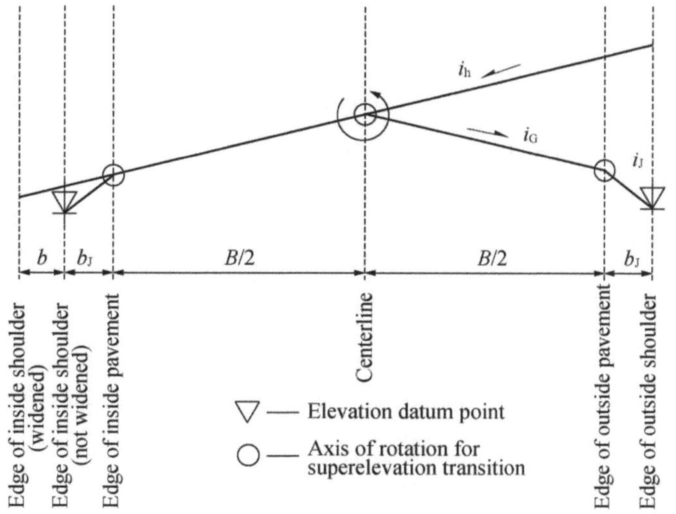


**Figure 4 - 12 Superelevation transition when superelevation rate equal to the cross slope**

centerline until the integral traveled way forms a straight cross slope; then, the integral traveled way revolves about the inside-edge profile until the superelevation rate is reached.

② Revolve about the centerline

The second method, as shown in Figure 4 - 13(b), let the traveled way revolve about the centerline. First, the outside traveled way revolves about the centerline until the integral traveled way forms a straight cross slope; then, the integral traveled way revolves about the centerline until the superelevation rate is reached.

③ Revolve about the outside-edge profile

The third method, as shown in Figure 4 - 13(c), let the traveled way revolve about the outside-edge profile. First, the outside traveled way revolves about the outside-edge profile, and the inside traveled way descends at the same pace until the outside and inside traveled ways form a straight cross slope; then, the integral traveled way revolves about the outside-edge profile until the superelevation rate is reached.

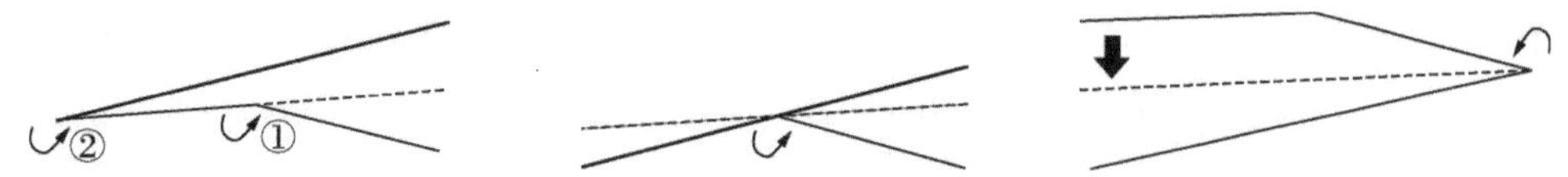


(a) About the inside-edge profile (b) About the centerline (c) About the outside-edge profile

**Figure 4 - 13 Diagrams of superelevation transitions of roads without the median**

(2) Superelevation transition of roads with the median

① Revolve about the centerline of the medial separator

First, the outside traveled way revolves about the edge profile of the medial separator until the cross slopes of the inside and outside traveled ways are equal; second, the integral cross section revolves about the centerline of the medial separator until the superelevation rate is reached. With this method, the medial separator is inclined, as shown in Figure 4 - 14(a).

(a) About the centerline of the medial separator    (b) About the edge profile of the medial separator    (c) About the centerline of both traveled ways

**Figure 4 - 14 Superelevation transition of roads with the medial separator**

② Revolve about the edge profile of the medial separator

The inside and outside traveled ways revolve about the edge profile of the medial separator, which make them the independent unidirectional superelevated section. For this method, the medial separator remains in its original shape, as shown in Figure 4 - 14(b).

③ Revolve about the centerline of both traveled ways

The inside and outside traveled ways revolve about their respective centerlines so that they become independent unidirectional superelevated sections. For this method, one side of the medial separator is raised and the other side is lowered, as shown in Figure 4 - 14(c).

(3) A crown line provided for highways with six or more lanes

When the edge profile of the medial separator is used as the rotation axis for the freeway and Class Ⅰ highway, even if the superelevation gradient is more than 1/330, the transverse drainage of the traveled way may not be sufficient due to the wider cross section with a gentle grade. In order to mitigate the adverse effect, measures such as reducing the length of the superelevation transition section, increasing the superelevation gradient, and attaining superelevation within a certain section of spirals may be adopted. Besides, a crown line could be used in the middle of the traveled way to reduce the runoff distance and improve the drainage conditions, so as to mitigate water accumulation on the road surface.

(4) Superelevation transition of the paved shoulder

The cross slope and superelevation rate of the paved shoulder should generally be consistent with the traveled way; if necessary, different cross slopes may be used. When the superelevation rate is less than or equal to 5% on the curve, the rate and direction of the cross slope of the paved shoulder shall be the same as those of the adjacent lane; when the superelevation rate is greater than 5% on the curve, the cross slope of the paved shoulder shall not be greater than 5%, and the direction shall be the same as that of the adjacent lane.

The superelevation rotation axis of the paved shoulder is located at the inside-edge profile of the paved shoulder. When the superelevation rate of the paved shoulder is the same as that of the adjacent lane, its superelevation transition section shall be the same as that of the traveled way, and the same superelevation relative gradient shall be used. When the superelevation rate of the paved shoulder is less than that of the adjacent lane, the cross slope of the paved shoulder shall first be transitioned to the same rate as that of the traveled way and then both the paved shoulder and traveled way perform transition together until the paved shoulder reach its maximum superelevation rate.

**3) Length of superelevation transition section**

When a road changes from the normal crown to the superelevated slope, the superelevation transition section shall be offered to ensure smooth driving, elegant road alignment, and smooth drainage. The minimum length of the superelevation transition section of highways is calculated as follows:

$$L_c=\frac{\Delta_i \cdot B}{P} \tag{4-12}$$

Where:

$L_c$—length of superelevation transition section, m.

$\Delta_i$—algebraic difference, %; between the superelevated slope and normal cross slope; when revolving about the inside-edge profile, $\Delta_i=i_h$; when revolving about the centerline, $\Delta_i=i_h+i_G$.

$B$—width from the superelevation rotation axis to the outside-edge profile of the traveled way, m; if there is a marginal strip, it is to the outside-edge profile of the marginal strip.

$P$—superelevation relative gradient, the relative slope between the rotation axis and the outside-edge profile of the traveled way; if there is a marginal strip, it is between the rotation axis and the outside-edge profile of the marginal strip. The maximum superelevation relative gradient is specified in the **Specification (2017)**.

The calculated length of the superelevation transition section shall be rounded to 5 m, and shall not be less than 10 m.

The longer the length of the superelevation transition section, the better the driving comfort. However, when the grade is small, the longer the superelevation transition section, the more difficult the road drainage. Therefore, the length of the superelevation transition section needs to be comprehensively considered in horizontal alignment design.

(1) In general, when determining the length of the spiral, the minimum length of the superelevation transition section has been considered, and $L_c = L_s$.

(2) If $L_c > L_s$ by calculation, the horizontal alignment shall be modified to make $L_c \leqslant L_s$; if the horizontal alignment could not be modified, the start point of the superelevation transition section may be moved forward.

(3) If $L_c < L_s$ by calculation, $L_c = L_s$ is still acceptable as long as the superelevation relative gradient $p \geqslant 1/330$.

(4) Freeways and Class Ⅰ Highways generally have long spirals so that the superelevation transition may be carried out only in a certain section of the spirals. Otherwise, the superelevation transitions may be carried out in two sections separately. The demarcation point is where the superelevation transition develops to the same slope of the road crown.

(5) When the spiral is not used for the Class Ⅳ highway, the superelevation transition section shall be applied on the tangent section. If the length of the tangent section is insufficient, it is allowed to allocate the superelevation the transition section in the tangent section and circular curve section half by half.

In the superelevation transition section, the road with a cross slope of zero (flat cross slope) shall be altered, drainage analysis shall be carried out, and corresponding measures shall be taken.

**4) Superelevation calculation**

When the superelevation of the horizontal curve is determined, the elevation differences $h$ between the road centerline, inside-edge and outside-edge of roadway, and the design elevation shall be calculated and listed in the "Subgrade Design Table" for construction.

(1) Revolving about the inside-edge profile for roads without a median

The attainment of superelevation for roads without a median revolving about the inside-edge profile is shown in Figure 4 - 15 and the corresponding calculation formula is shown in Table 4 - 11.

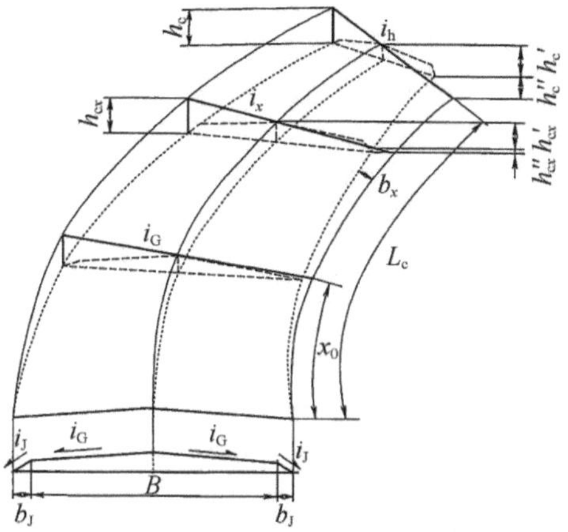

**Figure 4 - 15 Diagram of superelevation transition revolving about the inside-edge profile**

*Notes*: Taking the point on the shoulder edge profile as the elevation datum point.

$i_x$—cross slope at distance $x$;

$b_x$—roadway widening value at distance $x$;

$x_0$—distance between the start point of the superelevation transition section to the point that has the straight cross slope of the crown, namely the critical cross section;

$h_c$—maximum elevation of the outside-edge of roadway;

$h_c'$—maximum elevation of the centerline;

$h''_c$—maximum elevation of the inside-edge of roadway;

$h_{cx}$—elevation of the outside-edge of roadway at distance $x$;

$h_{cx}'$—elevation of the centerline at distance $x$;

$h_{cx}''$—elevation of the inside-edge of roadway at distance $x$.

As revolving about the inside-edge profile, the relationship between the elevation increase of the outside-edge of roadways on the transition section and the distance from the start point of the transition section is shown in Figure 4 - 16.

**Table 4 - 11 Calculation formula of the superelevation revolving about the inside-edge profile**

| Superelevation position | | Calculation formula $x \leqslant x_0$ | Calculation formula $x > x_0$ | Note |
|---|---|---|---|---|
| On the circular curve | Outside-edge $h_{cx}$ | $b_J i_J + (b_J + B) i_h$ | | 1. Widening value at distance $x$: $b_x = \frac{x}{L_c} b$ |
| | Centerline $h_c'$ | $b_J i_J + \frac{B}{2} i_h$ | | |
| | Inside-edge $h_c''$ | $b_J i_J - (b_J + b) i_h$ | | 2. Distance between the start point of transition section and the critical cross section: $x_0 = \frac{i_G}{i_h} L_c$ |
| On the transition section | Outside-edge $h_{cx}$ | $b_J (i_J - i_G) + \frac{x}{x_0} (2b_J + B) i_G$ | $b_J i_J + (b_J + B) \frac{x}{L_c} i_h$ | |
| | Centerline $h_{cx}'$ | $b_J i_J + \frac{B}{2} i_G$ | $b_J i_J + \frac{B}{2} \cdot \frac{x}{L_c} i_h$ | 3. Cross slope at distance $x$ ($x \geqslant x_0$): |
| | Inside-edge $h_{cx}''$ | $b_J i_J - (b_J + b_x) i_G$ | $b_J i_J - (b_J + b_x) \frac{x}{L_c} i_h$ | $i_x = \frac{x}{L_c} i_h$ |

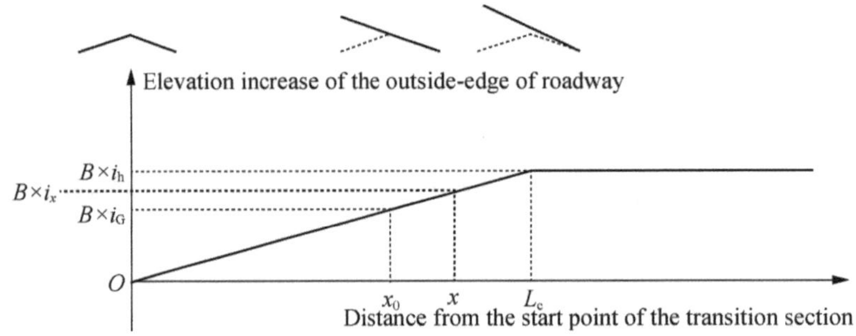


**Figure 4 - 16 Diagram of elevation increase of the outside-edge of roadway**

(2) Revolving about the centerline for roads without a median

The attainment of superelevation for roads without the median revolving about the centerline is shown in Figure 4 - 17 and the corresponding calculation formula is shown in Table 4 - 12.

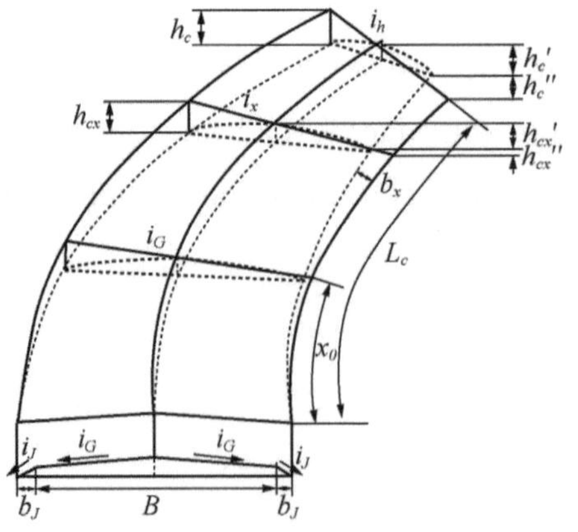


**Figure 4 - 17 Diagram of the superelevation transition revolving about the centerline**

The symbolic meanings are the same as Figure 4 - 15.

**Table 4 - 12 Calculation formula of superelevation revolving about the centerline**

| Superelevation position | | Calculation formula $x \leq x_0$ | Calculation formula $x > x_0$ | Note |
|---|---|---|---|---|
| On the circular curve | Outside-edge $h_c$ | $b_J(i_J - i_G) + \left(b_J + \frac{B}{2}\right)(i_G + i_h)$ | | 1. Widening value at distance $x$: $b_x = \frac{x}{L_c} b$<br>2. Distance between the start point of transition section and the critical cross section: $x_0 = \frac{2i_G}{i_G + i_h} L_c$<br>3. Cross slope at distance $x$: $i_x = \frac{i_h + i_G}{L_c} \cdot x - i_G$ |
| | Centerline $h'_c$ | $b_J i_J + \frac{B}{2} i_G$ | | |
| | Inside-edge $h''_c$ | $b_J i_J + \frac{B}{2} i_G - \left(b_J + \frac{B}{2} + b\right) i_h$ | | |
| On the transition section | Outside-edge $h_{cx}$ | $b_J(i_J - i_G) + \left(b_J + \frac{B}{2}\right)(i_G + i_h)\frac{x}{L_c}$ | | |
| | Centerline $h'_{cx}$ | $b_J i_J + \frac{B}{2} i_G$ | | |
| | Inside-edge $h''_{cx}$ | $b_J i_J - (b_J + b_x) i_G$ | $b_J i_J + \frac{B}{2} i_G - \left(b_J + \frac{B}{2} + b_x\right)\left(\frac{i_h + i_G}{L_c} \cdot x - i_G\right)$ | |

As revolving about the centerline, the relationship between the cross slope on the transition section and the distance from the start point of the transition section is shown in Figure 4 - 18.

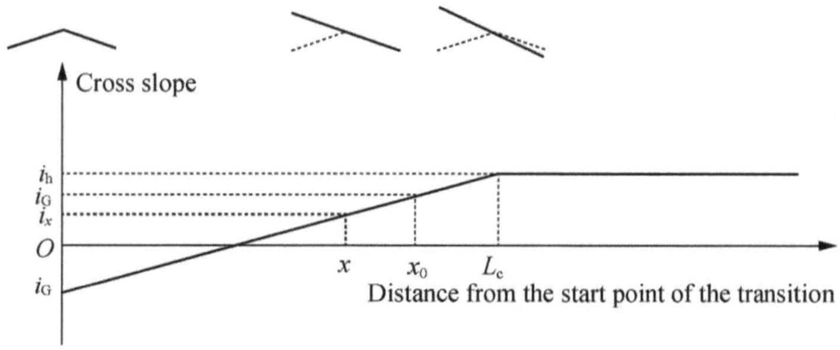


**Figure 4 - 18 Diagram of cross slope development on the transition section**

**Example 1**: A two-lane highway in a mountainous region is designed with a design speed of 60 km/h, a traveled way width of 7 m, and a shoulder width of 0.75 m. The stake of intersection point $JD_8$ is K87+952.00, the deflection angle $\alpha_{right}$ is 46°37′, the radius of the circular curve $R$ is 160 m, the cross slope of the traveled way $i_G$ is 0.02, and the cross slope of the shoulder $i_J$ is 0.04. Please complete the design and calculation of horizontal curve widening (linear) and superelevation (about the axis of centerline).

**Solution:**

(1) Determine whether to set superelevation and widening and relevant cross section parameters

$$i_h = V^2/127R - \mu = 60^2/(127\times160) - 0.12 = 0.06$$

In the **Specification (2017)**: widening value $b = 0.70$ m, superelevation gradient rate $p = 1/175$.

Accordingly:

$$L_c = B(i_G + i_h)/2p = 7\times(0.02+0.06)/(2\times1/175) = 49(\mathrm{m})$$

$$L_s = \max(L_c, 0.035\times V^3/R, V/1.2, 15\times b) = \max(49, 47.25, 50, 10.5) = 50(\mathrm{m})$$

Therefore $L_c = L_s = 50$ m.

(2) Calculation of stake of main points

① Calculation of horizontal curve elements (Figure 2 - 8 of this book).

$$q = \frac{L_s}{2} - \frac{L_s^3}{240R^2} = \frac{50}{2} - \frac{50^3}{240\times160^2} = 24.98(\mathrm{m})$$

$$p = \frac{L_s^2}{24R} - \frac{L_s^4}{2688R^3} = \frac{50^2}{24\times160} - \frac{50^4}{2688\times160^3} = 0.65(\mathrm{m})$$

$$\beta_0 = \frac{L_s}{2R}\cdot\frac{180}{\pi} = \frac{50}{2\times160}\times\frac{180}{\pi} = 8.95° = 8°57'00''$$

$$T = (R+p)\tan\frac{\alpha}{2} + q = (160+0.65)\tan\frac{46°37'}{2} + 24.98 = 94.21(\mathrm{m})$$

$$L = (\alpha - 2\beta_0)\frac{\pi}{180}R + 2L_s = (46.62 - 2\times8.95)\times\frac{\pi}{180}\times160 + 2\times50 = 180.19(\mathrm{m})$$

$$E = (R+p)\sec\frac{\alpha}{2} - R = (160+0.65)\sec\frac{46°37'}{2} - 160 = 14.93(\mathrm{m})$$

$$J = 2T - L = 2\times94.21 - 180.19 = 8.23(\mathrm{m})$$

② Stake of main points:

$$ZH = \mathrm{K}97+952.00 - 94.21 = \mathrm{K}87+857.79$$

$$HY = \mathrm{K}87+857.79 + 50 = \mathrm{K}87+907.79$$

$$QZ = \mathrm{K}87+857.79 + \frac{180.19}{2} = \mathrm{K}87+947.89$$

$$YH = \mathrm{K}88+037.98 - 50 = \mathrm{K}87+987.98$$

$$HZ = \mathrm{K}87+947.89 + 180.19 = \mathrm{K}88+037.98$$

Verification: $JD = QZ + \frac{J}{2} = \mathrm{K}87+947.89 + \frac{8.23}{2} = \mathrm{K}87+952.00$

(3) Calculation of superelevation and widening of the horizontal curve

To summary, the results are shown in Table 4 - 13:

**Table 4 - 13 Superelevation and widening calculation results**

| Stake number | Note | Distance $x$/m | Subgrade elevation/m | | | Subgrade widening/m | |
|---|---|---|---|---|---|---|---|
| | | | Left | Middle | Right | Left | Right |
| — | Normal crown | | 0.00 | 0.10 | 0.00 | — | — |
| K87+857.79 | Start point (ZH) | 0.00 | 0.02 | 0.10 | 0.02 | — | 0.00 |
| K87+867.79 | — | 10.00 | 0.08 | 0.10 | 0.01 | — | 0.14 |
| K87+877.79 | — | 20.00 | 0.15 | 0.10 | 0.01 | — | 0.28 |
| K87+887.79 | — | 30.00 | 0.22 | 0.10 | −0.03 | — | 0.42 |
| K87+897.79 | — | 40.00 | 0.29 | 0.10 | −0.11 | — | 0.56 |
| K87+907.79 | End point (HY) | 50.00 | 0.36 | 0.10 | −0.20 | — | 0.70 |
| Superelevated and widened section (circular curve) | | | | | | | |
| K87+987.98 | End point (YH) | 50.00 | 0.36 | 0.10 | −0.20 | — | 0.70 |
| K87+997.98 | — | 40.00 | 0.29 | 0.10 | −0.11 | — | 0.56 |
| K88+007.98 | — | 30.00 | 0.22 | 0.10 | −0.03 | — | 0.42 |
| K88+017.98 | — | 20.00 | 0.15 | 0.10 | 0.01 | — | 0.28 |
| K88+027.98 | — | 10.00 | 0.08 | 0.10 | 0.01 | — | 0.14 |
| K88+037.98 | Start point(HZ) | 0.00 | 0.02 | 0.10 | 0.02 | | 0.00 |
| — | Normal crown | 0.00 | 0.00 | 0.10 | 0.00 | — | 0.00 |

Among them, elevation of the road centerline: $0.04\times0.75+0.02\times3.50=0.1$(m).

Elevation of the outside-edge on the start of the spiral: $(0.04-0.02)\times0.75\approx0.02$(m).

Elevation of the inside-edge at K87+897.79: $0.10-(4.25+0.48)\times\left[\frac{40}{50}\times(0.06+0.02)-0.02\right]\approx-0.11$(m).

Elevation of the outside-edge at the end of the spiral: $0.10+4.25\times0.06\approx0.36$(m).

## 4.5 Sight distance

### 4.5.1 Type of sight distance

A driver's ability to see ahead is of utmost importance in the safe and efficient operation of a vehicle on a highway. Sight distance is the length of the roadway ahead that is visible to drivers. For safety on highways, the designer shall provide sight distance of sufficient length so that drivers can control the operation of their vehicles to avoid striking an unexpected object in the traveled way. Whether sight distance is sufficient is one of the important indicators to evaluate road design and operation quality. Sight distance can be divided into the following categories.

(1) Stopping Sight Distance: the minimum distance required for the driver to safely stop from the instant the driver sees an object necessitating a stop.

(2) Meeting Sight Distance: the minimum distance required for two vehicles to travel from the instant the driver sees the vehicle in front to the instant they safely meet.

(3) Give-way Sight Distance: on a two-lane road without clear lane separating lines, the minimum distance required for two opposing vehicles to decelerate and give way safely when they meet each other.

(4) Passing Sight Distance: on a two-lane road, when the following vehicle overtakes the preceding vehicle, the minimum distance required to complete a normal passing maneuver in which the passing driver can determine that there are no potential conflicting vehicles ahead before beginning the maneuver.

(5) Decision Sight Distance: the shortest distance required for a driver to detect an unexpected or otherwise difficult-to-perceive information source or condition in a roadway environment that may be visually cluttered, recognize the condition or its potential threat, select an appropriate speed and path, and initiate and complete the maneuver safely and efficiently.

Among them, the stopping sight distance is the shortest and shall be assured under any circumstances. Meeting sight distance, give-way sight distance, and passing sight distance are specific to two-lane roads, among which passing sight distance is the longest. Sight distance adequate for passing shall be encountered frequently on two-lane highways. Each passing section along a length of roadways with sight distance ahead equal to or greater than the minimum passing sight distance shall be as long as practical. The frequency and length of passing sections for highways principally depend on the topography, the design speed, and cost; for urban roads, the spacing of intersections is the principal consideration. Decision sight distance is the sight distance required at the entrances and exits of roads.

## 4.5.2 Sight distance requirement

**1) Sight distance requirement for the freeway and Class Ⅰ highway**

Stopping sight distance shall be adopted for the freeway and Class Ⅰ highway. For general road sections, the stopping sight distance of each lane shall not be less than that stipulated in Table 4 - 14.

**Table 4 - 14 Stopping sight distance of the freeway and Class Ⅰ highway**

| Design speed/km • $h^{-1}$ | 120 | 100 | 80 | 60 |
|---|---|---|---|---|
| Stopping sight distance/m | 210 | 160 | 110 | 75 |

*Note*: This table refers to the **Specification (2017)**.

**2) Sight distance requirement for Class Ⅱ, Class Ⅲ and Class Ⅳ highways**

Meeting sight distance shall be adopted for Class Ⅱ, Class Ⅲ and Class Ⅳ highways. Stopping sight distance may be adopted for road sections where lane separation measures are taken due to the restriction of terrain or other conditions, and it shall not be less than that stipulated in Table 4 - 15.

**Table 4 - 15 Stopping sight distance and meeting sight distance of Class Ⅱ, Class Ⅲ and Class Ⅳ highways**

| Design speed/km • $h^{-1}$ | 80 | 60 | 40 | 30 | 20 |
|---|---|---|---|---|---|
| Meeting sight distance/m | 220 | 150 | 80 | 60 | 40 |
| Stopping sight distance/m | 110 | 75 | 40 | 30 | 20 |

*Note*: This table refers to the **Specification (2017)**.

**3) Passing sight distance requirement**

For two-lane highways of Class Ⅱ, Class Ⅲ and Class Ⅳ highways, sections that meet passing sight distance shall be provided at intervals. Class Ⅱ highways of arterial functions should provide a passing section that meets the requirement of passing sight distance within 3 minutes of traveling time, and shall conform to the provisions in Table 4 - 16.

**Table 4 - 16 Minimum passing sight distance**

| Design speed/km · h$^{-1}$ | | 80 | 60 | 40 | 30 | 20 |
|---|---|---|---|---|---|---|
| Minimum passing sight distance/m | General value | 550 | 350 | 200 | 150 | 100 |
| | Limit value | 350 | 250 | 150 | 100 | 70 |

*Note*: This table refers to the **Specification (2017)**.

**4) Sight distance requirement for downhill sections**

For downhill sections of the freeway and Class Ⅰ highway, as well as Class Ⅱ and Class Ⅲ highways with a high proportion of large vehicles, the stopping sight distance of trucks shall be used to examine these sections. The stopping sight distance of trucks on the downhill section of highways at all classes shall not be less than that stipulated in Table 4 - 17.

**Table 4 - 17 Stopping sight distance of trucks on the downgrade** unit: m

| Design speed/km · h$^{-1}$ | | 120 | 100 | 80 | 60 | 40 | 30 | 20 |
|---|---|---|---|---|---|---|---|---|
| Grade/% | 0 | 245 | 180 | 125 | 85 | 50 | 35 | 20 |
| | 3 | 265 | 190 | 130 | 89 | 50 | 35 | 20 |
| | 4 | 273 | 195 | 132 | 91 | 50 | 35 | 20 |
| | 5 | — | 200 | 136 | 93 | 50 | 35 | 20 |
| | 6 | — | — | 139 | 95 | 50 | 35 | 20 |
| | 7 | — | — | — | 97 | 50 | 35 | 20 |
| | 8 | — | — | — | — | — | 35 | 20 |
| | 9 | — | — | — | — | — | — | 20 |

*Note*: This table refers to the **Specification (2017)**.

**5) Decision sight distance requirement at the entrance and exit of the road**

All types of exits such as interchanges, service areas, parking areas, and passenger car stops of highways at all classes shall meet the requirement of decision sight distance as well as the following requirements:

(1) Decision sight distance at different design speeds should conform to the provisions in Table 4 - 18.

**Table 4 - 18 Decision sight distance**

| Design speed/km • $h^{-1}$ | 120 | 100 | 80 | 60 |
|---|---|---|---|---|
| Decision sight distance/m | 350(460) | 290(380) | 230(300) | 170(240) |

*Notes*: (1) The value in parentheses shall be taken when the operation environment is complicated and there are many prompt messages at the roadside exit.

(2) This table refers to the **Specification (2017)**.

(2) For road sections restricted by terrain, geology and other conditions, 1.25 times of the stopping sight distance may be used as the decision sight distance, but necessary speed limit control and management measures shall be taken.

## 4.5.3 Sight distance requirement on curves

On the horizontal curve of a road, there may be trees, houses, slopes, road facilities, etc. inside the curve that obstruct the driver's line of sight, which is called a "blind curve". In the design, the "blind curve" needs to be checked in terms of sight distance. If the sight distance does not meet the requirements, the obstacles obstructing the sight shall be cleared.

The method of checking sight distance is to draw the border of sight along the trajectory of the driver's viewpoint to form a continuous envelope, as shown in Figure 4 - 19. An obstacle in the envelope shall be cleared.

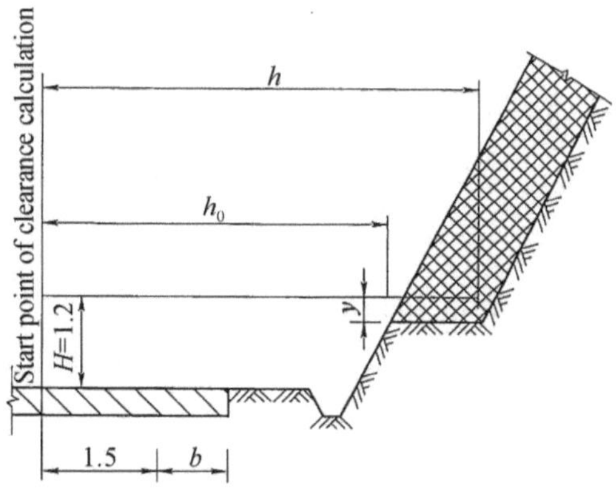


**Figure 4 - 19 Diagram of horizontal clearance**

### 1) Sight distance curve and horizontal clearance

The sight distance curve refers to a series of outer border lines tangent to the viewpoint drawn at certain intervals along the driver's viewpoint trajectory. As shown in Figure 4 - 20, curve *AB* is the driver's viewpoint trajectory, multiple lines of sight with a length of the sight distance $S$ (1 - 1′, 2 - 2′, 3 - 3′, …, 10 - 10′, etc.) are determined, and the envelope curve tangent to these lines of sight is the so-called sight distance curve.

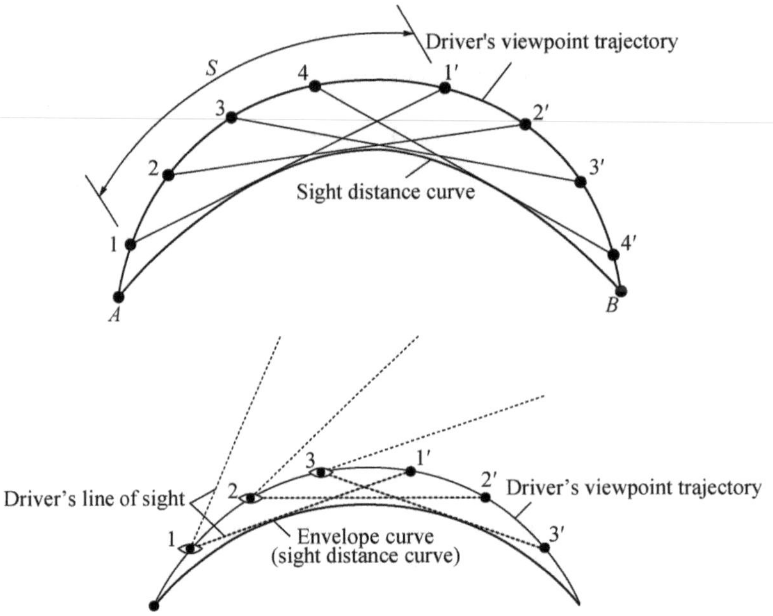


**Figure 4 - 20 Diagram of the sight distance curve**

The driver's viewpoint is generally 1.20 m above the ground and 1.5 m away from the inside edge of the roadway (excluding widening). The horizontal clearance $h$ at each point of the horizontal curve is the maximum distance between the driver's viewpoint trajectory and the sight distance curve; and $h_0$ is the distance from the obstacle to the driver's viewpoint. If horizontal clearance $h$ of a cross section is less than $h_0$, the requirement of the sight distance is met; and vice versa.

**2) Calculation of horizontal clearance**

Calculation of horizontal clearance can be divided into two scenarios: with and without a spiral.

(1) Calculation of horizontal clearance without a spiral

When the horizontal curve does not have a spiral, the calculation of horizontal clearance is as follows:

① If horizontal curve length $L$>sight distance $S$

$$h = R_S - R_S \cos\frac{\gamma}{2} \tag{4-13}$$

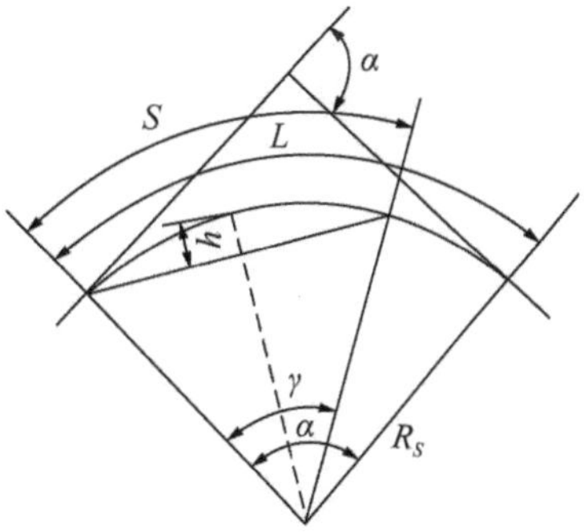


**Figure 4 - 21 Horizontal curve length $L>$sight distance $S$**

$$\gamma=\frac{180S}{\pi R_S} \tag{4-14}$$

where: $\alpha$—deflection angle, °;

$h$—horizontal clearance, m;

$R_S$—curve radius of viewpoint trajectory when vehicle drives on the inside of a curve, m, $R_S=R-\frac{B}{2}+1.5$; $B$—pavement width before widening, m;

$\gamma$—central angle corresponding to the sight distance, °;

$S$—sight distance. m.

② If horizontal curve length $L\leqslant$sight distance $S$

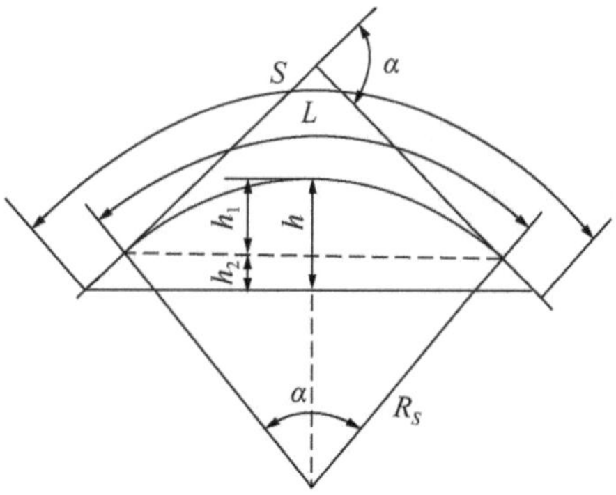


**Figure 4 - 22 Horizontal curve length $L\leqslant$sight distance $S$**

$$h=R_S-R_S\cos\frac{\alpha}{2}+\frac{S-L}{2}\sin\frac{\alpha}{2} \tag{4-15}$$

$$\alpha=\frac{180L}{\pi R_S} \tag{4-16}$$

where: $\alpha$—deflection angle of the road, °;

$L$—horizontal curve length, m.

(2) Calculation of horizontal clearance with a spiral

① If circular curve length $L'>$ sight distance $S$

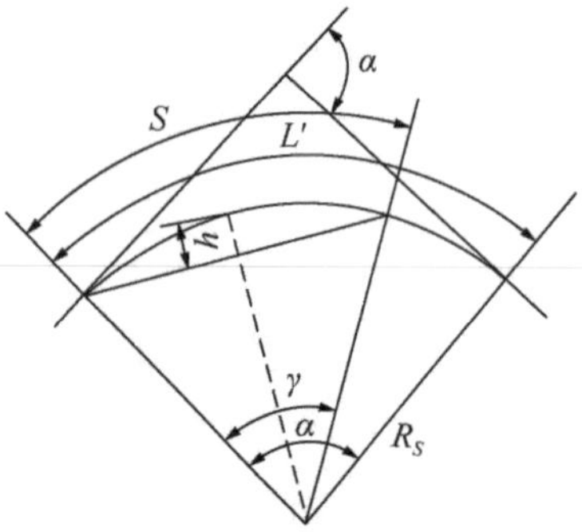


**Figure 4 - 23 Circular curve length $L'>$sight distance $S$**

In this case, the calculation of horizontal clearance is the same as that for the calculation of horizontal clearance without a spiral when the horizontal curve length $L$ is greater than the sight distance $S$. And the circular curve length $L'$ can be used instead of the horizontal curve length $L$ by using the formulas (4 - 13) and (4 - 14).

② If circular curve length $L'\leqslant$ sight distance $S<$ horizontal curve length $L$

$$h=h_1+h_2=R_S-R_S\cos\frac{\alpha-2\beta}{2}+(l-l')\sin\left(\frac{\alpha}{2}-\delta\right) \qquad (4-17)$$

where: $\delta=\arctan\left\{\frac{1}{6R_5}\left[1+\frac{l'}{l}+\left(\frac{l'}{l}\right)^2\right]\right\}$;

$l$—transition curve length;

$\beta$—the spiral angle corresponding to the full length of the transition curve of the road centerline;

$l'=\frac{L-S}{2}$.

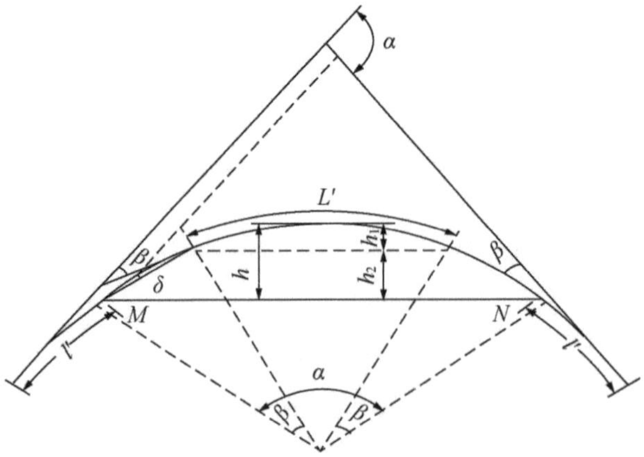


**Figure 4 - 24 Circular curve length $L'<$sight distance $S<$horizontal curve length $L$**

③ If horizontal curve length $L \leqslant$ sight distance $S$

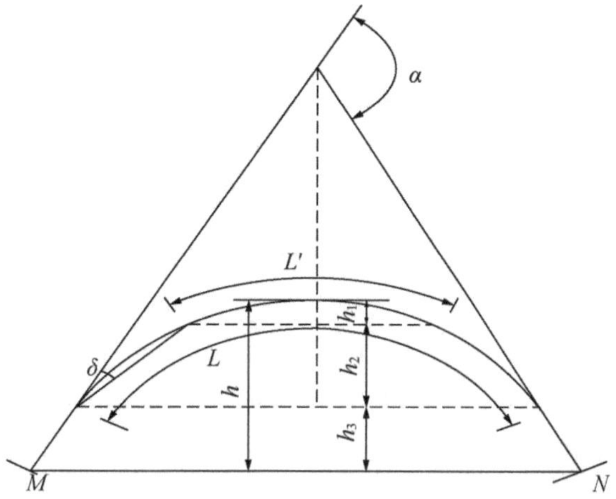


**Figure 4 - 25 Horizontal curve length $L \leqslant$ sight distance $S$**

$$h = h_1 + h_2 + h_3 = R_S - R_S \cos\frac{\alpha - 2\beta}{2} + l\sin\left(\frac{\alpha}{2} - \delta\right) + \frac{S - L_S}{2}\sin\frac{\alpha}{2} \quad (4-18)$$

where: $\delta = \arctan\frac{1}{6R_S}$.

It is worth noting that the horizontal clearance calculated according to the above formulas is the maximum one that needs to be cleared on a certain sight distance curve, and the horizontal clearance of different cross sections on the entire horizontal curve is different. If the obstacles are removed according to the maximum horizontal clearance in the entire horizontal curve, it will cause engineering waste. Therefore, the sight distance envelope method (Figure 4 - 20) is often used to complete the sight distance inspection. All objects within the spatial range of the viewpoint trajectory and sight distance curve are obstacles that block the line of sight and need to be cleared or implement certain measures. When the sight distance inspection is completed, additional parts of the cross section need to be plotted as a basis for the calculation of earthwork and construction.

## 4.6 Cross section design and deliverable

The main content of the cross section design is to determine the location of each component of cross section within the scope of road land. Highways and urban roads differ in steps in terms of cross section design.

Cross section layout shall be considered when determining the centerline and designing the vertical alignment of highways. During the cross section design, the layout shall be set according to the standard cross section diagram (see Figure 4 - 26). Occasionally, horizontal and vertical alignments are found to be inappropriate in the cross section design, and may be adjusted. In addition to the components within the roadway, the protection works such as the fill slope, cut, drainage ditch, intercepting ditch and retaining wall shall be also determined. Cross section design shall meet the requirements of environmental and ecological protection, land saving, and geological disaster reduction.

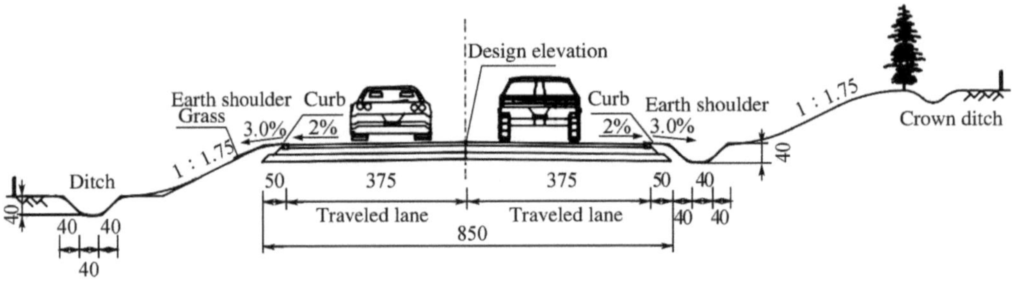


**Figure 4 - 26 Standard cross section of highways (cm)**

Cross section components of urban roads are mainly the layout according to the standard cross section diagram (see Figure 4 - 27) within the boundary of the road or the specified range. Compared with that of highways, the cross section of urban roads has higher requirements for the placement of various pipelines and the coordination with surrounding structures.

The cross section depicting the general layout of the whole road (or a certain section) is called the standard cross section.

A scale of 1 : 100 or 1 : 200 is usually adopted for the long-term and short-term cross section diagrams. The form and the requirements of each component in the cross section diagram shall be marked, such as the building boundary, width of the roadway, sidewalk, and greening, cross slope, slope direction, drainage method, etc.

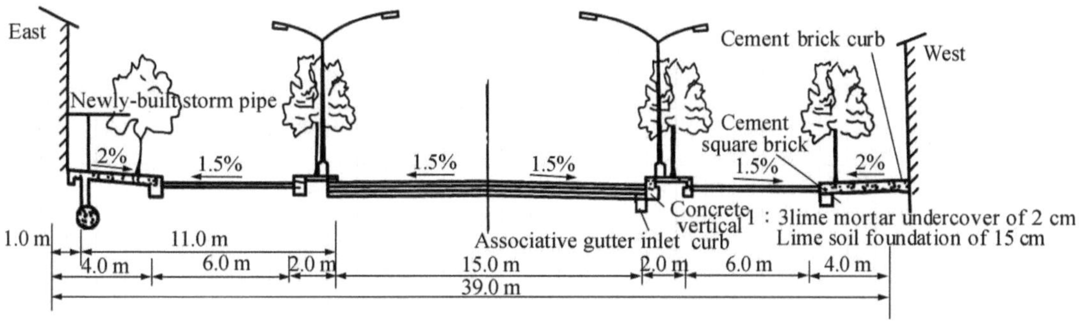


**Figure 4 - 27 Standard cross section of urban roads**

When reconstructing or expanding the existing roads, it is necessary to obtain the cross section diagram of the existing road to show the information on each component of the road, the type and thickness of pavement structure, the layout of the aboveground poles and underground pipelines, and pay attention to the relative position of the centerline of current cross sections and the short-term and long-term standard cross sections. The standard cross sections of highways and urban roads are shown in Figures 4 - 26 and 4 - 27.

## 4.6.1 Subgrade design table

The subgrade design table is one of the main technical documents in road alignment design, which is compiled by integrating the design data of horizontal, vertical and cross sections of the route. In the table, there is horizontal and vertical alignments' information, such as the center stake, horizontal alignment parameters, vertical curve parameters, elevation of the center stake, design elevation, construction height, etc., as well as cross section information, such as the subgrade width, crown slope, superelevation, widening on the small radius curve, etc. The subgrade design table can basically replace the horizontal, vertical and cross design drawings, and can be used on the construction site conveniently. Table 4 - 19 is the subgrade design table.

## 4.6.2 Cross section design drawing

Subgrade cross section diagram mainly describes the relationship between the ground line of center stakes and the design line. The marked information includes the ground height, design height, fill and cut height, fill and cut area, the distance and elevation from center stakes at specific positions, as well as the protection and drainage facilities on both sides of the road. The cross sections of each stake are arranged from left to right, from bottom to top, as shown in Figure 4 - 28.

## Table 4 - 19 Subgrade design table

| Horizontal curve | Rate and length of grades, stake and elevation of VPIs | Vertical curve | | Stake | Ground elevation/m | Design elevation/m | Cut or fill height/m | | Distance between each point of the cross section and the design line/m | | | | | | Elevation difference between each point of the cross section and the design elevation/m | | | | | | Note |
|---|---|---|---|---|---|---|---|---|---|---|---|---|---|---|---|---|---|---|---|---|---|
| | | | | | | | | | Left | | | Right | | | Left | | | Right | | | |
| | | Crest | Sag | | | | Cut | Fill | $WB_1$ | $WB_2$ | $WB_3$ | $WA_3$ | $WA_2$ | $WA_1$ | $B_1$ | $B_2$ | $B_3$ | $A_3$ | $A_2$ | $A_1$ | |
| 1 | 2 | 3 | 4 | 5 | 6 | 7 | 8 | 9 | 10 | 11 | 12 | 13 | 14 | 15 | 16 | 17 | 18 | 19 | 20 | 21 | 22 |
| K0+011.722 | G4K0+000.000 | | | K0+000.000 | 327.61 | 327.61 | | | | 1.75 | | | 1.75 | | | 327.645 | | | 327.575 | | |
| | 327.610 | | | K0+010.000 | 328.223 | 327.96 | | 0.263 | | 2.247 | | | 1.75 | | | 327.926 | | | 327.986 | | |
| JD1(左)48°31′29.6″ | | | | K0+020.000 | 328.153 | 328.31 | 0.158 | | | 2.35 | | | 1.75 | | | 328.193 | | | 328.398 | | |
| R=50.000 A1=0.000 | | | | K0+030.000 | 328.661 | 328.66 | | | | 2.35 | | | 1.75 | | | 328.543 | | | 328.748 | | |
| A2=0.000 T1=22.536 | | | | K0+040.000 | 329.092 | 329.01 | | 0.082 | | 2.35 | | | 1.75 | | | 328.893 | | | 329.098 | | |
| T2=22.536 L=42.346<br>J=2.727 E=4.844 | | | | K0+050.000 | 330.894 | 329.36 | | 1.534 | | 2.35 | | | 1.75 | | | 329.243 | | | 329.448 | | |
| K0+054.068 | | | | K0+060.000 | 331.639 | 329.71 | | 1.929 | | 1.994 | | | 1.75 | | | 329.652 | | | 329.761 | | |
| | | | | K0+070.000 | 327.719 | 330.06 | 2.341 | | | 1.75 | | | 1.75 | | | 330.07 | | | 330.05 | | |
| | | | | K0+080.000 | 320.808 | 330.41 | 9.602 | | | 1.75 | | | 1.75 | | | 330.445 | | | 330.375 | | |
| | | | | K0+090.000 | 320.976 | 330.76 | 9.785 | | | 1.75 | | | 1.75 | | | 330.795 | | | 330.725 | | |
| K0+122.303 | 190.00 \| 3.50% | | | K0+100.000 | 320.898 | 331.11 | 10.212 | | | 1.75 | | | 1.75 | | | 331.145 | | | 331.075 | | |
| JD2(左)29°04′30″ | | | | K0+110.000 | 325.707 | 331.46 | 5.753 | | | 1.75 | | | 1.75 | | | 331.435 | | | 331.486 | | |
| R=50.000 A1=0.000 | | | | K0+120.000 | 333.649 | 331.81 | | 1.838 | | 2.443 | | | 1.75 | | | 331.664 | | | 331.915 | | |
| A2=0.000 T1=7.779 | | | | K0+130.000 | 339.931 | 332.16 | | 7.771 | | 2.65 | | | 1.75 | | | 331.975 | | | 332.283 | | |
| T2=7.779 L=15.224 | | | | K0+140.000 | 340.874 | 332.51 | | 8.364 | | 2.427 | | | 1.75 | | | 332.367 | | | 332.613 | | |
| J=0.335 E=0.992 | | | | K0+150.000 | 339.111 | 332.86 | | 6.251 | | 1.75 | | | 1.75 | | | 332.836 | | | 332.884 | | |
| | | | | K0+160.000 | 336.036 | 333.21 | | 2.825 | | 1.75 | | | 1.75 | | | 333.245 | | | 333.175 | | |
| K0+137.526 | | K0+167.557 | | K0+170.000 | 335.871 | 333.553 | | 2.318 | | 1.75 | | | 1.75 | | | 333.588 | | | 333.518 | | |
| K0+193.218 | | SJD1 | | K0+180.000 | 335.503 | 333.728 | | 1.775 | | 1.75 | | | 1.75 | | | 333.763 | | | 333.693 | | |
| | K0+190.000 | R=425.173 | | K0+190.000 | 334.785 | 333.668 | | 1.117 | | 1.75 | | | 1.75 | | | 333.703 | | | 333.633 | | |
| JD3(左)4°24′22″ | 334.260 | T=22.443 | | K0+200.000 | 333.694 | 333.372 | | 0.322 | | 1.75 | | | 1.75 | | | 333.407 | | | 333.337 | | |
| R=250.000 A1=0.000 | 22.443 \| −7.06% | E=0.592 | | K0+210.000 | 332.877 | 332.842 | | 0.035 | | 1.75 | | | 1.75 | | | 332.877 | | | 332.807 | | |
| A2=0.000 T1=9.617 | K0+212.443 | K0+212.443 | | K0+212.443 | 332.676 | 332.676 | | | | 1.75 | | | 1.75 | | | 332.711 | | | 332.641 | | |
| T2=9.617 L=19.225 | 332.676 | | | | | | | | | | | | | | | | | | | | |
| J=0.009 E=0.185 | | | | | | | | | | | | | | | | | | | | | |
| | | | | | | | | | | | | | | | | | | | | | |
| K0+212.443 | | | | | | | | | | | | | | | | | | | | | |
| | | | | | | | | | | | | | | | | | | | | | |
| XXX highway | | | | Subgrade design table | | | | | Design | | | Review | | | Approve | | | Date | | Chart ID | |

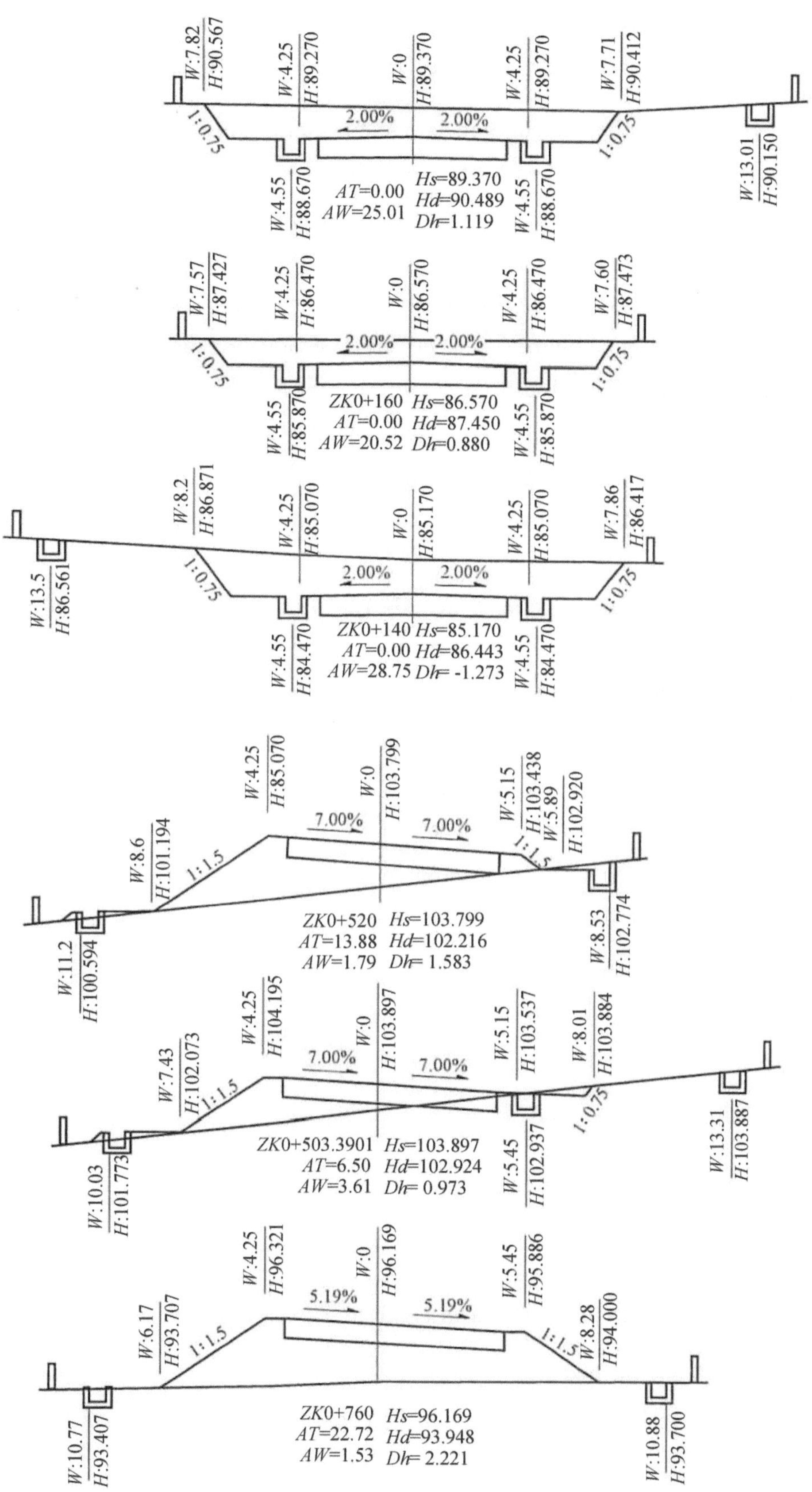


**Figure 4 - 28 Cross section design diagram of highways**

## 4.6.3 Calculation and allocation of earthwork

Earthwork quantity calculation is an important content of road design and one of the important indicators of road scheme comparison and selection. At present, the calculation principle of earthwork quantity mostly adopts the average cross section method, that is, assuming that the fill and cut area of adjacent stakes changes continuously, and then multiplying the average area by the distance to obtain the fill and cut volume, respectively. Obviously, this assumption is not rigorous, but considering the accuracy of the cross section ground line and the inherent error of earthwork allocation and other factors, the average cross section method is acceptable and widely used in engineering. Therefore, the calculation of earthwork quantity is converted to the calculation of the fill and cut area of each stake.

The commonly used methods include the product distance method and the coordinate method.

**1) Product distance method**

The principle of the product distance method is: first, divide the cross section into several small blocks of equal width (see Figure 4 - 29), which are generally the triangle, trapezoid or rectangle. The area of each small block is approximately the product of the center height and unit width: $A_i = b\times h_1 + b\times h_2 + b\times h_3 + \cdots + b\times h_n = b\times \sum h_i$. When $b$ is 1 m, the area $A$ is equal to the sum of the average heights of blocks.

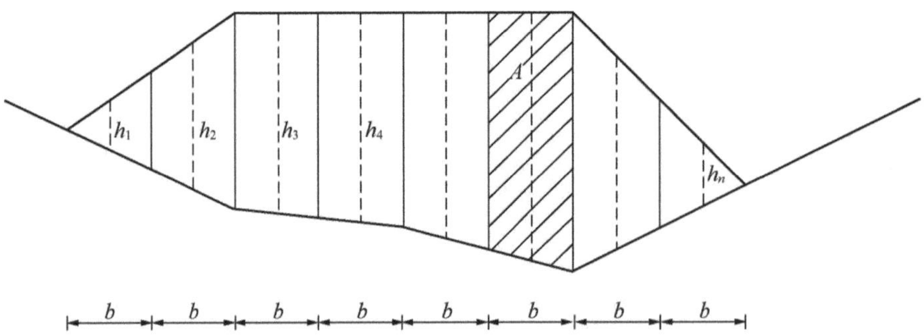


**Figure 4 - 29 Cross section area calculation (Product distance method)**

**2) Coordinate method**

As shown in Figure 4 - 30, given the coordinates of each vertex $(x_i, y_i)$ on the section, the area of the section is calculated by:

$$F = \frac{1}{2}\sum_{i=1}^{n}(x_i y_{i+1} - x_{i+1} y_i) \qquad (4-19)$$

The coordinate method has a higher precision, which is suitable for the computer.

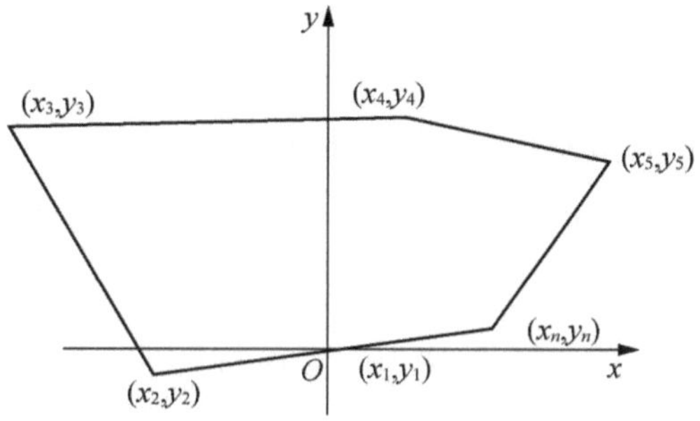


**Figure 4 - 30 Cross section area calculation (Coordinate method)**

If the two adjacent cross sections are both fill or cut with similar area, it can be assumed that it is a prism between the sections (Figure 4 - 31), and the calculation formula of the volume is as follows:

$$V = \frac{1}{2}(F_1 + F_2)L \qquad (4-20)$$

where: $V$—volume, i. e., earthwork quantity, $m^3$;

$F_1$, $F_2$—area of two adjacent cross sections, $m^2$;

$L$—distance between adjacent sections, m.

This method is easy to calculate and is commonly used. It is generally called "average cross section method".

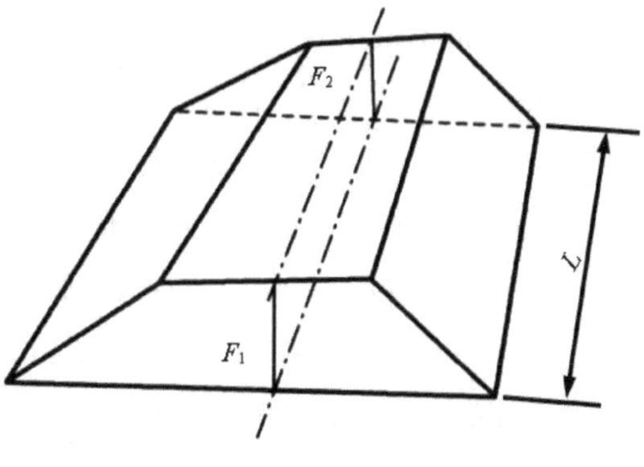


**Figure 4 - 31 Volume calculation**

If the difference between $F_1$ and $F_2$ is very large, it is closer to a pyramid.

The calculation formula is as follows:

$$V=\frac{1}{3}(F_1+F_2)L\left(1+\frac{\sqrt{m}}{1+m}\right) \tag{4-21}$$

where: $m=\frac{F_1}{F_2}$, and $F_2>F_1$.

The following problems shall be paid attention to in the calculation of the cross section area and earthwork quantity:

(1) Fill area and cut area shall be calculated separately.

(2) The area of fill shall be calculated separately according to rock fill, side slope reinforcement and earth fill.

(3) For sludge excavation, both the area of cut and fill shall be calculated at the same time. In this way, the area of the same place is counted twice because of the need to change the soil. Similarly, the area of bench excavation shall also be calculated twice.

(4) The quantity of earthwork between the start and end of large and medium bridges shall not be included in the quantity of subgrade earthwork.

The main functions of subgrade earthwork calculation and allocation are as follows:

(1) Determine the earthwork quantity of the whole road section and each individual section so as to determine the plans of fill and cut and the source of fill, as well as to select the appropriate construction methods and transportation tools.

(2) As the basis for the preparation of project budget and completion settlement.

(3) As the basis of the construction schedule, quality inspection, financial settlement, and salary payment.

After the calculation of subgrade earthwork quantity, the problem of earthwork allocation and transportation shall be considered, so as to determine the source of fill and the destination of spoil (waste earthwork), and the quantity and transportation volume of priced earthworks. It is expected to reasonably solve the problem of balance and utilization of earthworks in each road section via earthwork allocation. The earthwork excavated can be used to fill under the condition of economic and reasonable transportation so as to avoid unnecessary borrow and spoil of earthworks and minimize the amount of farmland occupied.

**1) Allocation principle**

(1) In the fill-cut section, it is necessary to first use cut for fill in this section to reach transverse balance and then conduct longitudinal allocation so as to reduce the total transportation volume.

(2) The influence of bridge and culvert location on transportation shall be taken into account when allocating. Generally, earth shall not be transported across large ditches. Meanwhile, attention shall be paid to the possibility and convenience of construction and to avoid or reduce earth transportation on the upgrade as much as possible.

(3) In order to make the earthwork allocation reasonable, it is necessary to select the appropriate transportation mode according to the terrain and construction conditions, and determine the reasonable economic transportation distance, so as to analyze whether the earthwork is obtained by longitudinal transportation or nearby borrow.

(4) Although the economic transportation distance shall be taken into account in earthwork allocation, it is not the only index considered. It is also necessary to comprehensively consider such issues as land occupation by borrow and spoil of the earthwork, compensation for young crops loss and impact on agricultural production. Sometimes, although the transportation distance is a bit longer than economical transportation distance and the transportation cost may be slightly higher, it will be still economic if less land and agricultural production are occupied and influenced.

(5) Different types of earth and rock shall be allocated according to the needs of the project to ensure the stability of the subgrade and the material supply of artificial structures.

(6) For the switch-back curve section on the hillside, priority shall be given to the longitudinal allocation of the earthwork in the up and down directions.

(7) For earthwork allocation, it is necessary to negotiate with local authorities in advance for proper spoil and borrow of the earthwork. The site of the borrow pit shall be selected according to the terrain and farmland planning, and the measures of returning the borrowed earth to farmland and land preparation shall be taken into consideration. The spoil shall not occupy the arable land; if possible, it should be leveled into arable land to prevent blocking

the river and damaging farmland.

**2) Allocation method**

There are many methods of earthwork allocation, such as the cumulative curve method, the allocation chart method, and the earthwork calculation table method. At present, the earthwork calculation table method is often used in production, which can be directly allocated on the earthwork table without drawing the cumulative curve and allocation diagram. It has the advantages of simple practice, clear deployment and sound accuracy. The specific steps are as follows:

(1) Allocation is based on the calculation and review of earthwork quantity. Before allocation, the location of bridges and culverts, steep slopes and large ditches that may affect transportation shall be marked in the table for reference.

(2) Be clear of the fill and cut conditions of the subgrade among different stakes and make transverse balance first; clarify the quantities of the utilized, insufficient, and available distant earthwork of certain stakes.

(3) Before longitudinal allocation, the reasonable economic transportation distance shall be determined according to the construction method and the possible transportation mode.

(4) According to the distribution of the earthwork shortage and available distant earthwork, combined with the profile grade and natural conditions of routes, the allocation scheme shall be specifically formulated based on the principles of economy and supporting agriculture. The method is to transport earthwork to sections of earthwork shortage either by nearby longitudinal allocation or by transportation far from the neighboring section, stake by stake, section by section. The specific allocation orientation and quantity are indicated in the longitudinal utilization-allocation column with arrows, as shown in Table 4 – 20.

(5) After longitudinal allocation, if there is still earthwork shortage or left distant earthwork, the locations of the borrow pit or spoil bank shall be determined through consultation with the local authorities. And the quantity of borrow or spoil and the distant earthwork shall be filled in the column, respectively.

(6) After the allocation of the earthwork, it shall be checked according to the following formulas.

## Table 4 - 20 Calculation table of subgrade earthwork quantity

| Stake | Area of cross section/m² | | | Average area/m² | | | Distance /m | Classification and quantity of cut/m³ | | | | | | | | | | | | | Quantity of fill/m³ | | Quantity of utilization units/m³ and longitudinal allocation | | | | | | | Note |
|---|---|---|---|---|---|---|---|---|---|---|---|---|---|---|---|---|---|---|---|---|---|---|---|---|---|---|---|---|---|---|
| | Cut | Fill | | Cut | Fill | | | Total Quantity | Earth | | | | | | Rock | | | | | | | | Utilization of this stake | | Shortage of fill | | Surplus of cut | | Diagram of longitudinal allocation for long-distance transportation utilization | |
| | | | | | | | | | I | | II | | III | | IV | | V | | VI | | | | | | | | | | | |
| | | Earth | Rock | | Earth | Rock | | | % | Quantity | % | Quantity | % | Quantity | % | Quantity | % | Quantity | % | Quantity | Earth | Rock | Earth | Rock | Earth | Rock | Earth | Rock | | |
| 1 | 2 | 3 | 4 | 5 | 6 | 7 | 8 | 9 | 10 | 11 | 12 | 13 | 14 | 15 | 16 | 17 | 18 | 19 | 20 | 21 | 22 | 23 | 24 | 25 | 26 | 27 | 28 | 29 | 30 | 31 |
| K0+000.000 | 4.1 | 0.2 | | | | | | | | | | | | | | | | | | | | | | | | | | | | |
| | | | | 5.1 | 0.2 | | 10.0 | 51.0 | 100.0 | 51.0 | | | | | | | | | | | 2.3 | | | | 2.3 | | 51.0 | | | |
| K0+010.000 | 6.1 | 0.3 | | | | | | | | | | | | | | | | | | | | | | | | | | | | |
| | | | | 4.9 | 0.3 | | 10.0 | 49.3 | 100.0 | 49.3 | | | | | | | | | | | 2.6 | | | | 2.6 | | 49.3 | | | |
| K0+020.000 | 3.7 | 0.2 | | | | | | | | | | | | | | | | | | | | | | | | | | | | |
| | | | | 6.0 | 0.1 | | 10.0 | 59.8 | 100.0 | 59.8 | | | | | | | | | | | 1.2 | | | | 1.2 | | 59.8 | | | |
| K0+030.000 | 8.2 | | | | | | | | | | | | | | | | | | | | | | | | | | | | | |
| | | | | 8.9 | 0.1 | | 10.0 | 89.4 | 100.0 | 89.4 | | | | | | | | | | | 0.9 | | | | 0.9 | | 89.4 | | | |
| K0+040.000 | 9.6 | 0.2 | | | | | | | | | | | | | | | | | | | | | | | | | | | | |
| | | | | 15.2 | 0.1 | | 10.0 | 151.9 | 100.0 | 151.9 | | | | | | | | | | | 0.9 | | | | 0.9 | | 151.9 | | | |
| K0+050.000 | 20.7 | | | | | | | | | | | | | | | | | | | | | | | | | | | | | |
| | | | | 21.8 | 0.3 | | 10.0 | 218.1 | 100.0 | 218.1 | | | | | | | | | | | 3.5 | | | | 3.5 | | 218.1 | | | |
| K0+060.000 | 22.9 | 0.7 | | | | | | | | | | | | | | | | | | | | | | | | | | | | |
| | | | | 11.4 | 8.6 | | 10.0 | 114.4 | 100.0 | 114.4 | | | | | | | | | | | 86.4 | | | | 86.4 | | 114.4 | | | |
| K0+070.000 | | 16.6 | | | | | | | | | | | | | | | | | | | | | | | | | | | | |
| | | | | | 100.5 | | 10.0 | | 100.0 | | | | | | | | | | | | 1,005.4 | | | | 1,005.4 | | | | | |
| K0+080.000 | | 184.5 | | | | | | | | | | | | | | | | | | | | | | | | | | | | |
| | | | | | 187.7 | | 10.0 | | 100.0 | | | | | | | | | | | | 1,877.4 | | | | 1,877.4 | | | | | |
| K0+090.000 | | 191.0 | | | | | | | | | | | | | | | | | | | | | | | | | | | | |
| | | | | | 157.2 | | 10.0 | | 100.0 | | | | | | | | | | | | 1,571.9 | | | | 1,571.9 | | | | | |
| K0+100.000 | | 123.4 | | | | | | | | | | | | | | | | | | | | | | | | | | | | |
| | | | | | 98.4 | | 10.0 | | 100.0 | | | | | | | | | | | | 983.8 | | | | 983.8 | | | | | |
| K0+110.000 | | 73.4 | | | | | | | | | | | | | | | | | | | | | | | | | | | | |
| | | | | 12.9 | 36.7 | | 10.0 | 129.5 | 100.0 | 129.5 | | | | | | | | | | | 366.8 | | | | 366.8 | | 129.5 | | | |
| K0+120.000 | 25.9 | | | | | | | | | | | | | | | | | | | | | | | | | | | | | |
| | | | | 88.1 | | | 10.0 | 880.5 | 100.0 | 880.5 | | | | | | | | | | | | | | | | | 880.5 | | | |
| K0+130.000 | 150.2 | | | | | | | | | | | | | | | | | | | | | | | | | | | | | |
| | | | | 149.5 | | | 10.0 | 1,494.8 | 100.0 | 1,494.8 | | | | | | | | | | | | | | | | | 1,494.8 | | | |
| K0+140.000 | 148.7 | | | | | | | | | | | | | | | | | | | | | | | | | | | | | |
| | | | | 135.2 | | | 10.0 | 1,352.0 | 100.0 | 1,352.0 | | | | | | | | | | | | | | | | | 1,352.0 | | | |
| K0+150.000 | 121.7 | | | | | | | | | | | | | | | | | | | | | | | | | | | | | |
| | | | | 105.7 | | | 10.0 | 1,056.7 | 100.0 | 1,056.7 | | | | | | | | | | | | | | | | | 1,056.7 | | | |
| K0+160.000 | 89.7 | | | | | | | | | | | | | | | | | | | | | | | | | | | | | |
| | | | | 69.5 | | | 10.0 | 694.8 | 100.0 | 694.8 | | | | | | | | | | | | | | | | | 694.8 | | | |
| K0−170.000 | 49.3 | | | | | | | | | | | | | | | | | | | | | | | | | | | | | |
| | | | | 44.9 | | | 10.0 | 449.3 | 100.0 | 449.3 | | | | | | | | | | | | | | | | | 449.3 | | | |
| K0−180.000 | 40.6 | | | | | | | | | | | | | | | | | | | | | | | | | | | | | |
| | | | | 30.9 | | | 10.0 | 309.1 | 100.0 | 309.1 | | | | | | | | | | | | | | | | | 309.1 | | | |
| K0−190.000 | 21.2 | | | | | | | | | | | | | | | | | | | | | | | | | | | | | |
| | | | | 14.1 | 0.0 | | 10.0 | 140.9 | 100.0 | 140.9 | | | | | | | | | | | | | | | | | 140.9 | | | |
| K0−200.000 | 6.9 | 0.0 | | | | | | | | | | | | | | | | | | | | | | | | | | | | |
| | | | | 5.6 | 0.4 | | 10.0 | 55.6 | 100.0 | 55.6 | | | | | | | | | | | 3.6 | | | | 3.6 | | 55.6 | | | |
| K0−210.000 | 4.2 | 0.7 | | | | | | | | | | | | | | | | | | | | | | | | | | | | |
| | | | | 4.2 | 1.0 | | 2.4 | 10.4 | 100.0 | 10.4 | | | | | | | | | | | 2.3 | | | | 2.3 | | 10.4 | | | |
| K0−212.443 | 4.3 | 1.2 | | | | | | | | | | | | | | | | | | | | | | | | | | | | |
| Total of this page | | | | | | | | 7,307.5 | 7,307.5 | | | | | | | | | | | | 5,909.0 | | | | 5,909.0 | | 7,307.5 | | | |
| Consecutive accumulation | | | | | | | | 7,307.5 | 7,307.5 | | | | | | | | | | | | 5,909.0 | | | | 5,909.0 | | 7,307.5 | | | |
| XXX highway | | | | | | | | Calculation table of earthwork quantity | | | | | | | | | | | Design | | | Review | | | Approve | | Date | | Chart ID | |

$$V_{\text{transverse transportation}}+V_{\text{longitudinal transportation}}+V_{\text{borrow}}=V_{\text{fill}} \tag{4-22}$$

$$V_{\text{cut}}+V_{\text{borrow}}=V_{\text{spoil}} \tag{4-23}$$

$$V_{\text{cut}}+V_{\text{borrow}}=V_{\text{fill}}+V_{\text{Spoil}} \tag{4-24}$$

The above inspection is generally carried out page by page. If there is cross-page allocation, its quantity shall be taken into account. Whether there are errors in the allocation and calculation process can be identified via review. After verification, the quantity, transportation distance and volume of the priced earthwork can be calculated respectively to provide data for the preparation of budgets.

**3) Issues in allocation and calculation**

(1) Economic transportation distance

The sources of earthworks for fill section are as follows: first, by longitudinal transportation; second, nearby borrow outside the road. In general, it is more economical to transport the cut earthwork to fill section within a shorter distance. However, if the transportation distance is too long, the transportation cost exceeds the cost of borrowing the earthwork near the fill section. Then nearby borrow will be more economical. Therefore, there is a limit distance whether to "transport" or "borrow", which is the so-called "economic transportation distance". The calculation formula is as follows:

$$L_E=\frac{B}{T}+L_F \tag{4-25}$$

Where: $L_E$—economic transportation distance, km;

$B$—unit price of the borrowed earthwork, yuan/m$^3$;

$T$—unit price of distant transportation, yuan/(m$^3$ · km);

$L_F$—free transportation distance, km.

It can be seen that the economic transportation distance is an index to evaluate the earthwork borrowing or transportation. When the transportation distance is less than the economic transportation distance, the longitudinal transportation is economic; and vice versa.

(2) Average transportation distance

The transportation distance of earthwork allocation generally refers to the average transportation distance, which is the distance from the center of gravity of cut volume to that of fill volume. In order to simplify the calculation, the average transportation distance can be calculated by the distance from the center of the cut section to that of fill the section.

In longitudinal allocation, when the average transportation distance exceeds the free transportation distance stipulated in the quota, the earthwork transportation volume shall be calculated according to the excess transportation distance.

(3) Transportation volume

The earthwork transportation volume is the product of average transportation distance and earthwork allocation quantity, and the unit of transportation volume is "unit-cubic meter". In the engineering quota, the average transportation distance is divided into a transportation unit every 10 m, namely a "unit". 20 m is divided into two transportation units, which are called "2 units", and so on. In the calculation of the earthwork, it can be indicated by symbols ① and ②. When it is less than 10 m, it is still calculated according to ① or rounded. The formula is as follows:

Total transportation volume = volume of earthwork allocation $\times n$

where, $n$—units of average transport distance unit, and its value is:

$$n=\frac{L-L_F}{10} \tag{4-26}$$

where, $L$—average transportation distance, m;

$L_F$—free transportation distance, m.

(4) Priced earthwork

In the allocation of the earthwork, all cuts, whether "discarded" or "transported", shall be priced; but fills shall be determined according to the earthwork origination. If the earthwork is borrowed from outside the road, it shall be priced; if it is longitudinally transported for fill, it shall not be priced. Therefore, the quantity of priced earthworks shall be determined after the allocation of earthworks, and the quantity is

$$V_{\text{priced}}=V_{\text{cut}}+V_{\text{borrow}} \tag{4-27}$$

**Questions:**

1. What are the basic forms of the cross section of highways and urban roads?
2. What are the forms of widening transition?
3. What is the superelevation? Why set the superelevation?

4. Are the length of the spiral, superelevation transition section and widening transition section the same? What are their functions and relationships?
5. What are the ways to set the superelevation? What is the design method? What shall be considered in the design?
6. The design speed of a class Ⅲ highway is 40 km/h, the pavement width is 7.0 m, the shoulder width is 0.75 m, the cross slope of the crown section is 2%, the cross slope of the shoulder is 3%, the sideway force coefficient is 0.05, the deflection angle $\alpha$ of a horizontal curve is $82°17'26''$, the radius $R$ is 150 m, and the spiral length $L_S$ is 40 m.
   (1) Does the curve need to be widened and superelevated? Why?
   (2) If it is needed to set the superelevation, please calculate the rate of the superelevated cross slope and the length of the superelevation transition section.
   (3) If the widening value of the curve is 1.5 m, please calculate the design elevation at the start of the horizontal curve and the subgrade width at the middle point of the horizontal curve, respectively.
7. A two-lane highway is with a design speed of 60 km/h, and the subgrade width and pavement width are 8.5 m and 7.0 m. The radius of a horizontal curve is 150 m, the spiral length $L_S$ is 60 m, and the deflection angle $\alpha$ is $48°52'36''$. An obstacle near the inside center of the curve is 2.8 m away from the edge of the subgrade. Please check whether the curve can meet the requirements of stopping sight distance and passing sight distance. If not, what is the maximum width to be cleared and where is the position? Please draw a diagram to illustrate.
8. How to calculate and allocate the quantity of the subgrade earthwork?

# Chapter 5 Route Selection and Alignment Location

【**Objectives**: 1. Master the concept of road design. 2. Be familiar with the overall design principles and main contents. 3. Understand the relationship between the overall design, route selection, and alignment location. 4. Understand the contents of route selection, alignment location and the dynamic changes of their relationship. 5. Be familiar with the main contents and methods of route selection in the level and mountainous terrain, and have the preliminary ability to use computer-aided tools for route selection and alignment location on (digital) maps.】

## 5.1 Introduction

Route selection and alignment location of roads are the whole process of determining a technically feasible and economically reasonable route centerline through comprehensive comparison, which is based on the determination of the basic trend, construction scale, and technical standards of the route. In this process, safety, environmental protection, land use, and other factors are considered in combination with the topographic and geological conditions. In addition, the subgrade and pavement, bridges and culverts, tunnels, intersection hubs, drainage protection, and other engineering works have been taken into account.

Traditionally, route selection and alignment location need to go through three stages from global to local: route scheme design, route belt design and specific alignment location. Route scheme design mainly determines the start and end points, trends and important points of the route, and usually compares and selects several schemes, which is a global work. On the basis of outlining the basic route orientation, route belt design refers to the addition of some detailed control points according to the topographic, geological, hydrological and other

© The Author(s), under exclusive license to Springer Nature Singapore Pte Ltd. 2025
B. Yu and J. Cheng, *Road Survey and Design*,
https://doi.org/10.1007/978-981-97-4836-5_5

conditions, commonly known as the route layout, which belongs to local optimization work. According to the technical standards and route scheme, the specific alignment location is the comprehensive horizontal, vertical and cross section design based on the above two steps, which is the final stage and final result of route selection and alignment location.

What is the relationship between route selection and alignment location?

Route selection usually has several meanings. Firstly, it refers to the determination of the road route scheme in the early stage of road design, performed in the pre-feasibility study, project feasibility study or preliminary design stage. Secondly, it refers to the comparison of multiple schemes. Thirdly, it refers to the adjustment of local alignments of the tentative route scheme. The task of alignment location is to determine the final road centerline, accompanied by field survey. Route selection and alignment location are two complementary stages. With the rapid development of survey technologies and computer-aided road design technology, the contents and relationship of the two stages have also changed as follows:

(1) In the 1980s and before the total station is not widely used, route selection is carried out on paper topographic maps. Because the scale of topographic maps is usually not very high and the information update is infrequent, alignment location needs to supplement a lot of on-site survey work. In addition, the surveying and mapping equipment for locating the road centerline include theodolite, level gauge and electronic tachymeter. Much work is devoted to determining road centerline accurately and effectively. Accordingly, the road centerline is adjusted and optimized on-site to determine the ultimate one.

(2) The total station is put into use in the 1990s and electronic topographic maps and GPS equipment are prevalent in the 2000s, which greatly improves the quality and efficiency of route selection and alignment location on paper (digital) topographic maps. The coordinate method can be directly used for field layout, which leads to the efficiency improvement of field survey and workload reduction.

(3) After 2010, with the rapid development of LiDAR and oblique photography technology, the quality and efficiency of obtaining ground points and building 3D models have been greatly improved. On the other hand, the continuous improvement of computer-aided design technologies, including BIM

technology, has greatly boosted the level of indoor route selection and alignment location. Route design results are close to the field effect so the on-site survey and design work is mainly to supplement the survey of key positions such as bridges, tunnels and hubs. The present route selection and alignment location technology can basically achieve "which route to select is what alignment to get".

## 5.2 New concept and overall road design

### 5.2.1 New concept of road design

"The idea of design is the soul!"

The concepts of road design are closely related to the level of social and economic development. China's road design concepts have experienced the following three stages. From the 1950s to the 1970s, the guiding ideology is "more, fast, good, cheap, safe and economic". From the 1980s to the late 1990s, it becomes "fast, safe, comfortable and economic". Since the 2000s, under the premise of "people-oriented and safety first", road design and survey has formed new concepts:

"Adhere to the people-oriented principle and establish the concept of safety first.

Adhere to harmony between human and nature and establish the concept of respecting nature and protecting environment.

Adhere to sustainable development and establish the concept of saving resources.

Adhere to quality first and establish the concept of public satisfaction.

Adhere to reasonable selection of standards and establish the concept of design and creation.

Adhere to the idea of system theory and establish the concept of life cycle cost analysis."

Similarly, the concept of context-sensitive design widely followed in the world pursues not only the smoothness and safety "inside the road", but also the coordination with the natural and cultural environments "outside the road", which reflects the design concept of "thinking beyond road". Its core idea is consistent with the six new concepts in China.

### 5.2.2 Overall design principles

The overall design is a comprehensive design of scheme formulation, design conception, and coordination between disciplines inside and outside the project on the premise of considering the construction scale and design standards. The overall design of roads at all classes is mainly based on the highway function, highway class and its role in the road network. This section describes the corresponding design principles for roads in level-rolling areas and mountainous areas.

**1) Overall design principles of roads in level-rolling areas**

Level-rolling areas have the features of plain terrain, dense towns, the large population and developed agriculture. Natural obstacles frequently encountered in this area include villages and towns, farmland, rivers, lakes, etc. The overall design of roads in level-rolling areas shall generally follow the following principles:

(1) Follow the requirements of the overall layout of regional road network planning

The overall scheme and layout of the route shall properly handle the relationship between the proposed project and the arterial road network as well as its planning. It shall reasonably select the location of the collector-distribution center of traffic to fully play the role of roads, so as to provide sustainable development conditions for the area where the route passes.

(2) Consider the relationship between the route and towns

The route shall be close to but not into towns and away from but not neglecting towns. Therefore, the overall plan of the route shall be arranged in combination with the current situation of cities and towns along the route. A reasonable relationship between the route and towns is beneficial to harvest the best operation benefits of roads and promote the economic development of towns alongside the route.

(3) Consider the relationship between the route and agriculture

It is not allowed to occupy a large area of good farmland by unilaterally pursuing a straight route, nor is it necessary to strictly control the land occupation by lowering the alignment standard. In the design, sufficient connections shall be provided for the production and life of residents along the route by reasonably arranging bridges, interchanges, channels and culverts. If possible, the route shall be conducive to farmland creation and protection.

(4) Pay attention to coordination between the route and bridges

Due to the large number of rivers and lakes in level-rolling areas and the large quantities of bridges and culverts, the smoothness of the route shall be maintained in both horizontal and vertical planes as far as possible when crossing rivers.

(5) Pursue short and straight alignment

The ground grade of level-rolling areas is relatively gentle, so it is possible to pursue a short and straight route. Within the scope of reasonable project cost, various indicators in the specification shall be flexibly selected to achieve safety, comfort, efficiency and economy.

(6) Mediate conflict between the route and land occupation and demolition

In order to reduce the demolition, the overall design of the route shall strive to make a detour around important facilities such as power, telecommunications and national defense optical cables and public facilities such as factories and schools. In addition, the scale of land acquisition shall be minimized to protect the increasingly scarce land resources on which local people depend.

(7) Emphasize coordination with social and ecological environments

The overall design shall reduce damages to the ecological environment and cultural landscape and pay attention to the balance, continuity, and coordination of route indicators. Great importance shall be attached to the environmental protection design and the prevention of water and soil loss as well as noise disturbance. Elegant road design is pursued.

**2) Overall design principles of roads in mountainous areas**

Undulating terrain, complex geological conditions and changeable climate are characteristics of mountainous areas, so there are only a few suitable lands for engineering projects and road corridors. Because the horizontal, vertical and cross section designs are greatly limited, the technical indicators often adopt lower standards. Route selection in mountainous areas mainly includes the stream (river) line, over-ridge route, and ridge route. The overall design of roads in such terrain shall generally comply with the following principles.

(1) Take topographic route selection as basis, geological route selection as warrant, and ecological route selection as priority.

Topographic route selection means that the selected route shall conform to terrain, avoid large cut and fill, control the project scale, reduce project cost and protect the regional ecological environment. Geological route selection refers to studying the geological conditions in the route corridor, reasonably placing the route, determing detour plan about large areas of adverse geological zones, fundamentally improving the ability to resist natural disasters, ensuring construction and operation safety, and reducing project risks. Ecological route selection means that when comparing and selecting route schemes, it shall not only focus on the route and project scheme itself, but also pay attention to ecological and environmental protection.

(2) Pay attention to people-orientation and traffic safety design

The principle of "people-orientation, prevention first, combination of fault tolerance and protection" is advocated, and the scheme of the gentle grade, balanced alignment indicators and driving safety is desired. Alignment design shall strengthen the inspection, improve the combination of adjacent sections, reduce the allowable speed difference of adjacent sections, and improve the continuity and consistency of alignment design. Climbing lanes, emergency escape ramps and safety protection facilities shall be used to improve traffic safety and eliminate potential hazards.

(3) Utilize route corridor resources reasonably

Resources of the route corridors in mountainous areas are limited, and the proposed road is often located in the same corridor as the existing railway, road, pipeline and other projects. Therefore, it is necessary to conduct overall planning, propose rational layout, combine short-term and long-term needs, and perform comprehensive utilization. For example, in addition to leaving enough clearance, the needs of reconstruction and expansion of electrification shall also be considered for the overpass bridge crossing with the railway. Maintenance channels shall be set when crossing with pipelines.

(4) Consider relationship between road construction and natural and cultural landscapes

A beautiful ecological environment forms an attractive natural landscape, which is a good place to relax and travel. Therefore, the overall design of roads shall respect the importance of natural and cultural landscapes. A route shall not only coordinate with the surrounding environment and emphasize aesthetic feelings, but also reasonably provide parking areas, service areas, observation platforms, and other facilities along the route, which is conducive to the development of local tourism.

(5) Handle the contradiction between road construction and land occupation and demolition

The land available for agricultural cultivation in mountainous areas is scarce. Most of the highly cultivated land is distributed in plains or river valleys of mountains, and residence is also located herein. These areas are often places for superior route corridors. Therefore, the overall design shall minimize the occupation of high-yield fields, crop fields or forest gardens, so as to protect land resources on which the local people depend. In addition, the overall design shall also comprehensively compare schemes of land occupation, demolition, route detour and structure increase, so as to reasonably determine the implementation of land reclamation and residents' relocation.

(6) Consider the relationship between the route and the water source

There are abundant water resources in mountainous areas because of their unique terrain and ecological environment, which are often the water source for downstream residents. Road design shall protect water sources and avoid water and soil loss.

(7) Consider the balance of cut and fill

Earthwork quantity is an important indicator of road construction in mountainous areas. The reasonable planning and design of the route, proper use of technical indicators, and "replacing roads by bridges and tunnels" can help strive for the balance of cut and fill.

### 5.2.3 Main contents of the overall design

Route scheme selection is the pioneer of overall design of roads, and the realization of the overall design goal is under the lead of route discipline. However, in this process, it is also necessary to coordinate the relationships among the subgrade, bridge and culvert, tunnel, route intersection, traffic engineering, and facilities along the route, and to clarify the relevant design boundary and interface. It aims to form a complete system engineering and achieve the overall goal of safety, environmental protection, and sustainable development. The main contents of overall design include the determination of road functions and technical standards, construction scales and schemes, traffic engineering and environmental protection, design inspection and safety evaluation. Due to different construction conditions and characteristics of road projects, contents of overall design are slightly different. With the deepening of the road engineering construction stage, the contents of overall design are continuously refined.

**1) Determine road functions and technical standards**

Highway classes and technical standards shall be determined according to the highway function, traffic composition, vehicle type composition, terrain and geology along the highway, and other factors and parameters. It includes the following contents:

(1) The technical class of highways is comprehensively demonstrated according to the function of highways, combined with traffic volume and construction conditions. The same highway project may adopt different technical standards in different sections according to the changes in functions and traffic volume.

(2) The design vehicle shall be determined according to the road function, traffic composition, and vehicle type composition.

(3) For the freeway and Class Ⅰ highway, the lane number in the basic segment shall be determined according to the highway function and design traffic volume. And the lane number shall be increased by an even number.

(4) For highways of different classes, design speed may be selected by sections according to the changes of terrain, geology and natural conditions along the highway.

(5) The main control indexes such as horizontal and vertical planes, sight distance, superelevation and widening shall be reasonably determined according to the design speed of the road section, terrain, geology, environment, traffic demand, and other factors along the highway.

(6) Composition and width of highway cross sections shall be

comprehensively determined according to the technical class of highways, design traffic volume, environment along the highway and functions of each component of cross sections.

(7) The reconstructed highway shall follow the design highway technical standards and indexes. For sections reusing the original highway, the increment of design speed may induce engineering geological distress, increase project cost or produce adverse effects on environmental protection and cultural relics. The original design speed and index may be maintained for the local section after demonstration. It shall be noted that the length of such section should not be greater than 15 km for the freeway, and should not be greater than 10 km for Class Ⅰ and Class Ⅱ highways. And the technical class shall not be further reduced.

**2) Determine project scale and construction scheme**

(1) Project scale: It is necessary to comprehensively investigate the layout and planning of the railway, waterway, aviation, pipeline, and other comprehensive transportation systems in the route corridor, the current situation and development planning of industrial and mining enterprises, and the development and utilization of natural resources. On this basis, the start and end points, main control points, route length, number of intersections, management and service facilities are determined according to the highway network planning and highway function so as to determine the construction scale. The construction mode of the project and whether to construct in phase shall be demonstrated and determined according to the overall construction scale of the project, construction conditions of the control project, traffic volume growth, and project fund-raising and other relevant factors.

(2) Construction scheme: The appropriate cross section form of highways shall be selected. To study the relationship between the highway and adjacent railways and pipelines, it is necessary to rationally determine the direction and location of railways and various pipeline facilities based on the investigation. The crossing mode between the highway project and other highways shall be comprehensively determined according to the highway function, class and traffic organization mode. Highway reconstruction shall follow the principle of combining utilization and reconstruction. On the basis of traffic safety evaluation

of the original highway, as well as the survey and evaluation of the original subgrade, bridge and tunnel, the original highway could be reasonably and thoroughly utilized, which is conducted with the demonstration of utilization principles and utilization schemes of the existing routes and structures.

**3) Traffic engineering and environmental protection**

(1) Traffic engineering: Traffic engineering and facilities along the highway shall be designed synchronously with the main works. The location, type, spacing and configuration scale of management and service facilities such as the highway toll station, service area and parking area shall be reasonably determined according to the needs of the highway function and class, traffic organization mode, safety and operation management. If necessary, the scheme of planning in one time and construction in phase can be demonstrated according to the development needs such as traffic volume.

(2) Environmental protection: Road construction shall strictly implement environmental impact assessment of engineering construction projects, preparation of water and soil conservation plans and environmental protection systems. In the overall design, the relationship between highway construction and environment shall be coordinated in combination with the actual situation of the project, and relevant environmental protection measures shall be implemented to reduce the adverse impact on environment.

**4) Design inspection and safety evaluation**

Freeways and Class Ⅰ and Class Ⅱ highways shall perform traffic safety evaluation during design, and other highways could also perform traffic safety evaluation when feasible. For traffic safety evaluation, the operating speed method is used to analyze and exam alignment design, geometric design and alignment combination, so as to evaluate the coordination and consistency of operating speed. Based on traffic safety evaluation, alignment design and geometric index shall be optimized, and traffic safety facilities and management measures shall be checked.

## 5.3 Route selection

The route is the skeleton of roads, and its pros and cons affect the realization of road functions and its role in road network. In addition to the influence of natural conditions, route design is also restricted by many social factors. Not only the economic constraints, but also the requirements of safety, environmental protection, speed and elegance shall be considered. In order to ensure the quality of route selection, survey and design and reduce project cost, it is necessary to comprehensively consider the schemes, from coarse to fine, from outline to concrete. Multiple schemes shall be analyzed and compared in stages so as to determine the most reasonable route. This section mainly introduces the relevant contents of highway route selection. For route selection of urban roads, it mainly depends on the urban arterial road network and red line planning.

### 5.3.1 General principles and requirements for highway route selection

Many factors shall be considered in route selection, with the basic principles as follows:

(1) Urban layout and economic development planning of the area through which the route passes shall be obtained. The route scheme shall coordinate with the urban planning along the highway so as to promote local economic development.

(2) According to the highway network planning, the route scheme shall consider the cooperation between various transportation systems in the corridor and between different classes of road networks, so as to give full play to the comprehensive transportation benefits of highways.

(3) Route selection shall be refined from areas to corridors and from corridors to lines. The route scheme shall be determined based on the investigation and survey of topography, geology, hydrology, climate and meteorology, environmental sensitive period, etc. When there are multiple feasible route schemes in the section with the same start and end points, they shall be comprehensively compared to obtain the optimal route scheme.

(4) Routes shall conform to the planning of farmland and water conservancy construction, mineral resource development and urban development as far as possible.

(5) Route selection shall make full use of the construction land and strictly

protect the agricultural cultivated land. It is necessary to consider the capital construction on farmland in route selection. Farmland shall be less occupied, and high-yield farmlands or economic forest gardens (such as rubber forests, tea forests and orchards) shall not be occupied as much as possible. For the farmlands that must be occupied along the highway, schemes of land reclamation and land return shall be designed in accordance with the relevant national regulations.

(6) The route shall make a detour about immovable cultural relics, water sources and nature reserves as far as possible.

(7) The route shall keep a certain safe distance from flammable, explosive and other hazard sources and pollution sources.

(8) The highway reconstruction project shall pay attention to resource saving, adhere to the principle of combining utilization in reconstruction, and make rational use of the original project.

(9) During route selection, the opinions of the local government and people along the highway shall be consulted.

Highway route selection shall meet the following requirements:

(1) Engineering geology and hydrogeology of the area, corridor and route shall be deeply investigated to find out its impact on highway engineering. In the case of sections with adverse engineering geology, schemes of bypass, avoidance and crossing shall be demonstrated according to the impact on the route.

(2) Various sensitive points and mineral resources along the highway shall be investigated to study their impact on the route scheme, which is conducive to the rational selection of route location.

(3) Connections between the freeway, Class Ⅰ highway and traffic generation sources along the highway shall be realized through a regional road network or new connective roads.

(4) On the premise of conforming to the overall function and direction of the project, Class Ⅱ and Class Ⅲ highways shall avoid crossing cities and towns as far as possible, so as to reduce interference between traffic and the lives of urban residents and ensure operation safety.

(5) Geometric location relationship with relevant roads, railways, transmission lines, oil and gas pipelines, etc. shall be comprehensively considered

to make rational use of corridor resources and save the land.

(6) It is necessary to consider the position and elevation relationship of bridges, tunnels, interchanges, service areas, and other structures.

The above route selection principles and requirements are suitable for roads at all classes, but they shall be applied flexibly subject to project conditions.

### 5.3.2 Steps and methods of route selection

When the start and end points of a route are determined, there are many paths connecting them. Many factors affect route selection, some of which are contradictory and some restrict each other. The importance of each factor varies with the scenario, so it is impossible to determine the ideal scheme at one time. The most effective way is to start from a large area and gradually optimize the route scheme from areas to lines and from frames to refinements. Finally, the specific location of the route is determined through comprehensive economic and technical comparisons. Route selection is generally divided into the following three steps according to working contents.

**1) Route scheme selection**

Route scheme selection mainly studies the start and end points and important control points of the proposed highway and frame the general direction of the route. This work usually searches various possible schemes within a large area on a small-scale (1 : 50,000 – 1 : 100,000) topographic map. Then a preliminary comparison of possible schemes is conducted through the collected relevant data, based on which all possible route alignment schemes are proposed. According to the field survey, important control points and route trends are basically determined after the comparison of multiple schemes. When it is a lack of a topographic map, surveys or reconnaissance could be carried out to collect data on site for scheme comparison. When there is complex terrain or a large area, data can be collected through aviation techniques (such as aerial photography, remote sensing, etc.) for scheme comparison.

**2) Route corridor selection**

Based on the preliminary determination of the start and end points, important control points and the trend of routes in the pre-feasibility study stage, some detailed control points are determined according to the terrain, geology, hydrology and other natural conditions, which constitute the route corridor, also known as the route belt or route layout. By comparing different route corridor schemes and local route schemes and estimating the project scale, the final route corridor is basically determined. The determination of the route corridor shall generally be carried out on the topographic map with a scale of 1 : 5,000 - 1 : 10,000. Only sections with simple terrain and clear schemes may be directly selected on site.

**3) Alignment location**

Alignment location is conducted based on the above two steps. According to the technical standards and the route scheme, horizontal, vertical and cross section designs are performed in the favorable route corridor to determine the road centerline. There are two methods of alignment location: map location (including digital map) and field location.

## 5.3.3 Route selection in the pre-feasibility study stage

The screening, comparison, and optimization of route schemes follow throughout all stages of the project. In the pre-feasibility study stage, it mainly starts from the function and role of construction projects in the highway network and puts forward route schemes to ensure their function and role through analysis, research and demonstration. In addition, according to the requirements of highway network planning, the relationship between the route of the proposed project and relevant projects shall be coordinated to give full play to the overall function of highway network. The route scheme shall properly handle the connective relationship with towns along the route and other transportation modes. The location of super-large bridges and long tunnels that may occur when crossing rivers or mountains shall be determined in advance. All possible route schemes shall be listed to basically determine the important control points and the route trends.

Route schemes shall be determined according to the three stages: basic data investigation and collection, screening of possible route schemes and comprehensive comparison of schemes. The steps and methods are shown in Figure 5 - 1.

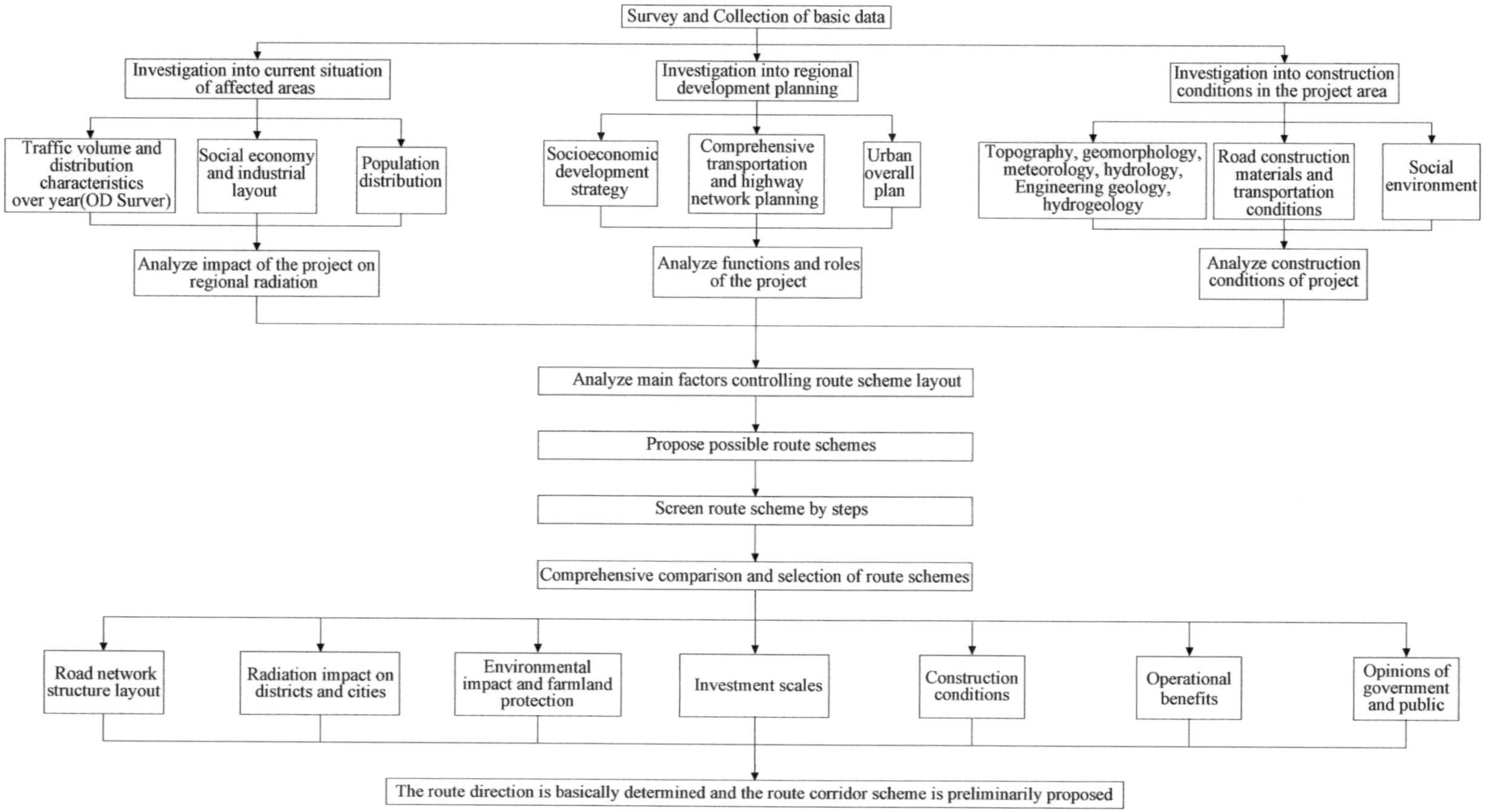


**Figure 5 - 1 Flow chart of route determination in the pre-feasibility study stage**

**1) Basic data investigation and collection**

(1) Investigation into the present situation of project-affected areas. Traffic volume and traffic distribution characteristics (including traffic volume origination and destination (OD) survey), economic development status, industrial layout characteristics, population distribution, and other data need to be collected. Traffic characteristics of the project-affected area and the current traffic volume distribution of road network need to be analyzed. In addition, traffic volume development trends of the proposed project and regional road network need to be predicted. The impact scope of different route schemes on regional radiation and the degree of stimulating local economic development and meeting regional traffic demand need to be analyzed.

(2) Investigation into regional development planning affected by the project. The social and economic status and development planning, comprehensive transportation development planning, highway network planning, urban overall planning, and land use planning of the project-affected areas need to be collected. The role of the proposed project in the comprehensive transportation network and highway network needs to be analyzed. The rationality of different route schemes in the highway network and their coordination with urban planning also need to be analyzed.

(3) Investigation into construction conditions in the project area. Topographic, geomorphological, meteorological, hydrological, engineering geology and hydrogeology, unfavorable geology and special geotechnical and other characteristic data need to be collected. Sources of road construction materials and transportation conditions of the project need to be investigated as well. Social and environmental analysis and investigation shall be performed.

In addition to the collection of the above data, route scheme selection shall fully consider the development needs of local economy along the highway, and consult the opinions of local governments and relevant administrative departments (including urban planning, transportation, farmland, water conservancy, environmental protection, railway, tourism, cultural relics, waterway, etc.) on the proposed highway route scheme. Their opinions and suggestions on the start and end points, main control points, route direction, connections with urban access roads and other roads of the proposed highway shall be considered.

**2) Screening of possible route schemes**

Selection of start and end points of the route shall be based on the nodes determined in the approved highway network planning. Government and administrative transportation department shall propose the preliminary node location or town name in the letter of authorization for feasibility study of the proposed project. Through demonstration, comparison from the aspects of technology and economy, different start-end connection schemes shall be put forward according to the requirements of road network connection and traffic conversion.

Selection of important control points shall consider the influence of many factors such as the urbanization level, population distribution, resource distribution, productivity layout, natural geographical conditions, and so on. Selection of important control nodes needs to meet the demands of transportation, promote the development of regional economy and urbanization, attract traffic flow to the greatest extent, and improve the efficiency of transportation channels. Selection of the route direction shall be coordinated with the overall region planning.

**3) Comprehensive comparison of schemes**

Both qualitative and quantitative methods shall be employed for the comprehensive comparison of route schemes. Evaluation by individual indicators, such as engineering quantities (economy), shall be avoided. Multi-objective analysis and demonstration shall be carried out from the following aspects:

(1) The route scheme shall meet the requirements of highway network planning. The road network structure shall be reasonable, and connection between the scheme and urban road network planning along the highway shall be coordinated.

(2) The route scheme shall promote the development of regional economy and form an effective radiation influence range.

(3) The route scheme shall coordinate with natural and social environment.

(4) The route scheme shall compare project quantity and investment, and analyze the construction conditions of control projects such as large structures, so as to reduce project cost, save investment and facilitate construction.

(5) The route scheme shall evaluate and compare the topographical and geological conditions, adverse geological distribution, road construction materials, transportation conditions, construction site layout, construction access roads

local government support, etc.

(6) The route scheme shall meet regional traffic demand, attract local traffic, and give full play to the overall operation benefits of highways.

(7) The route scheme shall consider the opinions and suggestions of local governments and relevant departments.

After full comparison and demonstration, the main control points and route scheme are recommended.

### 5.3.4 Case of scheme comparison

**1) Scheme comparison of important control points**

Scheme comparison of a proposed freeway is as follows:

North Corridor: Short operation mileage, small proportion of bridges and tunnels, good basic conditions (the existing national highway), large service population and in line with the overall planning, which is conducive to opening the Sichuan-Tibet channel as soon as possible and forming a wider road network coverage.

**Figure 5 - 2 Schematic diagram of North-South Corridor schemes of the Sichuan-Tibet Freeway**

South Corridor: it can make up for the insufficient distribution of the southern road network, but the terrain is arduous, the construction scale and difficulty are large, the geological conditions are poor, and it is inconsistent with the planning.

The comprehensive comparison shows that it is more favorable to choose the

North Corridor.

**Table 5 - 1 Comparison of North-South Corridor Schemes of the Sichuan-Tibet Freeway**

| No. | Items | Unit | North Corridor | South Corridor |
|---|---|---|---|---|
| 1 | Construction mileage | km | 208 | 247 |
| 2 | Operation mileage | km | 210 | 299.3 |
| 3 | Subgrade earthwork | $10^4$ $m^3$ | 2,981 | 3,510 |
| 4 | Protective engineering | $10^4$ $m^3$ | 123 | 175 |
| 5 | Pavement engineering | $10^4 m^2$ | 3,876 | 4,563 |
| 6 | Total bridge length | m/count | 49,666/93 | 99,783/158 |
| 7 | Total tunnel length | m/count | 36,983/11 | 18,409/26 |
| 8 | Bridge-tunnel percentage | % | 41.29 | 47.83 |
| 9 | Interchange | count | 7 | 6 |
| 10 | Engineering geological conditions | — | Geological disasters affected by North Corridor are less than those of South Corridor, so they can be avoided or treated, and treatment difficulty of North Corridor is less than that of South Corridor | Adverse geological phenomena are relatively developed. Geological disasters are densely distributed on both sides of the corridor. The river valley is narrow and mountains on both sides are steep. It is difficult for the route to bypass geologic hazards |
| 11 | Construction conditions | — | K-line is close to G317 highway, with convenient transportation and good construction conditions | The road class in the corridor is very low, and there are no roads in some sections, so construction conditions are arduous |
| 12 | Consistency to planning | — | It is consistent with National Highway Network Planning | It is inconsistent with National Highway Network Planning |

**Continued Table 5 - 1**

| No. | Items | Unit | North Corridor | South Corridor |
|---|---|---|---|---|
| 13 | Local government opinions | — | Consistent with regional planning, it is a corridor recommended by local governments at all levels | It is not recommended by local governments at all levels |
| 14 | Total investment | hundred million¥ | 419 | 452 |

**2) Scheme comparison among important local points**

Overpassing Xuejila mountain section:

There is a tunnel on K-line with a length of 5,670 m and a total bridge length of 14.7 km. The Gongjue and Qingnidong tunnels are convenient for vehicles to go up and down, and construction transportation is convenient. But the operating mileage is the longest (46.5 km).

Compared with K-line, D1-line has a mileage reduction of 14 km. The total bridge length is 8,342 m and the total tunnel length (2 tunnels) is 8,746 m. The geological conditions are slightly better, and a 13 km connective line needs to be built. The bridge and tunnel are large in scale, poor in alignment, difficult in construction, high in construction cost, low in alignment index, and poor in alignment combination.

D2-line has the shortest operating mileage, but the tunnel is too long (17,064 m).

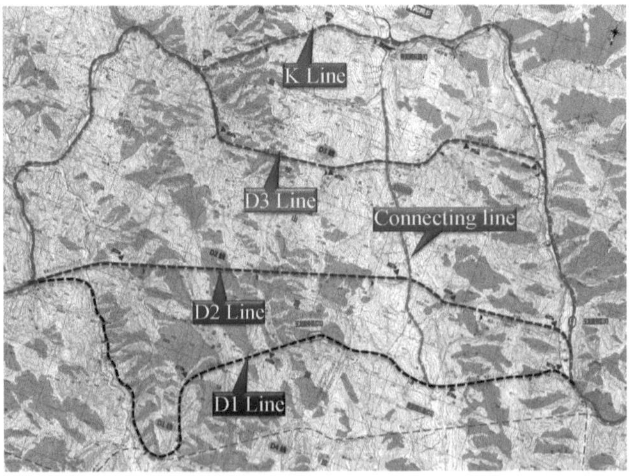


**Figure 5 - 3 Schematic plans for Xuejila section of the Sichuan-Tibet Freeway**

The mileage of D3-line is 2 km less than that of K-line. Although the height of the open line can be reduced, an additional tunnel (3,318 m) is required.

Comprehensive comparison suggests that a K-line corridor with lower cost is selected that may better serve Gongjue County.

### 5.3.5 Route selection in plain areas

The main feature of route selection in plain areas is to overcome plane obstacles, and the route plan shall be rationally laid out according to the function and nature of the proposed project. The main points are as follows:

(1) The terrain of plain areas has little restriction on the route. Route alignment shall be short and straight, with proper turning angles, matched curve lengths, and balanced horizontal and vertical technical indicators. When using lower technical indicators, attention shall be paid to the gradual transition of alignment. Long tangents and curves with small deflection angles are unfavorable.

(2) Routes shall adopt higher horizontal and vertical indicators as far as possible. When meeting the minimum fill height of the subgrade and the height of bridge and culvert construction, it shall adapt to the undulation of terrain, reduce the height of the subgrade as much as possible, and save project cost. With such practices, it is convenient to make full use of the original roadbed, bridge, and culvert projects when the road is upgraded in the future.

(3) For highway route selection and alignment location, corresponding environmental protection measures shall be taken according to the regional nature of social environment and ecological environment along the route. Residential drinking water source areas, habitats and growth areas of rare animals and plants shall be avoided. Main crop growing areas, orchards, nurseries and nature reserves should be avoided. If they could not be avoided, corresponding protection and management measures shall be taken; schools, hospitals, nursing homes and other sensitive areas shall be avoided, residential communities and densely housed villages and towns should be avoided. If it is impossible to bypass them, corresponding protection and prevention measures shall be implemented.

The setting conditions of bridges, culverts, crossings, channels, and other structures shall be comprehensively considered, and the favorable terrain shall be fully used to reduce the height of the subgrade and the quantity of the borrow earthwork. The borrow pit shall be selected on barren mountains and slopes as far as possible. If it is necessary to borrow the earthwork on both sides of the highway, reclamation and reconstruction designs shall be performed.

(4) Relationship between roads and agriculture shall be correctly handled. Route layout shall occupy as little arable land as possible, avoid cutting large plots of fertile land, and save land resources.

(5) The plane layout of the route shall be conducive to the traffic organization and the function of local road networks. For relatively developed and dense road networks, it can be summarized and sorted according to the class, traffic volume and importance of each road, and merged appropriately to reduce the number of intersections between the road network and the proposed project.

(6) Relationship between the route and towns shall be reasonably considered. There are more towns, villages, industries and other facilities in plain areas. The route shall bypass dense urban areas and produce less or no damage as much as possible, and adopt higher technical indicators to pass. The distance between the route and the periphery of towns shall be reasonable, which shall reserve enough space for the development of town and facilitate residents' travel simultaneously.

(7) As laying out of the route in the river network area, the river network shall be summarized according to the composition of irrigation canals, drainage canals, natural ditches, and rivers, as well as the characteristics of low gradient and slow flow rate, so as to distinguish the primary and secondary relationship and arrange the route location reasonably.

(8) Forms of intersections shall be reasonably determined. When two roads cross at-grade, the location of the route shall be selected according to the principle of priority for the main road; when the two roads are grade separation, the form of overpass or underpass shall be determined comprehensively according to the alignment in the front and back of the vertical profile section.

(9) When the route intersects or parallels with various pipe networks and pipelines, it shall meet the provisions of relevant industrial standards and specifications. The route shall bypass important power and telecommunications facilities as far as possible. When it must be close to or cross them, it shall

maintain sufficient distance and clearance and manage to dismantle no or less power and telecommunications facilities. When crude oil and natural gas transmission pipelines intersect with the freeway and Class Ⅰ highway, a special underground passage shall be used in the way of underpass. When they intersect with Class Ⅱ, Class Ⅲ and Class Ⅳ highways, the protective casing shall be embedded, and the buried depth shall not only meet the relevant industrial regulations, but also meet the relevant provisions of "General Code for Design of Highway Bridges and Culverts". The buried depth needs to be checked according to the vehicle load level of the crossed highway.

## 5.3.6 Route selection in mountainous areas

Mountainous areas have a distinct water system, and there are only two directions of mountainous highways: along the mountain and water and across the mountain. Generally, according to the landform and topographic characteristics along the highways, the route can be divided into four types: riverside (streamside) route, over-ridge route, on-ridge route, and hillside route (between riverside route and on-ridge route).

For one mountainous road, different route forms may be used for different segments. In this section, the layouts of the three routes are introduced, including the riverside (streamside) route, over-ridge route and on-ridge route.

**1) Riverside (streamside) route**

The riverside (streamside) route is the one laid out along the trend of rivers (stream), as shown in Figure 5 - 4.

**Figure 5 - 4 Riverside (streamside) route**

For rivers in the mountain, the bottom is generally not wide, and the terrace widths on both sides are different. There are a mixture of gentle slopes and steep slopes as well as shoals and cliffs along rivers. Most of the rivers are curved, with steep concave banks and gentle convex banks. Steep banks and gentle banks appear alternately if sticking to one side. Steep cliffs on both banks are next to valleys, and there are often wide terraces in open areas. Most of them are the few good cultivated land in mountainous areas. The geology of river valleys is complex, and there are often hazards such as landslides, rock piles, and mudslides. Valleys in cold regions often experience snow, avalanche, salivary ice, etc. due to the lack of sunshine. Rivers in the mountain usually have a small flow, but they may turn to rapids in case of rainstorms. The rapids carry sediments, gravels, trees, and other objects which scour riverbanks, destroy bridges and culverts, and inundate countryside, producing great damage.

The above-mentioned natural conditions have brought some difficulties to route selection. However, compared with the other routes in mountainous areas, the riverside (streamside) route has the advantages of the clear route direction and high horizontal and vertical alignment indexes. In addition, there are many residential areas along the river, which are conducive to serving industrial and agricultural production. Sources of building materials and water are sufficient and easy to obtain, which are convenient for construction and maintenance and could reduce project costs. It is necessary to make good use of favorable terrain and overcome unfavorable factors such as adverse geology and hydrology. Priority shall be given to the riverside (streamside) route when selecting routes in mountainous areas. As determining the riverside (streamside) route, the main issues to be resolved include: which side of the riverbank is used, how to determine route elevation, and where to set a bridge to cross the river.

(1) Riverbank selection

When the river valley is determined, a favorable riverbank for route location shall be selected in consideration of terrain, geology, hydrology, farmland, and urban distribution. When favorable riverbanks are distributed on both sides, attention shall be paid to choosing advantageous locations to cross the river. When it is necessary to conduct alignment development, it shall be located on the riverbank with a larger branch trench that is conducive to alignment development.

Advantageous conditions often appear on both sides of the river in a staggered manner. Therefore, thorough investigation, comprehensive comparison, and in-depth consideration shall be made during route selection. The following main factors shall be considered when choosing a riverbank.

① Geological, topographic and hydrological conditions. These are the main factors that affect the choice of the riverbank. An in-depth investigation is needed to find out their characteristics and patterns and place the design scheme in a safe and economical location. As shown in Figure 5 - 5, Plan B crosses the river to bypass two intermittent cliffs on the left riverbank and uses better topography on the right riverbank. However, after passing Xia Village, a steeper and longer cliff appears on the right riverbank and the route has to cross back to the left riverbank. Within 3 km, two middle bridges have to be built due to crossing the river twice. On the contrary, Plan A sticks to the left riverbank. Although it is necessary to excavate a stone section, it is more economical than building two middle bridges. Thus, it is not recommended to cross the river for the riverbank switch.

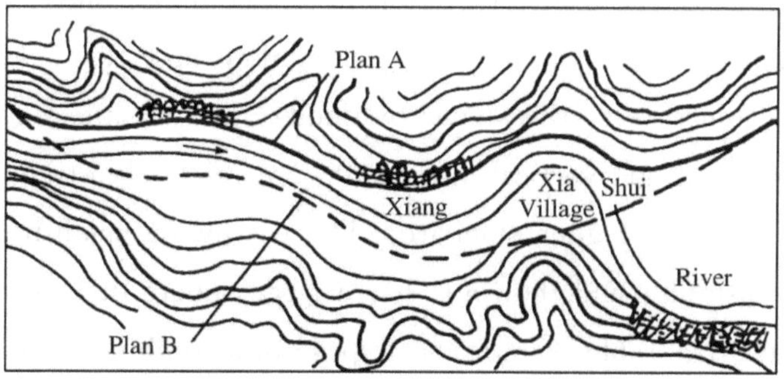


**Figure 5 - 5 Comparison of riverbank changes**

② Distribution of the farmland, city, industry, transportation, and water conservancy facilities. The scarcity and preciousness of land is the most prominent problem in the valley. Necessary measures are mandatory in route selection to occupy less or no farmland. Generally, the route shall be selected on the side with more residential areas, industrial, and mining enterprises and a more developed economy, so as to facilitate local service. However, in order to avoid the demolition of a large number of houses and facilitate urban development, it may also need to detour. In this case, comparison shall be carried out according to the specific situation. Highways and railways interfere with each other frequently which should be located on two riverbanks separately. If there is a main irrigation canal parallel to the route in the river valley, the highway should preferably be located above the main canal and keep a certain distance. Otherwise, the highway and the main canal are seperated on different banks.

③ Route selection in snow and frozen areas. The climates of the sunny slope and shady slope, the windward side and the leeward side in snowy and frozen areas differ greatly. On the premise of not affecting the overall route layout, the sunny slope and windward bank shall be selected as far as possible to reduce snow, salivary ice and other distresses. Sometimes even if the workload on the sunny slope is larger, the sunny slope plan shall be selected in light of driving time and safety.

(2) Height selection

The route height shall generally avoid direct flood erosion of the roadbed. According to the relationship between the height of the route and the design flood level along the river, it can be divided into two types: low route and high route.

The low route refers to the one that is not much higher than the design water level (including wave height and safety height), and the side slope of the subgrade near water is often threatened by floods. The advantage of the low route is that the horizontal and vertical alignments are relatively straight and gentle, and it is easy to win a higher standard. The earthwork quantity is small, the slope is low and stable, and the route has a large sphere of flexibility, which is convenient to use favorable terrain and bypass unfavorable terrain and geology. The route is convenient to cross the tributary and tractable when it must cross the mainstream. The disadvantage is that it is threatened by floods and demands numerous protective projects.

The high route refers to the one that is much higher than the designed water level and is basically free of flood threats. It is mainly adopted by using large sections of higher platforms, or by elevating the height of the low route when the low route close to mountains and rivers is vulnerable to snow or needs to bypass arduous projects. The advantage is that it is not affected by floods and the waste is easier to handle. However, since the high route is generally located on the hillside, it must bend with the mountain, leading to the sacrifice of alignment and an increase of in workload. In case of holes, it is often necessary to build a high retaining wall or other structures. The high route is difficult to avert unfavorable geology or river crossing.

The elevation of the riverside (streamside) route is selected according to the topographical, geological and hydrological conditions of riverbanks, combined with highway class and engineering economy. On one hand, the design elevation of the shoulder of the riverside (streamside) route shall ensure that it is higher than the design flood level of specified frequency. On the other hand, it shall prevent the route from being aloft on the hillside, which is unable to make full use of favorable terrain and difficult to cross the river. On the premise of flood level investigation, the low route is preferred as long as the design flood level of specified frequency is met.

Height selection relies on the full mastery of the characteristics of the river valley and the overall planning of profile design. As shown in Figure 5 - 6, the original route is to bypass a 1.7 km intermittent cliff along the river so that the scheme of the high route is adopted. The ascending section from the low route to the high route is very long, and the curve is sharp and steep, which jeopardizes driving safety. After local modification of the route, the grade has been improved but at the expense of the addition of a small radius curve and more curved alignment. Therefore, the low route is adopted by cutting through the cliff. The horizontal and vertical alignments are improved and the route length is shortened by 760 m.

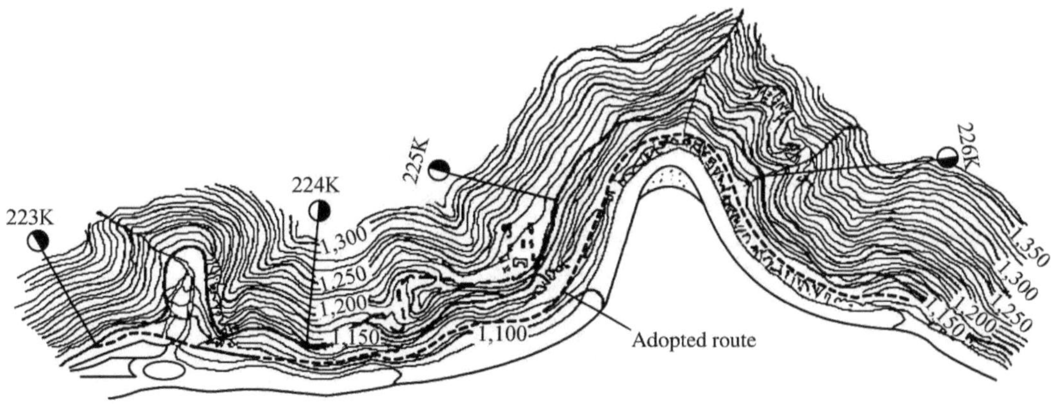


**Figure 5 - 6 Low route and high route of the valley route**

(3) Bridge location selection

According to the relationship between the route and river, there are two types of bridge locations: across the mainstream and across the tributary. Selection of bridge locations across the mainstream is normally a concern of route layout while selection of bridge location across the tributary is generally a concern of the local scheme.

The bridge location across the mainstream is often the control point that governs the route direction and shall be taken into consideration together with the riverbank selection. When the route needs to cross the river due to topography and geology conditions, an improper bridge location may lead to poor alignment of the bridgehead or an increase in construction workload of the bridge. Therefore, the selection of the riverbank shall meanwhile contemplate both the bridge location and bridgehead alignment. The following points shall be envisaged for the selection of bridge locations under the premise of meeting the relevant specifications: it should be selected on a river section with the straight course, stable river bed, free of tributary inflow near the upstream, and the narrow river bed; it locates on riverbanks of sound geology on both sides; it is suitable for connection with other routes; it negotiates with local short-term and long-term planning. The common situations are as follows:

① Cross the river at the waist of the S-shaped river section to strive for a large intersection angle between the bridge axis and the river, as shown in Figure 5 - 7. The example is a medium-small bridge so that the scheme of the skew bridge is more conducive to the coordination between the route and bridges.

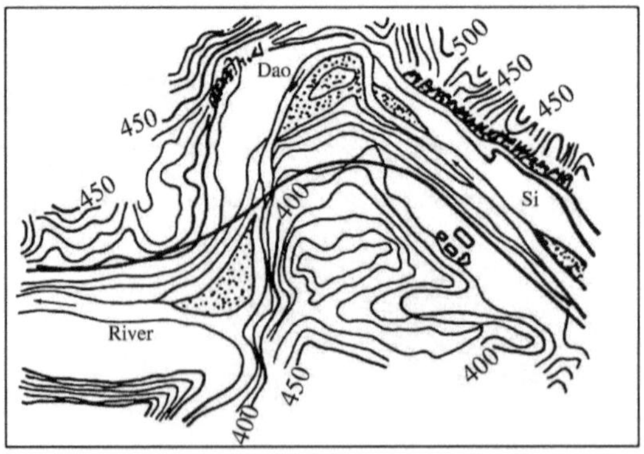


**Figure 5 - 7 Cross the river at the waist of the S-shaped river section**

② Cross the river near the river bend, as shown in Figure 5 - 8. Attention shall be paid to the impact of river flow on the bridge and corresponding protective measures shall be implemented.

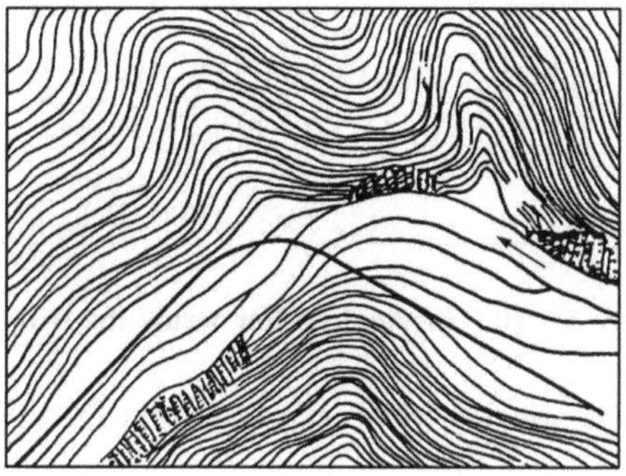

**Figure 5 - 8 Cross the river near the river bend**

③ As crossing the river in the straight reach, the alignment of the bridge approach shall be tackled properly. The bridge location shown in Figure 5 - 9(a) shall be avoided as far as possible. When it is necessary to cross the river in such a section, the medium-small bridge may be employed with the skewed shape to improve the bridgehead alignment. If the skew shape is not suitable for large bridges, the bridgehead alignment should turn into the dipper shape or a curved bridge approach should be planned, as shown in Figure 5 - 9(b). The curve at the bridgehead shall compete for a larger radius to facilitate driving.

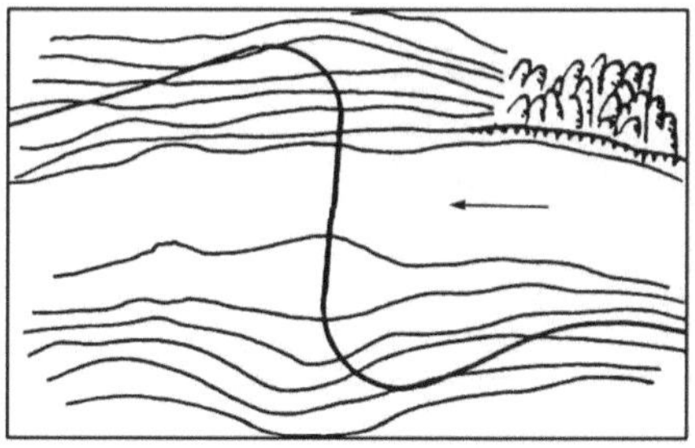

(a)

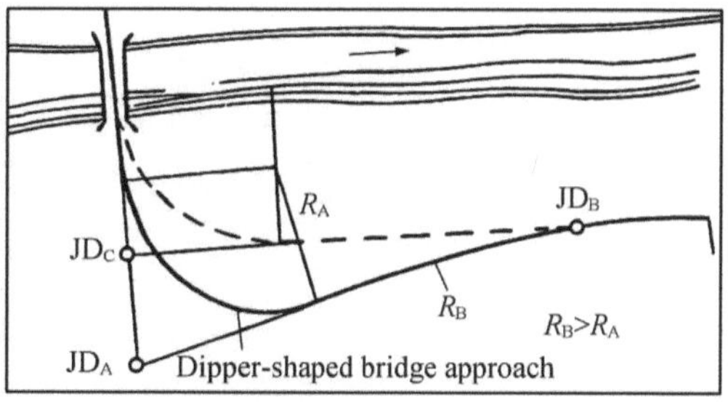


(b)

**Figure 5 - 9 Bridgehead alignment improvement**

The bridge location for the route spanning tributary has two alternatives: one is to span directly at the mouth of the tributary and the other is to cross the upstream of the tributary, as shown in Figure 5 - 10. The final plan shall be concluded after technical and economic comparisons based on the route class and the geological and topographical conditions of the bridge location.

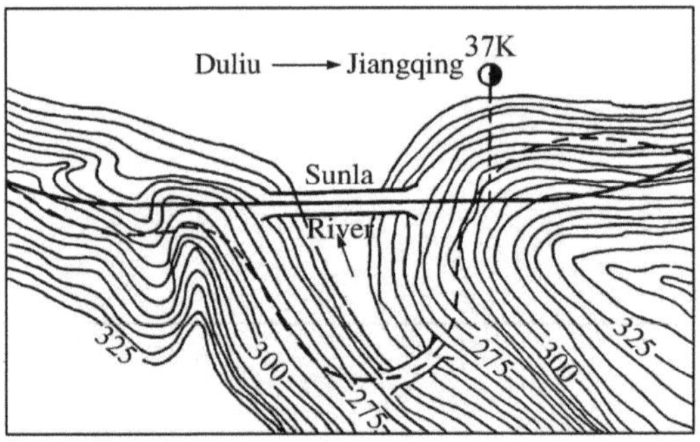


**Figure 5 - 10 Bridge location across the tributary**

(4) Determination of specific routes under several valley terrains

① When the river valley is relatively open, the cross slope is relatively gentle, the geology is superior, and most of the valley is farmland, as shown in Figure 5 - 11, there are three ways to lay out the route. The first way is to develop along the riverbank, as shown by the dashed line in Figure 5 - 11(a). In the first way, the grade is uniform and gentle with excellent alignment. Because the side slope is vulnerable to floods, protective works must be applied. It is a way that can be used. The second way is to lay out the route at the toe of the mountain, as shown by the solid line in Figure 5 - 11(b). In the second way, there

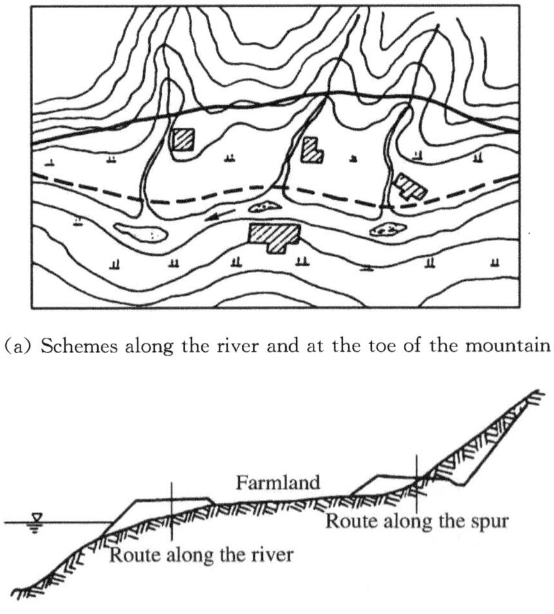


(a) Schemes along the river and at the toe of the mountain

(b) Cross sections along the river and at the toe of the mountain

**Figure 5 - 11 Route schemes on the open valley**

is a slight increase in route length, with ups and downs in the profile, but no or less fertile land is occupied. It is a way that can also be used. The third way is to lay out the route directly across farmland which has a high alignment standard but occupies the most farmland. In paddy fields, soil replacement is sometimes necessary to stabilize the roadbed. Except for freeways and Class-I highways, the third way is generally not recommended.

② When the river valley bends, schemes of circumvention or direct pass may be adopted subject to the actual situations of the spur or river bay.

When the route encounters the spur, as shown in Figure 5 - 12(a), there are two ways of route layout: the first way is to make a detour along the natural terrain of the spur. Due to the extension of the route, it is beneficial to change elevation in sections with limited grades (except for tunnels). However, it is susceptible to adverse geology and river erosion, and the route safety is discouraging. The second way is to pass directly by the cut or tunnels. The route is short, straight, and safe. Nevertheless, the construction cost is high with long tunnels. In general, when the workloads between direct pass and circumvention are limited, the former is adopted.

When the route encounters a river bend, as shown in Figure 5 - 12(b), there are three options: the first option is to make a detour along the river. The route is circuitous, the bank slope is steep, the current erosion is serious, the roadbed protection works are large, and the route safety is concerned. The second and third options are to build a bridge across the river and change the river course, respectively. Do the cutoff works, making the route short and safe. Regardless of which option to be used, it is decided according to the topography, geology, and hydrology conditions as well as farmland and water conservancy construction.

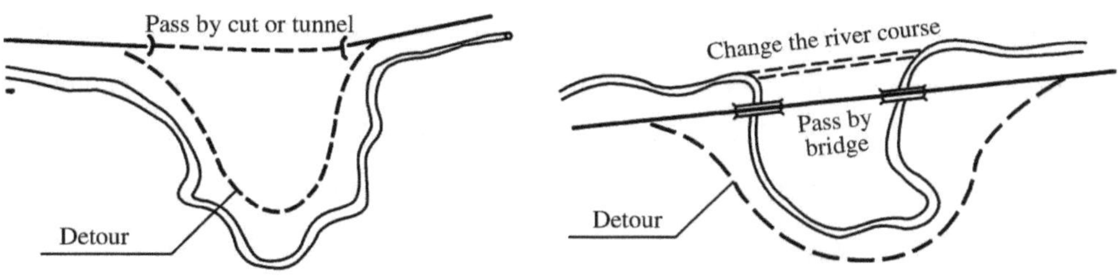


(a) Spur (b) River bay

**Figure 5 - 12 Route schemes of the spur and river bay**

For a specific large river bay with a wide river beach, the route can be placed on the river beach to improve the standard. As long as it is handled aptly, it can also protect and create farmland. Attention shall be paid to roadbed protection and reinforcement to hinder erosion and damage to the roadbed by water. For a specific protruding spur, the spur may be cut and the river bend may be filled. The balance between vertical fill and cut shall be taken care of to prevent a large amount of abandoned earth from being dumped on the beach and blocking the river course, as shown in Figure 5 - 13. When encountering the spur or river bend, whether to circumvent or pass directly shall be concluded in conjunction with the road class. Roads with the higher class should pass directly to attain a better alignment. Schemes for roads with the lower class shall be determined according to the technical and economic comparisons.

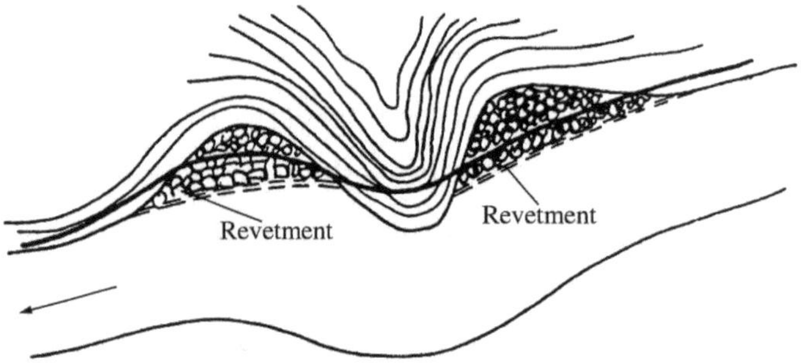


**Figure 5 - 13 Route schemes of cutting the spur and filling the river bend**

③ When the river valley is narrow, the cross slope is steep, and the geology is weak, route selection should avoid the hillside, and may compare with the scheme of shifting out to build bridges (bridges along the river). Steep cliffs often appear in mountain valleys, and the reach with steep cliffs on both banks is a valley. Generally, the river bed in a valley is narrow and the water flow is turbulent. The route may adopt two schemes of bypassing and crossing through this section. It shall be determined by comparison based on factors such as hydrology, geology conditions of the valley and highway class, technical standards, engineering workload, and construction conditions. As shown in Figure 5 - 14, the river valley is relatively tortuous and there are nearly 5 km cliffs, which produces difficulty in route layout. The Wayao Pass on the over-ridge route has a smoother direction, and the topography and geology conditions on both sides are better. It is advisable to bypass a valley by the over-ridge route.

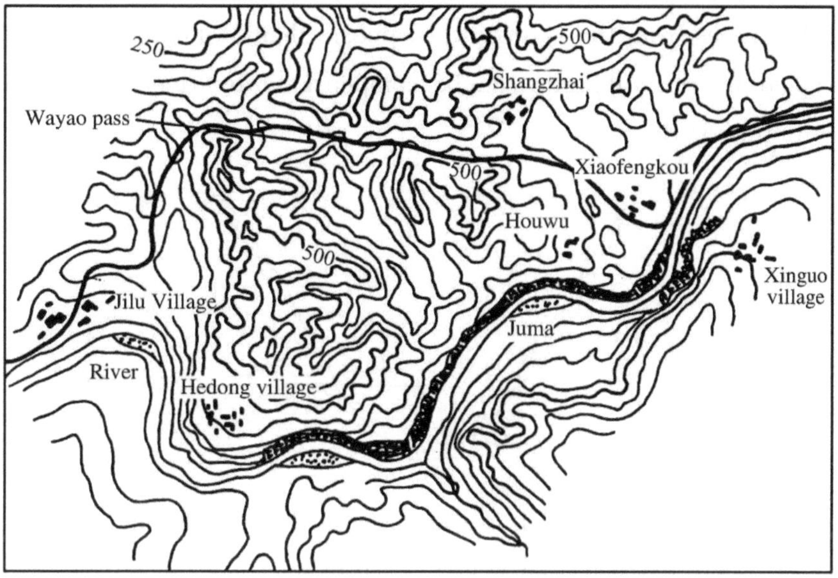


**Figure 5 - 14 Bypassing a valley by the over-ridge route**

For routes cutting through steep cliffs and valleys, the horizontal and vertical alignments are limited by the shape of the bank wall and the flood level, resulting in the constraint of route flexibility. It is advisable to adopt the low route. However, under the circumstances of high flood level or heavy snow accumulation, the low route is no longer recommended. For routes cutting through valleys, the following methods may be considered according to the width of the river bed, hydrological condition, and steepness of the bank:

Ⅰ. Compete with the river and occupy partial river bed. When the river bed is wide, the river is not deep, and the compression of partial river bed does not cause the excess elevation of flood level, a low route at the toe of the cliff may be used to pass through. According to the degree of compressibility of the river bed, there are two situations. First, the river bed is wider, the flood level is not raised much after compression, and the subgrade can be completely or mostly placed in water or the beach close to the toe of the cliff, which is constructed by borrowing rock or cutting stone cliffs moderately. Protective works shall be installed for the subgrade adjacent to water. Second, when the river bed is narrow and the flood level is greatly elevated after compression, a combination of road construction and river management is adopted, and the subgrade may also partially occupy the riverbed. "Cut" and "fill" are combined with fill dominating. The cut comes from the spur on the opposite bank, and the fill is mainly from the boulders from river bed and the stones from the spur on the opposite bank. Therefore, the drainage area occupied by the subgrade can be compensated by river bed cleaning, as shown in Figure 5 - 15.

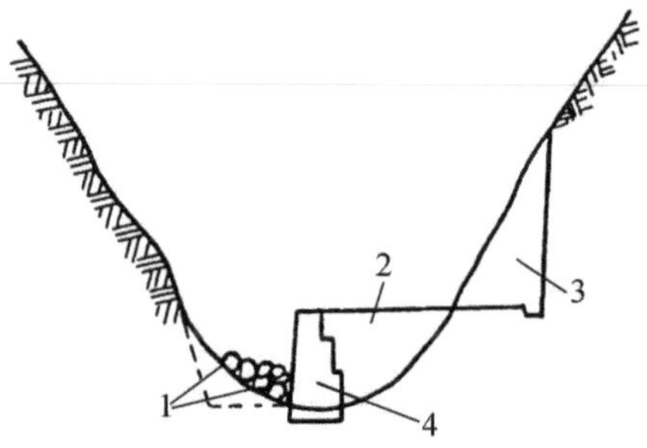


(1—Clean river bed; 2—Fill subgrade; 3—Cut rock cliffs; 4—Protection of the water-side subgrade)

**Figure 5 - 15 Partial subgrade occupying the river bed**

Ⅱ. Chisel the stone wall. When the cliffs on both sides of the bank are close and the river bed is too narrow to accommodate the river and the road in parallel, the rock wall can be cut through, as shown in Figure 5 - 16(a). The excavation situations are as follows: the subgrade is chiseled out on the stone wall, as shown in Figure 5 - 16(b). The excavated waste earthwork shall be properly handled and consumed as close as possible. The impact of earthwork scattered in the river on the water level shall also be regarded, and the route shall be appropriately elevated or the river course shall be cleared. If the bank wall is of good stone quality, a half tunnel may be chiseled out to reduce stone and waste, as shown in Figure 5 - 16(c). For specific gaps or sections that are not wide enough, a half-bridge or overhang abutment can be used. When the stone walls on both sides are very close and it is not suitable to open the subgrade, a bridge can be built along the river to pass.

**2) Over-ridge route**

The over-ridge route refers to the one that crosses over mountains. Its characteristic is to overcome a large height difference. The length and horizontal position of the route mainly depend on the grade of the route. During route selection, the grade must be deemed as the leading factor, and the layout of the

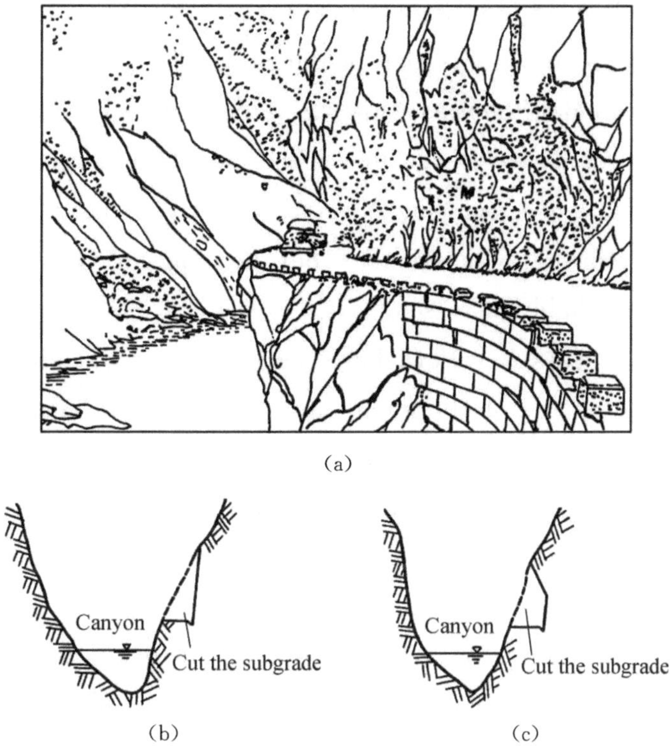


**Figure 5 - 16 Cutting the subgrade on the rock wall**

horizontal alignment and cross section shall be coordinated accordingly. Route selection of the over-ridge route mainly solves three issues, including pass selection, determination of ridge-crossing elevation, and route development on both sides of the pass. They are interrelated and affect each other, and the layout shall cope with the relationship between the three issues considering the hydrological and geological conditions.

(1) Pass selection

The pass is a saddle-shaped and evidently concave terrain on the ridge, and it is an important control point that reflects the scheme of the over-ridge route. It shall be identified within a larger range that basically conforms to the direction of the route. The location, elevation, geological conditions and route development conditions of the pass shall be fully judged to make the selection. Generally, an ideal pass is the one that basically is agreeable with the direction of the route, low in elevation, of good geological conditions, and with slopes on both sides that are conducive to route development.

① Selection of the pass location: The pass location shall be considered in association with the plan of route development on both sides on the premise that it basically conforms to the direction of the route. First to be considered is the pass that the elevation difference is small, and the route development may connect straightly to control points at the toe of the mountain as the slope descends, without invalid route development. Next to be considered are other passes that deviate slightly from the route direction, but the connection is smoother, and the mileage is not remarkably increased.

② Selection of pass elevation: Pass elevation and its height difference with control points at the toe of the mountain trigger a great impact on the route length, engineering quantity and operation condition. In alpine areas, especially snow and ice areas, the route with high altitude is unfavorable to driving. Sometimes a low pass, even if the direction deviates and the distance is far, shall also be compared. However, if the snow and ice are not too serious, the passes that are consistent with the route direction, of good condition for route development, of smooth connection, and sound geological conditions shall not be abandoned even if the altitude is slightly higher.

③ Selection of route development conditions of passes: The hillside route is the main component of the over-ridge route. The zigzag degree of hillsides, steepness of cross slopes, geological quality, and other conditions are directly related to the alignment index and engineering quantity. Therefore, pass selection must be determined in conjunction with the hillside alignment condition. If there is a hillside with good geology, gentle terrain, and conducive to route development downward, it shall be compared even if the pass is slightly deviated or higher.

④ Selection of geological conditions of passes: The geological structure of passes is generally weak with unfavorable geology existing. The stratigraphic structure shall be deeply investigated (Figure 5 - 17) to find out its nature and impact on the route. For the weak layer type, structural type and soft soil erosion type of passes, as long as the occurrence of rock stratum and the influence of water are taken care of, they may not produce negative impacts on the route. Generally, the fault fracture zone type and fault collapse type passes shall be bypassed as far as possible. When the route has to cross over such passes, the size and degree of the fracture zone shall be identified so that the favorable parts could be selected. And engineering measures (such as retaining wall and open cut tunnel) shall be put forward to ensure subgrade stability. For the passes with poor geological conditions, if the safety could not be warranted by local route adjustment or engineering measures, they shall be abandoned.

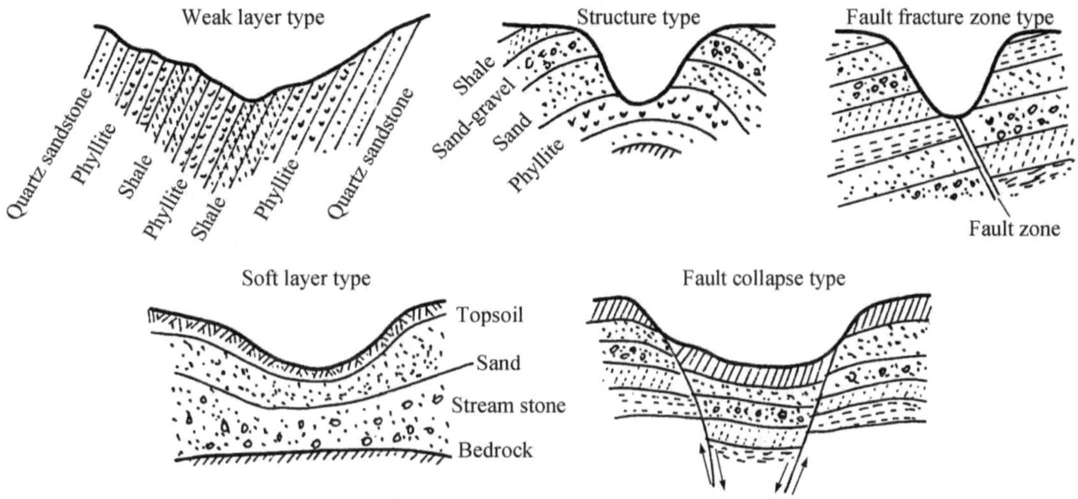


**Figure 5 - 17 Stratigraphic structures of passes**

(2) Determination of ridge-crossing elevation

When the route crosses ridges, cuttings or tunnels can be adopted. The lower the ridge-crossing elevation is, the shorter the route will be. However, with lower ridge-crossing elevation, cuttings or tunnels become deeper or longer, and engineering quantity also increases. Therefore, ridge-crossing elevation shall be selected through technical and economic comparison in association with factors such as the route class, pass terrain, geology, route development scheme on both sides and ridge-crossing method. These factors affect each other, and potential alternative schemes shall be comprehensively analyzed to make a reasonable choice. There are mainly the following methods to cross ridges.

① Shallow excavation and low fill: For the wide and gentle passes, some may reach several kilometers. In case of occasional swamps, shallow excavation and low fill should be adopted to cross ridges. The elevation to cross ridges is basically the elevation of passes.

② Deep excavation: When passes are thin, deep excavation is often used to cross ridges. For the deep excavation of passes, although the earthwork works are concentrated, the total quantity may not increase due to the reduction of ridge-crossing elevation and the shortening of route development. Even if they are increased, they can be compensated by improving driving conditions and saving operating expenses. The excavation depth of passes shall be determined according to the terrain, geology, climate and pass elevation required by route development. If the geology is sound, the excavation depth is generally within 30 m. The thinner the pass is, the deeper it digs. Since the workloads of deep excavation of passes are concentrated, a large amount of waste rock needs to be treated. And the construction condition is constrained, which affects the construction duration. Moreover, there are many slope distresses with weak stability during operation, which needs to be fully evaluated when selecting the ridge-crossing elevation.

As shown in Figure 5 - 18, the route crosses over the pass. There are three possible schemes due to different excavation depths. The excavation depth of scheme A is 9 m, and two switch-back curves are used. The excavation depth of scheme B is 13 m, and one switch-back curve is used. The excavation depth of scheme C is 20 m, the route may develop naturally along with the mountain without a switch-back curve. By comparison, scheme C has the best alignment and the shortest route, which is conducive to driving safety and saving operating cost.

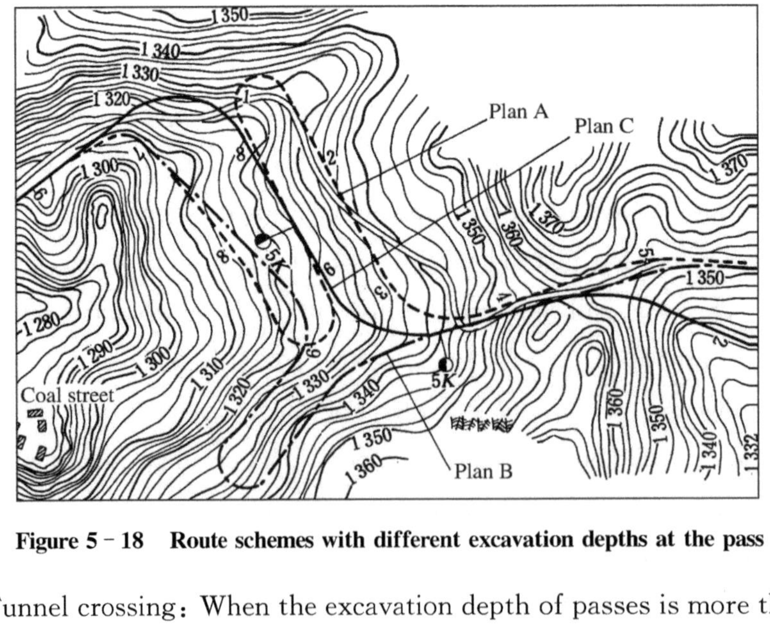


**Figure 5 - 18 Route schemes with different excavation depths at the pass**

③ Tunnel crossing: When the excavation depth of passes is more than 30 m, technical and economic comparison shall be conducted with the scheme of tunnel crossing. If the passes are thin, tunnels could reduce the route height, shorten the mileage, improve the alignment index and mitigate the impact of snow and ice. Generally, the lower the tunnel elevation is, the shorter the route is, the higher the technical index is, and the more beneficial the operation is. However, if the elevation is lower, the tunnel becomes longer, the construction period is longer, and the cost is higher. Therefore, the selection of tunnel elevation shall be based on the geological conditions of the over-ridge section and refer to the critical elevation. The critical elevation is the certain ridge-crossing elevation with the smallest sum of tunnel cost and route cost. If the design elevation is higher than the critical elevation, the cost of extending the route will be greater than that of shortening the tunnel. If the design elevation is lower than the critical elevation, the cost of extending the tunnel will be greater than that of shortening the route. The reduction of design elevation can save operating cost, and the route with large traffic volume is the key factor to be considered. In addition to economic factors, the following factors shall be considered in the selection of tunnel elevation: geological and hydrological conditions are important factors in the selection of tunnels, and the tunnel shall be constructed in a better stratum as far as possible. The tunnel elevation shall be below the perennial freezing line and perennial snow line to ensure construction and driving safety. Construction period and technical conditions shall be considered. Without great project cost increases, tunnel elevation shall be reduced as much as possible for the sake of long-term development.

(3) Route development on both sides of the pass

Route development is a layout method that allows the route grade in mountain areas to meet technical standards and takes advantage of terrain to extend the route length in order to overcome height difference.

① Route development form: There are three main route development methods of the over-ridge route: natural route development, switch-back route development, and spiral route development.

(a) Natural route development: It is a layout method to extend the route around the spur and side ditch with a appropriate grade, which is along the natural terrain to overcome the height difference. The advantages of natural route development are that the direction conforms to the basic trend of the route, the traveling is consistent with grade fluctuation, the route is the shortest, the alignment is simple, and the technical indicators are generally high. In particular, the route does not overlap, which is beneficial to driving, construction, and maintenance. The disadvantage is that there is little freedom to avoid arduous projects or bypass adverse geology because the only way is to adjust grades. In the case of high cliffs, deep valleys or large-area geological distresses, it is difficult to bypass them, and other methods have to be adopted.

(b) Switch-back route development: The route extends along one side of the hillside, selects a suitable location to turn into the opposite direction by the switch-back curve, and then returns back to the original hillside. Its advantage is that it is easy to use favorable terrain and avoid unfavorable terrain, geology and difficult projects. The disadvantage is that the upper and lower routes overlap on the same slope, especially closer for the upper and lower routes near the switch-back curve. It is unfavorable to driving, construction and maintenance. Therefore, it can be adopted only when it is compulsory, as shown in Figure 5 - 19.

**Figure 5 - 19 Switch-back route development**

The location of switch-back has a great impact on the construction quantity and operation quality of the switch-back curve, so it shall be selected cautiously. The shape of the switch-back curve depends on the terrain of the switch-back location. Generally, the following three terrains are used: the hillock with the large diameter, gentle cross slope, and adjacent to the low saddle or flat ridge, as shown in Figure 5 - 20(a) and Figure 5 - 20(b); the gentle hillside with sound geology and hydrology, as shown in Figure 5 - 20(c); the gully or col with the open terrain and gentle cross slope, as shown in Figure 5 - 20(d) and Figure 5 - 20(e).

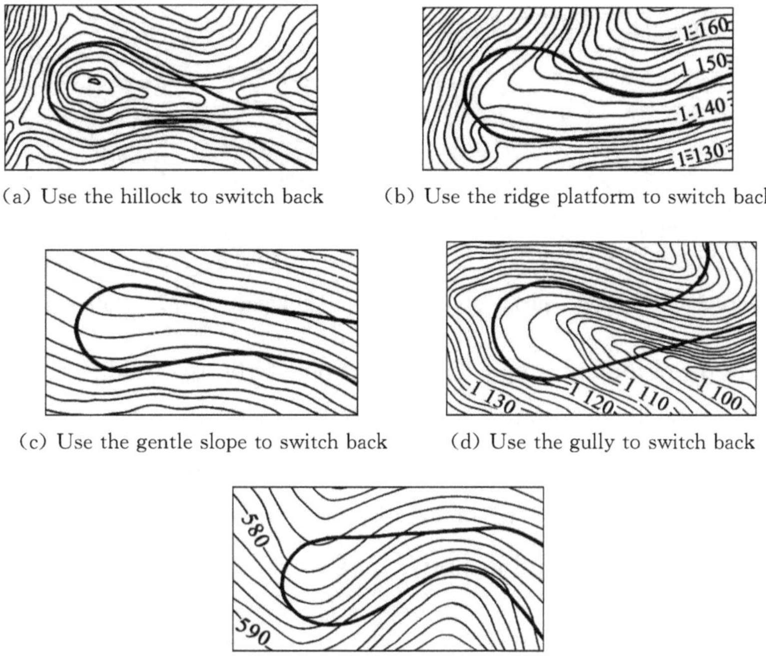


(a) Use the hillock to switch back (b) Use the ridge platform to switch back

(c) Use the gentle slope to switch back (d) Use the gully to switch back

(e) Use the col to switch back

**Figure 5 - 20 Favorable terrains suitable for the switch-back curve**

(c) Spiral route development: When the route is under constraint and it is necessary to raise or lower height at a certain place in order to make full use of favorable terrain or position nearby, the layout method of spiral route development could be adopted. Generally, spiral route development is realized by circling around the hillock on the ridge which cuts through with tunnels, as shown by the solid line in Figure 5 - 21, or by being circuitous locally in the valley which overpasses with bridges, as shown by the solid line in Figure 5 - 22. It can also rotate in the form of tunnels in the mountain. Spiral route development has the advantages of better alignment and avoiding route overlap as against switch-back route development, but it is not frequently used because of high cost of tunnels or bridges. When necessary, it shall be compared with the switch-back route development in detail according to the route's nature and task.

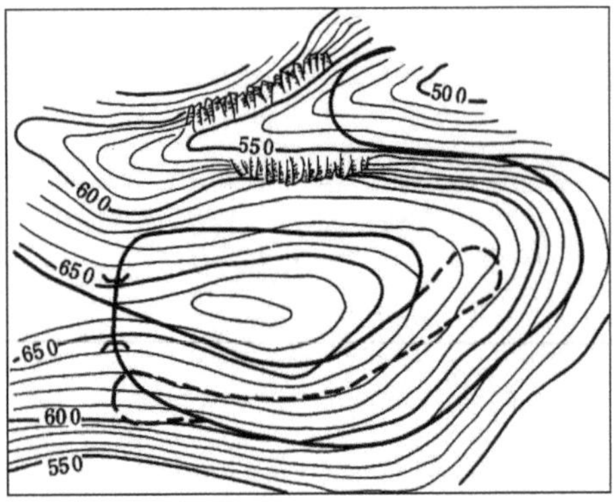


Figure 5 - 21 Ridge spiral

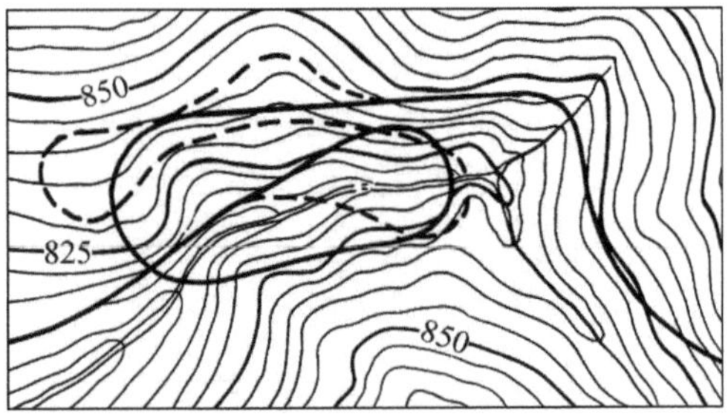


Figure 5 - 22 Valley spiral

② Route development step: The elevation of the over-ridge route is mainly overcome by route development on the hillsides on both sides of the pass. The route layout shall be dominated by grades, adjust grades reasonably and apply necessary switch-back curves to make use of favorable terrain and geology and bypass unfavorable terrain and geology. The working steps of route layout are as follows.

(a) Outlining the rough route. Conduct an extensive survey among the main control points determined in the inspection or survey stage, investigate the surrounding terrain and geological conditions, take a rough survey of grades with angular hand leveling as the guide, and outline the approximate route by using the favorable terrain and geology.

(b) Grading trial layout. The purpose of the grading trial is to further implement the possibility of the initially proposed route, discover and densify intermediate control points, find local comparative schemes, and formulate the route layout.

(c) Analyzing and implementing control points and determining the layout scheme. There are fixed and flexible control points: the first is that the position and elevation cannot be changed, such as the route of particularly difficult places and some severely restricted turning places, bridges that must be used, streets that must be passed, etc. The second is that the location is fixed and the elevation is flexible, such as passes, important bridge sites, etc. The third is that the location and elevation are both flexible, such as the side ditch crossing location, switch-back location of the wide and gentle hillside, etc. Most control points have room for flexibility, but the range of flexibility is different. There are two methods to adjust the flexible control points: for the switch-back location with high flexibility, it can be obtained from the intersection of the front and rear fixed control points with the appropriate grade; the non-switch-back control point between two fixed control points shall be adjusted within its movable range to make the grade between fixed control points as uniform as possible.

**3) On-ridge route**

The on-ridge route refers to the one that lies on the ridge of a mountain, which is often used as the local alternative to the riverside route or hillside route and the connective section of routes on both sides of the over-ridge route. The general alignment of the on-ridge route is mostly undulating and tortuous, and its undulating and tortuous degree varies depending on the ridge shape, the height difference between control passes and terrain. Generally, it has the advantages of small earthworks, good hydrological and geological conditions, fewer bridge and culvert structures, etc. The on-ridge route is relatively high and generally far away from residential areas, which is inconvenient to serve the industrial and agricultural productions along the route. Sometimes, construction is difficult due to a lack of construction materials and water. The terrain is high, the air is thin, and there are disadvantages such as clouds, snow and ice, which are unfavorable to driving and maintenance. The selection of the on-ridge route scheme shall mainly consider the following conditions: the direction of the ridge shall not deviate too far away from the general direction of the route, the plane of the ridge shall not be too tortuous, the height difference between passes on the profile shall not be too large, the geological condition of hillsides between control passes is good, the terrain is not too steep and messy, and the route going up and down the ridge shall have suitable terrain to be placed, which is one of the main conditions for whether the ridge route can be used. There are limited ridges that fully meet the above conditions, and there are few long on-ridge routes. The layout of the on-ridge route need to mainly solve the following issues: selection of control passes, selection of the side slope and grading trial layout.

(1) Selection of control passes

Each group of control passes represents a ridge on-route scheme, and the selection of control passes is the key to the selection of the on-ridge route. When the ridge direction is straight with little fluctuation, almost every pass can be temporarily determined as the control point. If the terrain is complex and the height of each pass varies greatly, the low pass between high passes is generally the control point of the route, and the prominent high pass can be omitted. When there are branches across, for several passes not far away and side by side, only one pass with a good connection with the front and rear shall be selected.

The selection of control passes shall also be comprehensively considered with the layout conditions on both sides of the ridge. The preliminarily selected control points shall be screened and implemented in the side slope selection and grading trial layout.

(2) Selection of the side slope

The side slope of the ridge is the main layout zone of the on-ridge route. The side slope with better layout conditions shall be selected to ensure good horizontal and vertical alignments, small construction quantity, and the stable subgrade. The sunny hillside with the uniform slope, gentle cross slope, good geological conditions and free of branches across is ideal. Except for a clear distinction of slopes, trade-offs shall be made on both sides. There may be different route schemes on the same slope, which can be determined by grading trial layout. For most of the primary selected control passes, the choice can be made in the process of side slope selection, and a few need to be investigated in grading trial layout.

(3) Grading trial layout

Layout shall be conducted between two fixed control points to strive for a short distance and gentle grade. On occasion, due to the large height difference between control points, the on-ridge route needs to be extended. Furthermore, to prevent the route from being too circuitous, fluctuating grades are used to shorten the length. The on-ridge route will inevitably have twists and turns, but it shall not be too sharp and frequent.

The horizontal and vertical curve indicators and sight distance shall be as higher as possible to facilitate driving. The on-ridge route layout often has the following three situations.

① The average grade between passes shall not exceed the limit value. When there are no major obstacles in topography and geology between two control passes, the layout shall be along the side slope with the uniform grade. If the average grade between control passes is relatively gentle, and there are obstacles or arduous projects, intermediate control points can be supplemented to modify the grade to bypass, and the intermediate control points and each pass are still laid out with the uniform grade.

② There are branches across control passes. As the route extends through branches, the appropriate pass shall be selected on the branch as the intermediate control point. The pass shall not make the route too tortuous. With excavation of reasonable deep, the grades of both wings of the route shall not exceed the limit values, and the route may proceed through sound terrain and geology areas. On occasions, although the control pass selected on the branch can meet the grade requirements, the alignment is too tortuous. In order to shorten the mileage, the control point may not be selected on the pass.

③ The average grade between passes shall be less than the limit value. According to the topographic and geological conditions, engineering measures such as the fill and cut, land bridges and tunnels are adopted to elevate the low pass and lower the high pass. The switch-back or spiral route development can also be adopted by using the favorable terrain of the side slope and ridge, as shown in Figure 5 - 23. For the route selection method, please refer to the over-ridge route in this section.

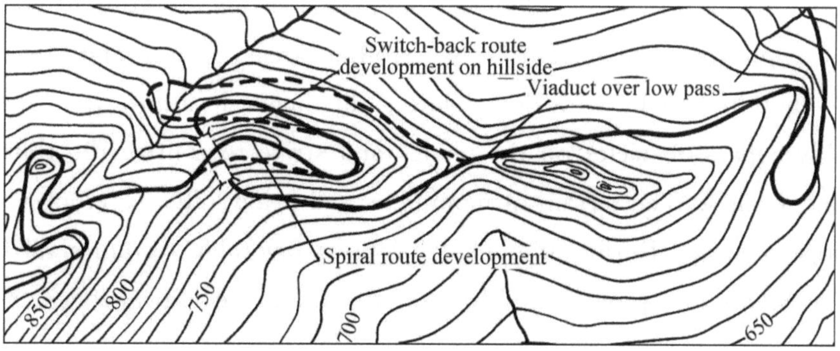


**Figure 5 - 23 Schematic diagram of route development of the on-ridge route**

## 5.4 Alignment location methods

Alignment location refers to the work of determining the centerline of roads within the scope of the established route corridor according to the technical standards. Alignment location shall be performed in association with topographic, geological and other conditions, and balance the horizontal alignment, vertical profile and cross section of the route. This is a work involving a wide range of disciplines and high technical requirements. Designers are required to have extensive knowledge, skilled alignment location skills and an attitude of perfection. It is difficult for the designer to determine an ideal route at one time. There are often multiple design schemes for research and comparison under complex conditions. Each scheme is a compromise of many mutually restrictive factors, and the ideal route can only be harvested by comparison. According to the different working objects, alignment location can be divided into map location, field location and aerial survey location. With the development of survey and design, most of alignment location in the design stage is carried out by using professional auxiliary tools on the basis of an electronic topographic map or even a three-dimensional model, which greatly improves the efficiency and quality of alignment location work. The following section briefly introduces the methods and steps of conventional map location.

Map location is a method to determine the position of the road centerline on a large-scale topographic map of 1 : 1,000 – 1 : 2,000. The topographic map has a wide range and broad field of vision. It is easy for alignment location personnel to determine a reasonable route indoors. Alignment location on different terrain has different emphases. The terrain of level-rolling areas is flat, and the route is generally not limited by elevation. During alignment location, it is mainly to reasonably bypass obstacles on the plane, and strive to control the route to be short and straight. The terrain of mountains and heavy hills is complex and the cross slope is steep. During alignment location, favorable terrain shall be fully used to avoid arduous projects and bypass unfavorable geological zones or ground features.

**1) Alignment location steps in level-rolling areas**

(1) Determine traverse points

Among the selected control points, points that can be crossed shall be close to and shall be avoided are analyzed and identified according to the key points of route layout in level-rolling areas. Then several intermediate traverse points are determined.

(2) Determine the traverse line

A traverse line of the route is initialized by connecting the traverse points.

(3) Initialize the horizontal curve

Read the intersection coordinates, calculate or directly measure the deflection angle and intersection distance, preliminarily determine the radius of the circular curve and the length of the spiral, and calculate the elements of the horizontal curve.

(4) Alignment location

Check whether all indicators meet the requirements of relevant specifications and adjust them appropriately until they are satisfactory.

**2) Alignment location steps for mountainous and hilly areas**

(1) Determine the traverse line

① Analyze the terrain and search various possible schemes. The terrain and geological conditions between the main control points determined in the stage of route layout shall be carefully studied. Favorable terrains, such as the gentle and straight hillside, open side ditch, location conducive to switch-back, etc. are selected to formulate various possible routes. As shown in Figure 5 - 24, the terrain on the left side of the figure is steep and the terrain on the right side is gentle. Points $A$ and $D$ are two control points, point $B$ is the available ridge platform, and point $C$ is the steep cliff that shall be bypassed. Then $a—b—c—d$ is a possible route, which must be determined by grading trial. The grading trial of map location is realized with a bipod gauge.

② Calculate the horizontal distance $a$ and determine the grading slope line. Based on the contour distance $h$ and the selected average grade $i_{mean}$ (5.0% - 5.5%), the horizontal distance $a$ between contour lines according to $a=h/i_{mean}$ is calculated depending on the tortuosity and elevation difference of the terrain. Then the opening of a bipod gauge is equal to $a$ (according to the scale of topographic map). Starting from a fixed point such as point $A$, the contour lines are intercepted along the proposed route to pinpoint points $a$, $b$, $c$, etc. Then it switches back near point $B$ (point $j$ as shown in the figure) and then intercepts to point $D$. When the position and elevation of the last point are close to point $D$, it indicates that the scheme is acceptable. Otherwise, the route scheme shall be modified (such as changing the switch-back position) or the $i_{mean}$ could be adjusted

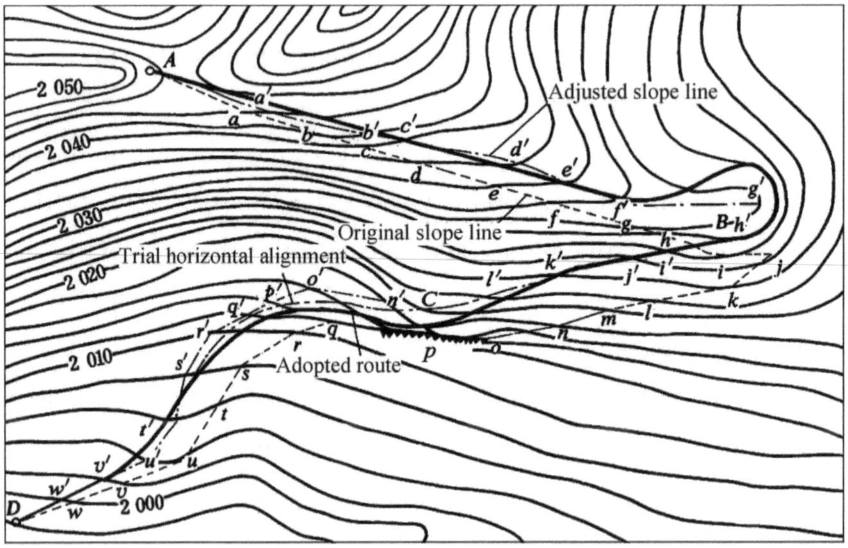


**Figure 5 - 24 Schematic plane chart of map location**

(within 5.0%–5.5%) until the scheme is proper. The connective line $A$—$a$—$b$—…—$D$ is a broken line with the average grade, which is called the slope line. It verifies the success of an alignment location scheme, and discovers some intermediate control points to provide a basis for future work.

③ Determine the intermediate control point, adjust the grade by sections and decide the traverse line. The slope line that uses terrain, bypass ground features, or adverse geological conditions is analyzed and the intermediate control points that shall be crossed or avoided are ascertained.

As shown in Figure 5 - 24, the favorable switch-back location at point $B$ is not used, and the steep cliff at $C$ is not avoided. If the grades before and after points $B$ and $C$ are adjusted (selected between the maximum and minimum grades, but the limit value shall not be adopted easily, and the reverse slope is prohibited), the steep cliff can be bypassed and the favorable switch-back location could be used. Points $B$ and $C$ may be confirmed as intermediate control points. Then the grades are adjusted by sections imitating grading trials to determine the uniform slope line. The line $A$—$a'$—$b'$—…—$D$ is formed by connecting the uniform slope lines of each section, which is called the traverse line. It is a broken line of which the grade is arranged by sections, makes use of favorable terrain, avoids adverse obstacles, and shows the approximate position of the route.

(2) Correct the traverse line

① Attempt to determine the plane and profile. The tangents and horizontal curves are proposed with reference to the traverse line, namely the plane trail line. The stakes and ground elevations are measured or red according to the characteristic points of topographic changes. The ground lines of the profile diagram are dot-plotted and the ideal grades are designed with reference to the ground lines and the grades determined in the previous sections. The rough design elevation of each stake is measured or red.

② Confirm the primary corrective traverse line. The purpose is to use the profile to correct the plane and avoid large fill and excavation on the profile. The points corresponding to the approximate design elevation in the cross section of each stake of the plane trail line are captured. Therefore, the connective line of these points is a broken line with the ideal grade, which is free of fill or cut in the centerline. It is called the primary corrective traverse line. When the fill and cut on the profile is too large, it shall be modified.

③ Confirm the secondary corrective traverse line. The purpose is to correct the plane with the optimal position in the cross section to avoid excessive transverse fill and excavation. A cross sectional view of each point of the primary correction traverse line is plotted, and the subgrade template is employed to search the most economical or optimal centerline position and its movable range by points, as shown in ② and ③ in Figure 5 - 25.

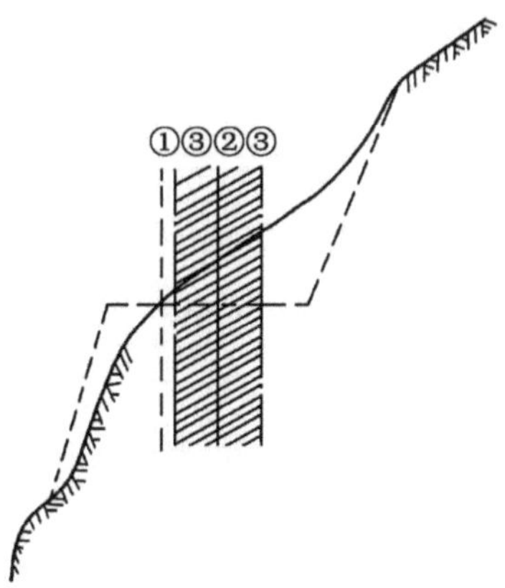


**Figure 5 - 25 Optimal location of the cross section**

According to the nature of the optimal positions, different symbolic points are adopted to locate these optimal positions on the plane view. The connective line of these points is a broken line with the ideal grade and optimal transverse positions on the plane, which is called the secondary corrective traverse line (It can be omitted when the optimal position is not displayed on the small-scale topographic map).

(3) Alignment location

Alignment location is continued based on the secondary corrective traverse line. The secondary corrective traverse line is a broken line on the plane, which does not meet the requirements of technical standards. It must be properly straightened by tangents and curved with horizontal curves to determine the exact position of the centerline. Alignment location shall be attempted repeatedly according to the nature and movable range of each characteristic point on the secondary corrective traverse line, so as to determine the centerline that meets technical requirements. The specific operation of alignment location may adopt the tangent alignment location method or the curve alignment location method.

Map location is a process of repetitive alignment trial and improvement. During the process, whether to modify the grade, move the centerline position or both shall be determined after a full study of horizontal, vertical and cross section aspects. To a certain extent, the more the repetition is, the better the final alignment is. It can be considered that map location work is not over until no further modification can be made to save project costs or improve alignment elegance. After the centerline is determined, the vertical profile, cross section, and the related content may proceed.

**Questions:**

1. What are the main contents of the overall design of roads?
2. What are the basic principles of the overall design of roads in plain areas and mountainous areas?
3. What are the main factors to consider in determining the function and technical standards of roads?
4. What are the typical engineering schemes in the plain area? What are their characteristics?
5. What is the basis of highway survey and design? What about the procedure?
6. What are the basic principles and general steps of route selection?

7. What are the main types of mountainous roads? What are the main aspects to consider when laying out various route types?
8. What kinds of route development are used for mountainous roads?
9. What are the terrain characteristics of the leveling area and the problems that shall be concerned in layout?
10. What are the basic steps of map location?

# Chapter 6 Intersection Design

【**Objectives**: 1. Understand traffic characteristics of intersections and their classification and application; 2. Understand conditions, principles, and provisions of intersection design; 3. Master intersection traffic management, including vehicles, pedestrians and non-motor vehicles; 4. Master plane design and sight distance design about intersections; 5. Master setting and calculation of the turning lane and auxiliary lane at intersections; 6. Master roundabout design.】

The methods and contents related to horizontal alignment, vertical alignment, and cross section design introduced above are mainly for ordinary road sections. As roads extend, they inevitably intersect with other facilities, such as the road-road intersection, road-railway intersection, road-pipeline intersection, etc., among which the road-road intersection is the most common.

A road intersection is defined as the general area where two or more highways join or cross, including the roadway and roadside facilities for traffic movements within the area. Each highway radiating from and forming part of an intersection is an intersection leg. The most common intersection at which two highways cross one another has four legs.

There are two general types of intersections, including the at-grade intersections and grade separation (grade separation without ramps and interchanges). An at-grade intersection (intersection hereafter) is the one that roads cross one another in the same plane; the grade separation refers to the one that roads cross one another in different planes, which manages the conflicting traffic flows on roads at different elevations to ensure traffic smoothness and safety.

Intersections are an important part of a highway facility because, to a great extent, the efficiency, safety, speed, cost of operation, and capacity of the facility depend on their design. Each intersection involves through- or cross-traffic movements on one or more of the highways and may involve turning

© The Author(s), under exclusive license to Springer Nature Singapore Pte Ltd. 2025
B. Yu and J. Cheng, *Road Survey and Design*,
https://doi.org/10.1007/978-981-97-4836-5_6

movements between these highways. Such movements reduce speed, block traffic and also induce traffic accidents, which may be facilitated by various geometric design and traffic management, depending on the type of intersections.

The complexity of intersections makes the design complicated, which is reflected in:

(1) Complexity of traffic. Traffic compositions at intersections include not only vehicles but also pedestrians, and there are traffic divergence and convergence from different ways. Therefore, in addition to geometric design, traffic management design is critical.

(2) Complexity of driving behavior. Different from a normal operation on ordinary sections, drivers at intersections need to decelerate and accelerate. To ensure driving safety, improve traffic capacity, and reduce vehicle exhaustive pollution, driving behavior at intersections shall be considered in intersection design.

(3) Complexity of geometric design. The plane, profile, and cross section, superelevation, sight distance, etc. on each intersection leg shall be designed in coordination. And the form, alignment, and elevation of intersections shall be coordinated with the surrounding environment.

Before the intersection design, it is necessary to analyze the traffic characteristics of intersections. The contents of the intersection design include the determination of intersection position and type, traffic management, geometry design, and so on.

## 6.1 Analysis of traffic characteristics at intersections

An intersection is the throat of roads where vehicles from different ways converge, cross and diverge. Thus, the traffic at intersections is more complex than those at ordinary road sections. Before the intersection design, traffic characteristics of intersections shall be known. There are three kinds of traffic movement at intersections, i. e. diverge, converge and conflict; so there are three kinds of traffic crossing points, i. e. divergence point, convergence point and conflict point. The spot where vehicles in the same way separate into different ways is called the divergence point, the spot where vehicles from different ways converge in the same way at a smaller angle is called the convergence point, and the spot where vehicles from different ways intersect with each other at a larger angle is called the conflict point. The existence of crossing points affects traffic safety and reduces traffic capacity, which is the most phenomenal characteristic of intersections compared with ordinary road sections, among which the conflict points have the largest impact on traffic. Figure 6 - 1 shows the crossing points at intersections without traffic management. The number of crossing points can be calculated by the following formula:

$$\text{Convergence point} = \text{Divergence point} = n(n-2) \tag{6-1}$$

$$\text{Conflict point} = \frac{n^2(n-1)(n-2)}{6} \tag{6-2}$$

where, $n$—number of intersection legs.

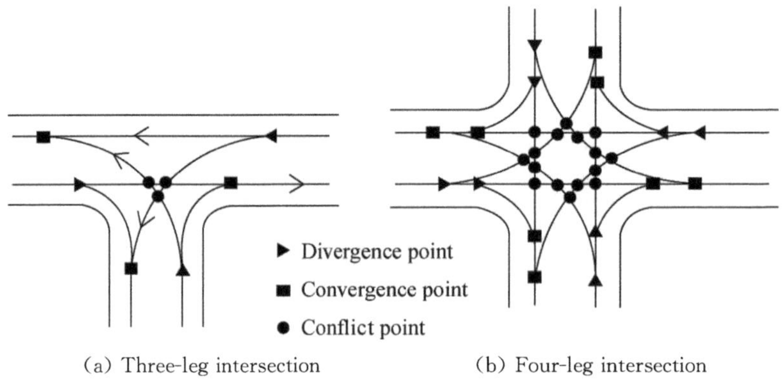


**Figure 6 - 1 Number of crossing points at intersections without traffic management**

To reduce crossing points, improve traffic capacity and ensure traffic safety, traffic management is often required at intersections. Figure 6 - 2 shows the number of crossing points at signalized intersections. Under this circumstance, the number of crossing points is greatly reduced. Table 6 - 1 shows the number of crossing points.

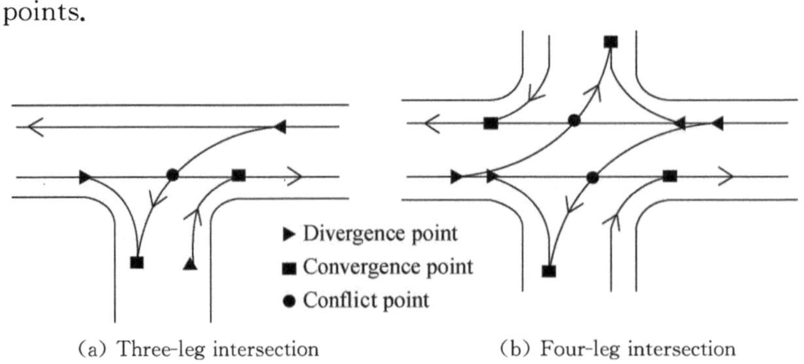


**Figure 6 - 2 Number of crossing points at signalized intersections**

**Table 6 - 1 Number of crossing points**

| Type of crossing points | Unsignalized Number of intersecting roads | | | Signalized Number of intersecting roads | | |
|---|---|---|---|---|---|---|
| | 3 legs | 4 legs | 5 legs | 3 legs | 4 legs | 5 legs |
| Divergence point | 3 | 8 | 15 | 2 or 1 | 4 | 4 |
| Convergence point | 3 | 8 | 15 | 2 or 1 | 4 | 4/6 |
| Conflict point | 3 | 16 | 50 | 1 or 0 | 2 | 4 |
| Total | 9 | 32 | 80 | 5 or 2 | 10 | 12 |

According to the analysis of Table 6 - 1, it can be summarized that the characteristics of crossing points at intersections are as follows:

(1) The number of conflict points caused by the left-turning or through traffic increases rapidly with the increase of intersecting legs;

(2) Most conflict points are caused by the left-turning traffic;

(3) With traffic management, the number of conflict points may reduce.

Therefore, the purpose of intersection design is to reduce or eliminate conflict points. There are three methods to reduce or eliminate conflict points:

(1) Traffic management. It uses traffic lights or is under the command of the traffic police, which staggers the passing phases of the through and left-turning vehicles at the intersection, that is, only one way of traffic is allowed to move through the intersection at one phase.

(2) Traffic organization and channelization. Reasonably arrange traffic islands at intersections, guide traffic flow to separate lanes, and transfer conflict points into divergence or convergence points, so as to reduce mutual interference of vehicles.

(3) Grade separation. Traffic flow of each way at intersections is allocated on lanes with different elevations at grade separation. However, grade separation covers a large area with high cost. Only at intersections with complex traffic, as well as the intersection of busy arterial roads and freeways, shall grade separation be adopted.

## 6.2 Type of intersection

Intersections can be classified according to different criteria. According to the traffic management mode, it can be divided into the priority intersection, non-priority intersection and signalized intersection; according to the shape of intersections, it can be divided into the T-shaped intersection, Y-shaped intersection, crossroad intersection, X-shaped intersection, roundabout, etc.; according to the different channelization modes, it can be divided into the intersection with widening corner, flared intersection, turning split intersection, etc.

## 6.2.1 Classified by the traffic management mode

The traffic management mode determines the geometric feature of intersections. Therefore, the traffic management mode of intersections shall first be determined according to the function, technical class and traffic characteristics of intersecting roads, and then the corresponding intersection type and geometric design details shall be decided. The traffic management mode is a prerequisite for intersection design. According to the function, class and traffic volume of intersecting roads, intersections may adopt three different traffic management modes: the priority intersection, non-priority intersection, and signalized intersection.

**1) Priority intersection**

A priority intersection is the crossing of the major road and minor road without signal control, in which vehicles on the major road claim priority while vehicles on the minor road shall stop or slow down to yield the right-of-way. In the priority intersection, traffic flow on the major road is usually continous. It is not required to stop, and the speed maintains the same at the intersection as that at the ordinary road sections. Traffic flow on the minor road needs to stop or slow down and observe at the entrance of the intersection; it have to wait for a gap of traffic flow on the major road to pass through the intersection.

If there are significant differences in the highway function, technical class and traffic volume between two highways or T-shaped intersections with large traffic volume, a priority intersection shall be adopted.

The priority intersection ensures the smooth passage of vehicles on the major road but vehicles on the minor road will experience delay as a result of yielding the right-of-way. The queue and delay are more serious when the traffic volume of the intersection is close to its capacity, so other traffic management modes shall be used under this circumstance.

**2) Non-priority intersection**

A non-priority intersection is the crossing of roads with very small traffic flow and vehicles in all ways waiting for gaps of traffic flow at intersections without any management measures.

When two or multiple intersecting roads are of low class and have small traffic volume, a non-priority intersection shall be adopted.

**3) Signalized intersection**

A signalized intersection is the crossing of roads whose traffic flow is dynamically controlled and regulated by means of traffic signals. Vehicles approaching the signalized intersections are required to queue and wait to pass according to the phase provided by signals.

The following intersections shall be implemented with signal control:

(1) The intersection of two highways with the high traffic volume, identical function and technical class, which is difficult to be managed by the priority intersection.

(2) Although the two intersecting highways have a difference in classes, the traffic volume of both roads is large (the two-way traffic volume of the major highway is greater than or equal to 750 vehicles/h, and the one-way traffic volume of the minor highway is greater than or equal to 300 vehicles/h). If the priority intersection is used, frequent traffic accidents and excessive traffic delays will be inevitable.

(3) The traffic volume in the major highway is rather large (the two-way traffic volume on the major highway is greater than or equal to 900 vehicles/h), whereas the volume on the minor road is small. If the priority is assigned to the major road, there will be unacceptable traffic delays on the minor highway due to difficulties in finding the available gap of the main flow, or will be dangers due to the risk of heading into the insufficient gap of the main flow.

(4) Despite the traffic volume in the two intersecting highways does not meet the above conditions, many pedestrians and non-motor vehicles crossing the intersection lead to traffic delays or even congestion and traffic accidents.

(5) There are excessive traffic delays at the entrance to a roundabout due to

high traffic volume.

(6) Intersections are located in urban areas. The implementation of signal control at intersections separates the conflicting traffic flows temporally, reduces the mutual interference between traffic in all ways, and improves the safety and efficiency of vehicle operation. Figure 6 - 3 shows the signalized intersection.

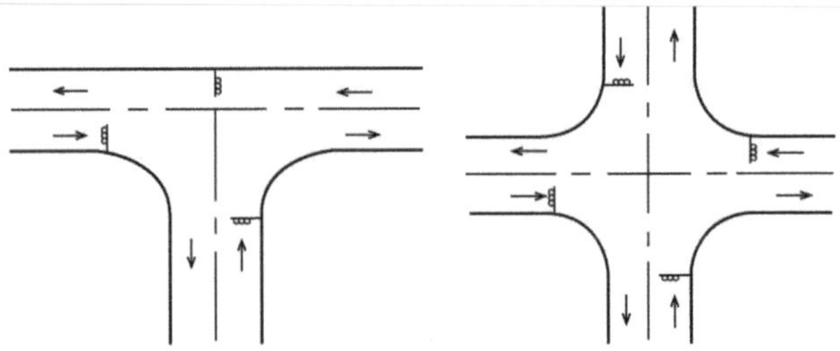

(a) Three-leg signalized intersection (b) Four-leg signalized intersection

**Figure 6 - 3 Signalized intersection**

## 6.2.2 Classified by number of intersecting roads

According to the number of intersecting roads, it can be categorized into three-leg intersections, four-leg intersections and multi-leg intersections, as shown in Figure 6 - 4.

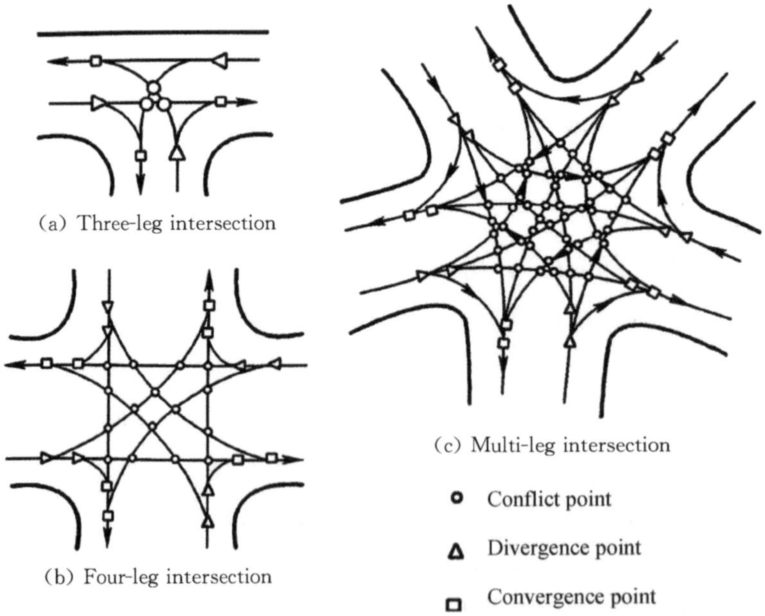


**Figure 6 - 4 Intersections classified by number of intersecting roads**

According to the above analysis of traffic characteristics of intersections, the more the number of intersecting roads, the more the number of crossing points (especially conflict points), and the more complex the traffic at intersections.

There shall not be more than four legs at intersections. When the number of legs is more than four, a roundabout shall be adopted. Besides, the number of legs in a roundabout should not be more than five.

## 6.2.3 Classified by the shape of intersection

Based on the shape of intersections, it can be divided into the T-shaped intersection, Y-shaped intersection, crossroad intersection, X-shaped intersection, dislocation intersection, multi-leg intersection and roundabout.

**1) T-shaped intersection**

A T-shaped intersection, or T, is the basic type of three-leg intersection, as shown in Figure 6 - 5(a). It is roughly orthogonal with an intersection angle of $90° \pm 15°$ and is the simplest type of intersections and one of the safe types. Usually, the through traffic claims priority while traffic on the other way shall yield the right-of-way.

**2) Y-shaped intersection**

A Y-shaped intersection is also one type of three-leg intersection with an intersection angle of less than 75° or greater than 105°, as shown in Figure 6 - 5 (b). This type of intersection is unfavorable to vehicle maneuver when the intersection angle is small, and the visual condition at the acute corner of intersections is poor.

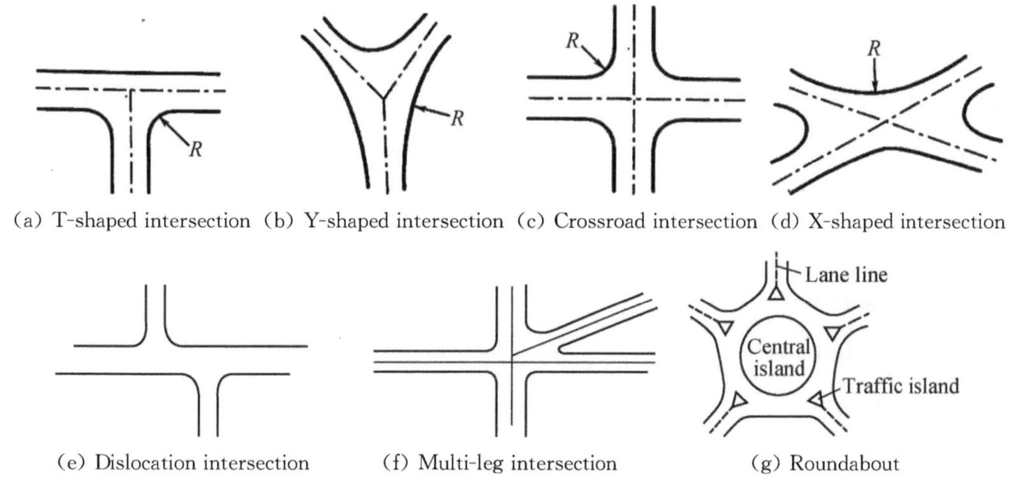


**Figure 6 - 5 Classified by shapes of the intersection**

**3) Crossroad intersection**

A crossroad intersection is the basic type of four-leg intersection. It is an orthogonal intersection with an intersection angle close to 90°, as shown in Figure 6 - 5(c). If the intersection angle is within the range of 90°±15°, it may also be designed as the crossroad intersection. It has the advantages of simple forms, less land occupation and low cost.

**4) X-shaped intersection**

An X-shaped intersection is also one type of four-leg intersection, which refers to an intersection angle of less than 75° or greater than 105°, as shown in Figure 6 - 5 (d). The area of X-shaped intersection is thin and long and the intersecting angle is small, which is unfavorable to vehicle maneuver. Thus it is used for special situations.

**5) Dislocation intersection**

A dislocation intersection refers to two offset T-shaped intersections which are very close to each other, as shown in Figure 6 - 5(e). For the highway with two lanes in two-way, if the traffic capacity is surplus and the priorities of intersecting highways are different, the dislocation intersection can be used to stagger the minor leg in order to prevent vehicles from the minor road moving through the principal one at high speed. The dislocation intersection should not be adopted when the through traffic volume of the minor road is large.

**6) Multi-leg intersection**

The intersection consisting of five or more roads is called the multi-leg intersection, as shown in Figure 6 - 5(f). The multi-leg intersection imposes a great challenge on traffic organization and management and shall be avoided as far as possible. When traffic safety is affected by numerous conflict points, a roundabout can be considered to simplify the traffic flow through the intersection and minimize potential conflict points.

**7) Roundabout**

A roundabout is an intersection type in which an island is located at the center of intersections, and all vehicles circulate around the central island in one way, as shown in Figure 6 - 5(g). A roundabout has no conflict points, continuous traffic flow, no signal control, and is convenient for landscape greening. It is suitable for intersections with multi-leg (5 roads or more), moderate traffic volume, and substantial turning vehicles (close to or exceeding the through traffic volume).

### 6.2.4 Classified by the channelization mode

Channelization design refers to the applications of the traffic sign, marking and traffic island at intersections to direct the traffic flow and pedestrians to travel along the specified way like water flows in the channel without interference.

Channelization design must be carried out for intersections of Class Ⅱ and above highways, channelization design shall be carried out for intersections of Class Ⅲ highways, and channelization design should be carried out for intersections of Class Ⅳ highways. Channelization design shall be conducted based on the intersection type, traffic management method, turning traffic volume, design speed, and other factors, and adopt the methods such as the intersection with the widening corner, flared intersection, and turning split intersection.

**1) Intersection with the widening corner**

An intersection with the widening corner refers to the one that connects the intersecting roads smoothly by the single circular curve of the appropriate radius or compound curve, as shown in Figure 6 - 6. This intersection has the advantages of simple forms, less land occupation, low cost and convenient design; however, it suffers from low capacity and slow traffic speed. Thus, it is suitable for Class Ⅲ and Class Ⅳ highways or local roads with low speed, small traffic volume, and few turning vehicles. If the skew of the intersection is not large, this intersection can also be applied when the major road with small turning traffic crosses the minor road. Finally, the key point of intersection design is the determination of the curve radius at the corner and the assurance of sight distance.

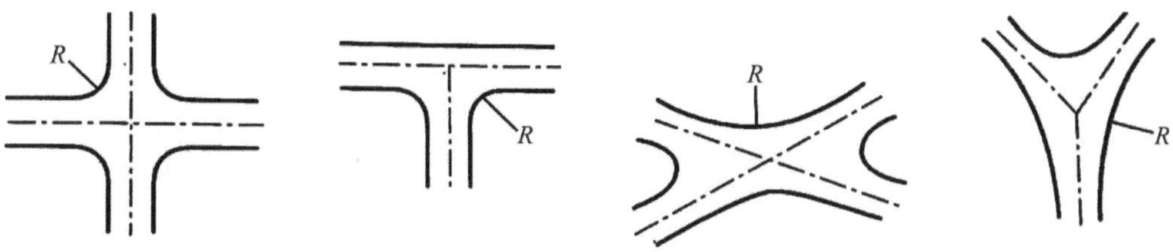


**Figure 6 - 6 Intersections with widening corners**

**2) Flared intersection**

A flared intersection is the one in which a speed-change lane and a turning lane are introduced at the junction of intersections in order to prevent the turning vehicles from affecting the normal movements of other vehicles. The right-turning or left-turning lanes may be added individually or concurrently. As a result, this intersection can reduce the interference of turning traffic on the through traffic, increase speed, decrease the accident rate, and enhance traffic capacity, but it requires more land use and a considerable investment. It is suitable for Class Ⅰ and Class Ⅱ highways and urban arterial roads with large traffic volume and many turning vehicles. Figure 6 - 7 shows an example of a flared T-shaped intersection, and Figure 6 - 8 shows an example of a flared crossroad intersection.

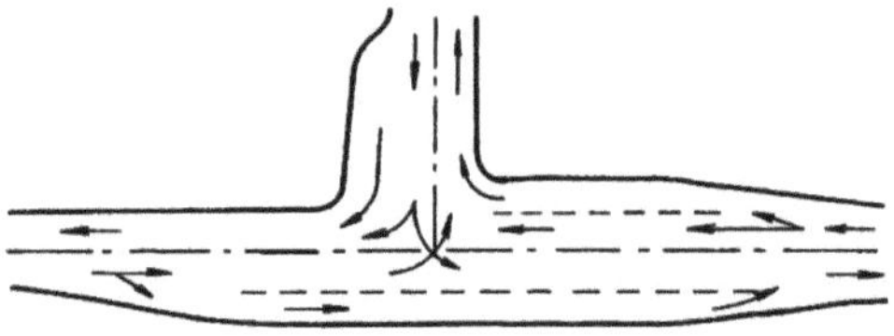

**Figure 6 - 7 Flared T-shaped intersection**

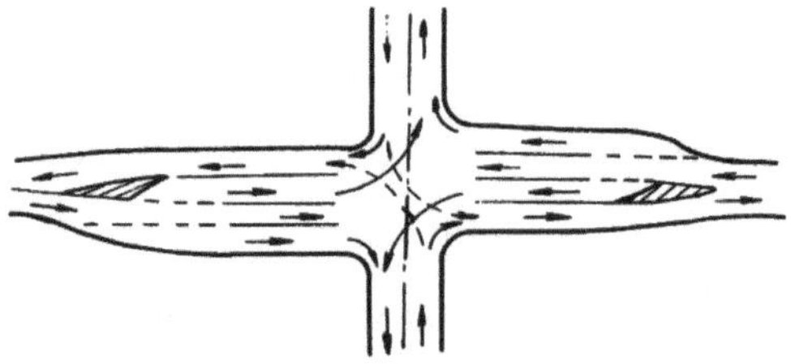

**Figure 6 - 8 Flared crossroad intersection**

**3) Split turning intersection**

A split turning intersection is the one in which the one-way right-turning or two-way left-turning and right-turning traffic flows operate separately with a large radius with the regulation of the divisional island, directional island, dividing lanes, etc., as shown in Figure 6 - 9. Therefore, the turning vehicles, especially the right-turning vehicles, can pass the intersection at high speed and the turning capacity is increased. It is readily for roads with high speed and many turning vehicles. The design focuses on the determination of the turning radius, the assurance of sight distance, as well as the requirements of the end radius of the divisional island.

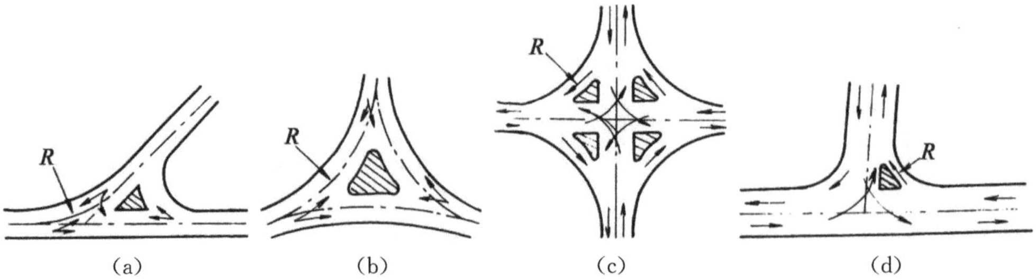


**Figure 6 - 9 Split turning intersections**

## 6.3 Principle and provision of intersection design

The number and spacing of intersections affect the efficiency of relevant sections of a highway. The more the intersection and the smaller the spacing, the higher the convenience for the connection of road networks and for the travel of villages and towns on both sides, but the greater the compromise on the efficiency of road sections. Meanwhile, with the increase in intersection density, it is detrimental to traffic safety and management. Therefore, in the specific project design, it is necessary to correctly comprehend the function and technical class of the project, and properly coordinate the balance between traffic efficiency and transportation serviceability along the route. The following conditions shall be met for the placement of intersections.

(1) Intersections shall be set reasonably according to the function and technical class of intersecting highways, the status and planning of regional road networks, as well as the topographical and geomorphological conditions of intersecting areas.

(2) When highways of different classes intersect each other, the introduction of intersections shall conform to the provisions in Table 6 - 2.

**Table 6 - 2 Requirements of highway intersections**

| Intersected highway | Highway mainline | | | | |
|---|---|---|---|---|---|
| | Class Ⅰ highway (arterial) | Class Ⅰ highway (collector) | Class Ⅱ highway (arterial) | Class Ⅱ highway (collector) | Class Ⅲ & Class Ⅳ highways |
| Class Ⅰ highway (arterial) | Prohibited | — | — | — | — |
| Class Ⅰ highway (collector) | Prohibited | Limited | — | — | — |
| Class Ⅱ highway (arterial) | Prohibited | Limited | Limited | — | — |

Continued Table 6 - 2

| Intersected highway | Highway mainline | | | | |
|---|---|---|---|---|---|
| | Class Ⅰ highway (arterial) | Class Ⅰ highway (collector) | Class Ⅱ highway (arterial) | Class Ⅱ highway (collector) | Class Ⅲ & Class Ⅳ highways |
| Class Ⅱ highway (collector) | Prohibited | Limited | Limited | allowed | — |
| Class Ⅲ & Class Ⅳ highways | Prohibited | Limited | Limited | allowed | allowed |

*Note*: This table refers to the **Specification (2017)**.

The following principles shall be followed in the design of highway intersections:

(1) Selection of the intersection location shall systemically consider the status and planning of highway network, terrain, ground features, geological conditions, economic and environmental factors, etc., and the location with flat terrain and wide vision should be selected.

(2) Selection of the intersection type shall systemically consider the function, technical class, traffic volume, traffic management, land use condition, engineering cost and other factors of intersecting highways. Besides, the selected intersection form shall enable smooth traffic in the major highway and mainstream, fewer conflict points, as well as the small conflict zone.

(3) Geometry design of intersections shall be combined with the traffic management method and consider the arrangement of relevant facilities.

(4) The technical indexes of alignments of intersecting roads within the intersection area shall meet the requirements of sight distance.

(5) Tangents should be used for intersecting highways within the intersection area; when curves are used, the radius should be greater than the radius of the circular curve without superelevation. The grade shall be as gentle as possible and meet the minimum radius of the vertical curve required by vision.

(6) Intersection design is based on the predicted traffic volume, and the traffic volume used in the design is the design hourly traffic volume.

(7) As for the pedestrian crossing facilities at intersections, such as the

crosswalk, pedestrian overcrossing, pedestrian passageway, etc., they are designed according to the pedestrian flow, highway technical class and traffic management method.

(8) Geometry design of intersections need to be considered together with signs, markings and signal facilities, and they are arranged as a whole. Small intersections with the insufficient sight distance can be augmented with reflectors according to specific conditions.

(9) In the reconstruction of intersections, in addition to collecting data about traffic volume, it is necessary to investigate the traffic delay and the condition of existing intersections such as the number, degree, and reason of traffic accidents.

(10) Intersection design shall satisfy the traffic requirements of corresponding design vehicles on the intersecting roads. In the case of special traffic demand, the traffic conditions of intersections shall be inspected according to the actual vehicle type.

The determination of the intersection angle and the number of legs shall meet the following requirements:

(1) The intersection angle should be the right angle. In case of skewed intersections, the acute angle shall not be less than 70°; when restricted by topographic conditions or other special circumstances, it shall be greater than 45°.

(2) The number of legs shall not be more than four; when the number of legs is more than four, a roundabout shall be adopted.

(3) The number of legs of the roundabout should not be more than five, and "yield-at-entry" shall be used for the roundabout if allowed.

(4) The newly built highway shall not be directly connected with the existing intersection with four legs or more.

The gap of intersections shall meet the following requirements:

(1) The gap of intersections shall be determined according to the highway function, technical class and its impact on operating safety, traffic capacity and vehicle delay.

(2) The minimum gap of intersections of Class Ⅰ and Class Ⅱ highways shall conform to the provisions in Table 6 - 3.

**Table 6 - 3 Minimum gap between intersections**

| Highway class | Class Ⅰ | | | Class Ⅱ | |
|---|---|---|---|---|---|
| Function | Arterial | | Collector | Arterial | Collector |
| | General value | Minimum value | | | |
| Gap/m | 2,000 | 1,000 | 500 | 500 | 300 |

*Note*: This table refers to the **Specification** (**2017**).

(3) When Class Ⅱ and Class Ⅲ highways serve as arterials, priority shall be given to ensure the smooth flow on the arterial, and measures shall be taken to eliminate the longitudinal and transverse interference. The gap between intersections shall maintain enough spacing, and grade separation can be employed if necessary.

(4) When Class Ⅰ and Class Ⅱ highways serve as collectors, the intersection shall be placed reasonably, and the number of intersections shall be reduced through branch merging and other measures.

## 6.4 Traffic organization at intersections

Traffic organization at intersections includes vehicle traffic organization, pedestrian traffic organization, and non-motor vehicle organization. The basic task of traffic organization design at intersections is to ensure the safety of vehicles and pedestrians, improve the traffic capacity of intersections, and make the traffic flow in all ways pass through intersections safely and efficiently.

### 6.4.1 Vehicle traffic organization

The common methods of vehicle traffic organization include application of exclusive lanes, traffic organization of left-turning vehicles, traffic channelization, signal control, and other traffic organization adjustment methods.

**1) Application of exclusive lanes**

By setting exclusive lanes, vehicles in different ways are organized to drive separately in the respective lanes without interference. According to the lane width and the traffic volume of left-turning, through and right-turning vehicles, multiple schemes of lane division can be proposed, as shown in Figure 6 - 10.

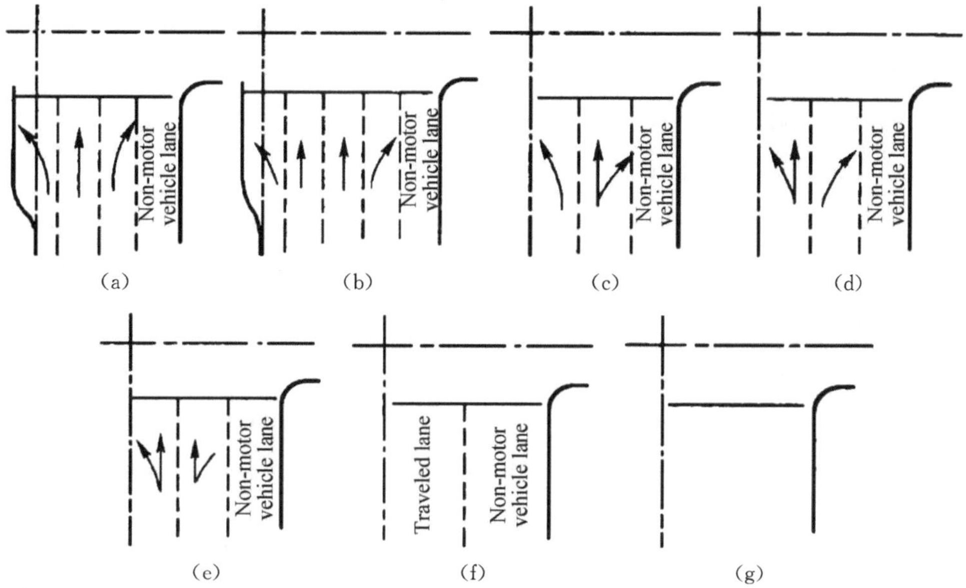


**Figure 6 - 10 Lane division at intersections**

Figure 6 - 10(a) shows the example of setting an exclusive lane for each way since there are almost equal through, left-turning and right-turning vehicles. In Figure 6 - 10(b), in the case of a large number of through vehicles and a certain number of left-turning and right-turning vehicles, the scheme of two through lanes, one left-turning lane and one right-turning can be adopted. In Figure 6 - 10(c), if there are more left-turning vehicles than right-turning vehicles, a left-turning lane can be provided separately, and the right-turning and through vehicles share one lane. On the contrary, in Figure 6 - 10(d), if the number of right-turning vehicles is more than that of left-turning vehicles, a right-turning lane can be provided as the exclusive lane, and the left-turning and through vehicles share one lane. In Figure 6 - 10(e), when there are few left-turning and right-turning vehicles, they can share a lane with the through vehicles, respectively. In Figure 6 - 10(f), if the traveled way is too narrow to offer the left-turning, through and right-turning lanes, it can be divided into the vehicle lane and non-motor vehicle lane. In Figure 6 - 10(g), it is impossible to divide lanes because the lane width is extremely narrow, or unconducive to lane utilization if the lane is divided. Thus, the lane division can be omitted.

**2) Traffic organization of left-turning traffic**

The left-turning vehicles are the main cause of traffic conflicts at intersections. It is an effective method to conduct reasonable organization of left-turning vehicles in order to ensure traffic safety and improve intersection capacity. The traffic organization of left-turning vehicles can be realized by the following methods.

(1) Application of left-turning exclusive lanes

As shown in Figure 6 - 10(c), within the traveled way, a lane is separated close to the centerline for the left-turning vehicles to avoid obstructing the through vehicles; as shown in Figure 6 - 10(a), if the width of the original traveled way is not enough, the left-turning lane may be appropriately extended to the left side of the centerline. After setting the exclusive left-turning lane, the left-turning vehicles need to wait for the signal phase to pass, or look for a gap to pass without affecting the through traffic.

(2) Traffic management

Left-turning within a specified time is prohibited through signal control or traffic police command.

(3) Changing left-turning to right-turning

As shown in Figure 6 - 11, traffic changes from left-turning to right-turning by means of the roundabout, detour and U-turn to reduce conflict points.

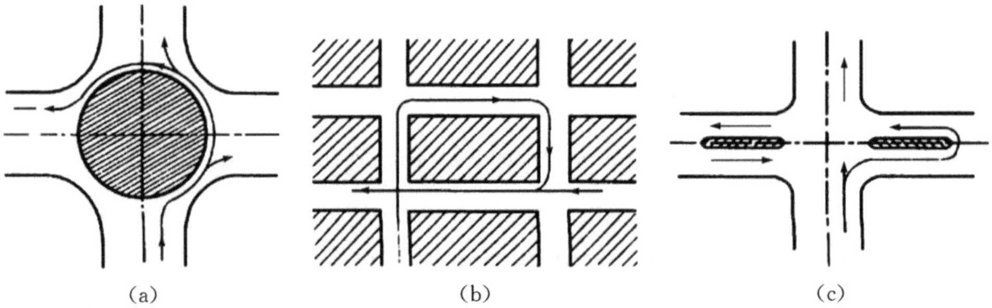


**Figure 6 - 11 Changing left-turning to right-turning**

As shown in Figure 6 - 11(a), the roundabout organizes all vehicles to travel anticlockwise around the central island. Left-turning movements change to right-turning movements so that the conflicting traffic flows become the diverging or converging ones.

As shown in Figure 6 - 11(b), block detour refers that vehicles turn right around the adjacent block instead of turning left directly. The trip is increased as a result of the detour. Moreover, this method is usually used only when the left-turning vehicles account for a small proportion and it is difficult to widen the old city road. Or it is used for intersections with a large grade of approach roads at the bridgehead, so as to prevent the right-angle turning accident when vehicles run downhill at high speed.

As shown in Figure 6 - 11(c), the U-turn is an indirect left-turning when a left-turning vehicle turns right at the intersection of which the left-turning is prohibited. After making right turns, vehicles turn back from the downstream median opening. This method simplifies the intersection signal timing, isolates the conflict area, and reduces the conflict points.

**3) Traffic channelization**

When the major road of T-shaped intersection is a Class Ⅱ highway, and in case of a small amount of through traffic volume and a considerable proportion of turning traffic volume at the intersection, channelization by divisional island on the Class Ⅱ highway can be used to prevent lane encroachment and traffic interference, as shown in Figure 6 - 12(a); contrarily, when the through traffic volume of the major highway is large, the channelized T-shaped intersection with divisional islands on both principal and minor highways can be adopted, as shown in Figure 6 - 12(b).

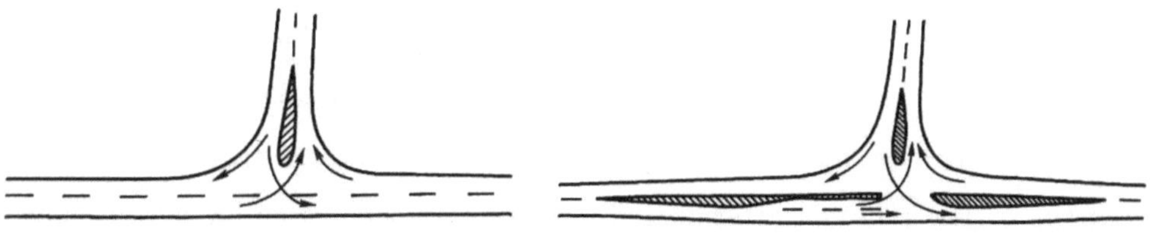

(a) (b)

**Figure 6 - 12 Channelized T-shaped intersection with divisional islands**

As shown in Figure 6 - 13, the channelized T-shaped intersection with the divisional island is adopted when the major highway has four lanes, or the Class Ⅱ highway is with a design speed greater than or equal to 60 km/h and a considerable proportion of turning traffic, or the two-lane highway connects directly with the interchange. Specifically, when the major highway has two lanes, certain channelization schemes shown in Figure 6 - 13(a), (b), (c) could be selected according to the left-turning and right-turning traffic volume; and the divisional island on the major highway should be a painted island. In addition, when the major highway has four lanes, the channelization scheme shown in

Figure 6 - 13(d) shall be adopted. The divisional island on the minor road can be treated referring to Figure 6 - 13(a), (b) and (c) according to the left-turning and right-turning traffic volumes respectively. The divisional island on the major highway shall be a solid island.

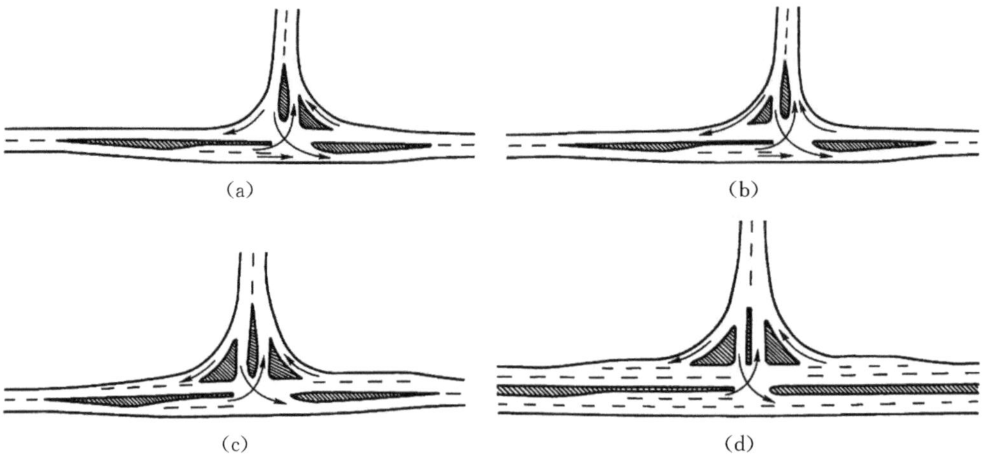

(a) (b) (c) (d)

**Figure 6 - 13 T-shaped intersection with divisional islands**

When highways of high-class or large traffic volume intersect, channelized intersections with vehicle tracks specified by divisional islands and divisional islands shall be adopted. The traffic speed and capacity of turning vehicles at the channelized intersection, especially the right-turning vehicles, are high. It is applicable to the arterial highway with high speed and many turning vehicles, such as the four-lane major highway or two-lane highway with a design speed of 80 km/h, or the regional arterial highway with two lanes and a design speed of 60 km/h. Figure 6 - 14 shows the channelized crossroad intersections.

**4) Signal control**

The traffic signal system with automatic control is adopted to improve the running speed and traffic capacity.

**5) Other traffic organization adjustment methods**

Occasionally, under special circumstances, the traffic organization mode may be adjusted by restricting vehicle movements, organizing one-way traffic, closing branches on the principal arterial roads appropriately, so as to improve traffic capacity of the whole road network.

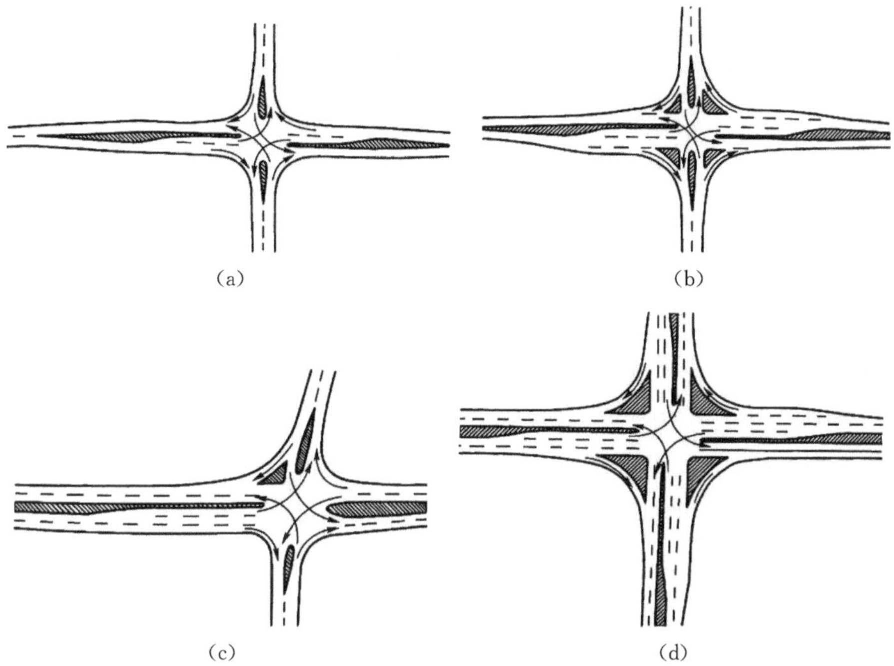

Figure 6 - 14 Channelized crossroad intersections

## 6.4.2 Pedestrian traffic organization

In urban road design, pedestrians and non-motor vehicles converge, turn, and cross streets at intersections. Hence it is necessary to consider the pedestrian traffic organization. In fact, pedestrian and non-motor vehicle organizations are seldom considered in highway design. However, it is an effective way to eliminate traffic congestion at intersections and ensure traffic safety by organizing the traffic of pedestrians and non-motor vehicles reasonably, especially on the highways with a high degree of land development on both sides, on the highways at the access of cities and on the urban roads with numerous pedestrians and non-motor vehicles.

The main contents of pedestrian traffic organization are two-fold: one is to organize pedestrians to walk on the sidewalks, and the other is to organize pedestrians to cross the street safely on the pedestrian crosswalk lines.

Sidewalks are usually symmetrically located on both sides of the traveled way, and generally extend from the normal road section. Sidewalks of the adjacent roads at intersections are connected, and their corners are widened properly to accommodate pedestrian flow concentration and the need to turn. In addition to the necessary road signs, traffic signals, lighting, and railings, it is not allowed to arrange other facilities on the sidewalk to ensure its effective width.

To enable pedestrians to cross the street safely and orderly (across the traveled way), pedestrian crosswalks shall be set at the appropriate positions of intersections, which are exhibited in the form of the zebra crossing. The sidewalk and crosswalk at intersections are interconnected to form a walkway network that can reach any way. The position and width of crosswalks are very important to the pedestrian traffic organization.

The key points of pedestrian crosswalk design at intersections are as follows:

(1) Crosswalks shall be located at a position easily observed by drivers, as close to the intersection as possible, and consistent with the natural way of pedestrian flow. Furthermore, to shorten the walking distance of pedestrians, crosswalks shall be perpendicular to the traveled way as much as possible.

(2) When the crosswalk is too long (greater than 15 m), a pedestrian refuge island shall be provided in the middle of the crosswalk, with a width of more than 1.5 m.

(3) Crosswalk width is related to the number of pedestrians and the signal phase. The crosswalk width of the major road should not be less than 5 m, and that of the branch road should not be less than 3 m. The width may be increased or decreased in units of 1 m.

(4) The crosswalk shall be located parallel to the extension line of the sidewalk of road sections, and shall be set back appropriately, with a distance no less than 1 m (as shown in part a of Figure 6 - 15). At intersections where the right-turning vehicles are prone to conflict with pedestrians, the impact of right-turning vehicles on the pedestrian of two adjacent entrances needs to be mitigated. Thus the crosswalk lines of adjacent entrances shall not intersect, and space shall be reserved for one right-turning vehicle at least. The setback distance should be 3.0 - 4.0 m (as shown in part b of Figure 6 - 15).

(5) The length of the sidewalk corner (as shown in part c of Figure 6 - 15) shall not be less than 6.0 m, which is the length of a passenger car. Guard rail and other isolation facilities shall be used.

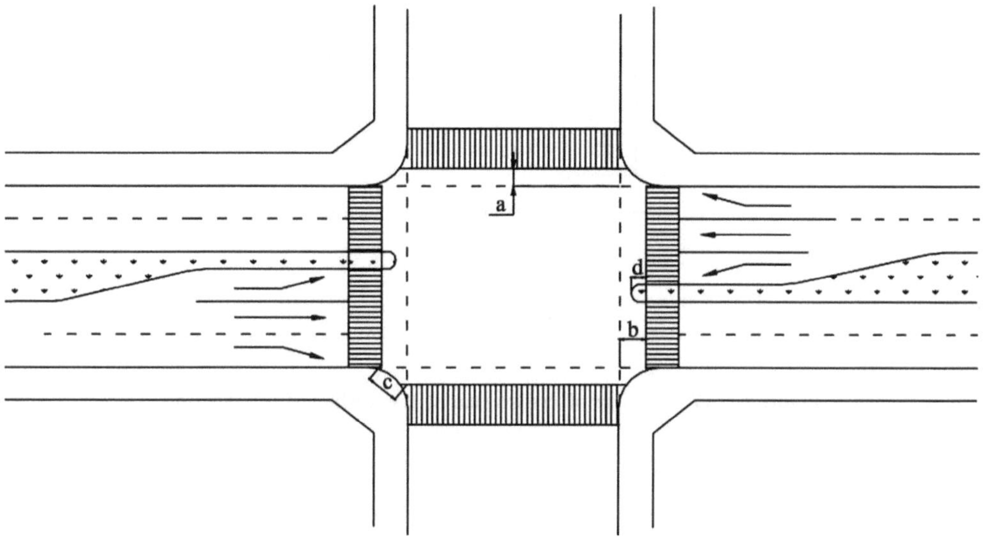


**Figure 6 - 15 Position of crosswalks**

(6) For the entrance lane with a medial separator, the crosswalk shall be set back by 1. 0 - 2. 0 m from the end of the separator. In addition, the medial separator shall provide safety protection for pedestrians crossing and stopping (as shown in part d of Figure 6 - 15).

(7) With the employment of the divisional island, the crosswalk may be applied at the Y-shaped intersection (as shown in Figure 6 - 16). If there are few pedestrians, crosswalk A is not required.

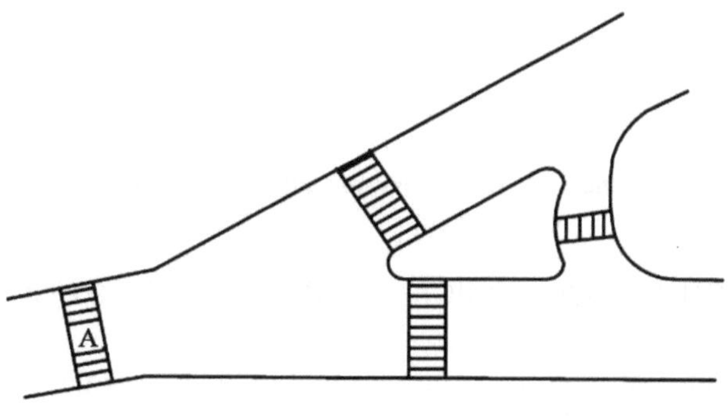


**Figure 6 - 16 Layout of crosswalks at the Y-shaped intersection**

(8) The layout of crosswalks at the T-shaped intersection should be designed as shown in Figure 6 - 17. In the case of few pedestrians, crosswalk A or B is not necessary.

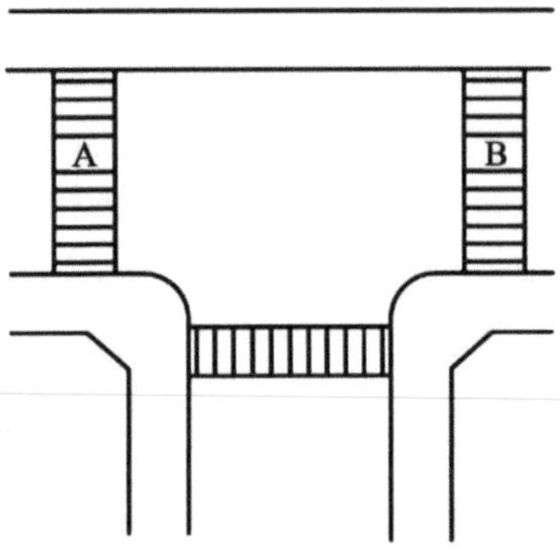


Figure 6 - 17 Layout of crosswalks at the T-shaped intersection

(9) When the viaduct pier is located near intersections, crosswalks shall be placed on the separator where the pier is located if it is under constraint (as shown in Figure 6 - 18).

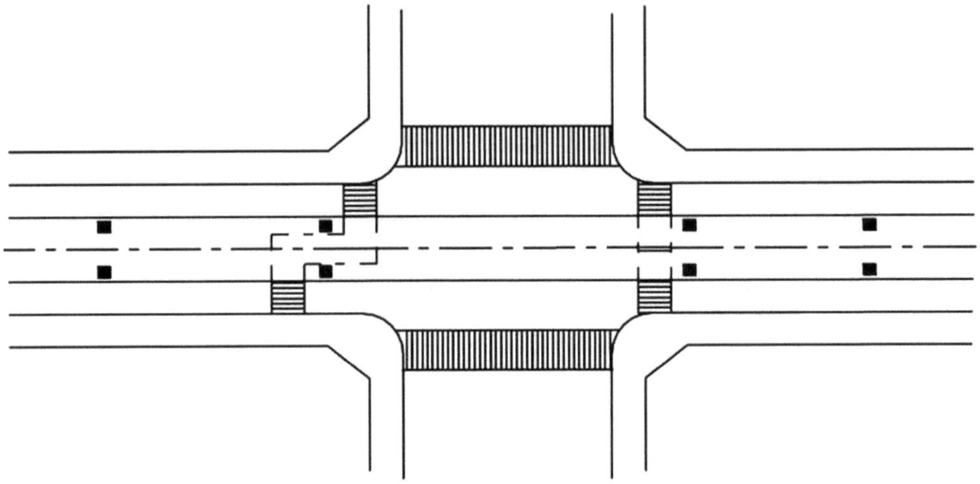

Figure 6 - 18 Crosswalk under the viaduct

### 6.4.3 Non-motor vehicle traffic organization

Mixture of non-motor vehicles and vehicles leads to many adverse effects on both. The competition of temporal and spatial resources between non-motor vehicles and vehicles at intersections increases the accident potential and reduces the traffic efficiency.

For the traffic organization design of an intersection, it shall be ensured that non-motor vehicles and vehicles can recognize each other clearly and discern the tracks and directions of each other. In addition, efforts shall be devoted to isolating non-motor vehicles from vehicles in time and space; if the isolation is not possible, appropriate space shall be reserved for lane separation. The conflict points between non-motor vehicles and vehicles shall be kept away from the conflict points between vehicles as far as possible.

The following methods can be adopted for the traffic organization of non-motor vehicles at intersections:

(1) Exclusive lane for right-turning vehicles

The existing road is used to set an exclusive lane for the right-turning non-motor vehicles. This method can be employed at intersections with high volume of non-motor vehicles.

(2) Left-turning waiting area

A left-turn waiting zone for non-motor vehicles is set in front of the non-motor vehicles' entrance lane at intersections. When the green light is on, left-turning non-motor vehicles first move with the through non-motor vehicles to the opposite left-turning waiting zone, and then move forward when the green light in the other way is on, that is, it turns left-turning into two through movements.

(3) Advance stop line

The stop line of non-motor vehicles is drawn in front of vehicles. When the green light is on, non-motor vehicles enter the intersection first to avoid crowding with vehicles. This method is beneficial to improve the traffic capacity and traffic safety of intersections, and is also suitable for the situation of large left-turning non-motor vehicles.

(4) Double green light

Stop lines of vehicles and non-motor vehicles are still in the same position. However, considering the characteristics that non-motor vehicles start quickly and always pass through the intersection in groups, the green light of the non-motor vehicles' traffic signal turns on first to allow the non-motor vehicle group to enter the intersection first, and then the green light of the vehicles' traffic signal turns on. The time interval between the two green lights depends on the traffic volume and the geometric properties at intersections.

When the traffic flow is large and the interference between non-motor vehicles and vehicles is serious, grade-separated non-motor vehicle organization can be considered in combination with pedestrian overpasses or tunnels.

Stairways, ramps, or the mixed type could be used for the pedestrian overcrossing or underpass. Generally, pedestrians should use stairways while non-motor vehicles should use ramps. When there are many non-motor vehicles and ramps could not be adopted due to terrain or other conditions, the mix of stairways and ramps is optional.

## 6.5 Geometric design of intersections

### 6.5.1 Design factors of intersections

**1) Design speed**

For the intersection of highways, the provisions of design speed are as follows:

(1) The design speed of major highways within the intersection area should be consistent with that of the highway section.

(2) When the function and class of two intersecting highways are the same or the traffic volumes are approximating, the design speed of the through lane at intersections can be appropriately reduced, but it shall not be less than 70% of the section.

(3) The design speed can be appropriately reduced when the minor highway is realigned due to the intersection angle and other reasons, or when lower alignment indexes are adopted due to constrained conditions.

(4) The design speed of the turning lane shall be determined according to the design speed, traffic volume, intersection type, traffic management method, land use, and other factors.

**2) Design vehicle**

Intersection design shall meet the traffic requirements of corresponding design vehicles for the intersecting highways. If there is a special traffic demand, traffic conditions at the intersections shall be inspected according to actual vehicle types.

**3) Design traffic volume and level of service**

Intersection design shall be based on the predicted traffic volume, and the traffic volume adopted in the design shall be the design hourly traffic volume.

The design level of service at intersections of Class Ⅰ arterial highways shall not be lower than level three; the design level of service at intersections of Class Ⅰ and Class Ⅱ collector highways shall not be lower than level four. And the traffic capacity and level of service at intersections of Class Ⅲ or above highways shall be analyzed and inspected.

## 6.5.2 Horizontal alignment design of intersections

Horizontal alignment design of highway intersections shall meet the following requirements:

(1) The two intersecting highways within the intersection area shall be orthogonal or nearly orthogonal. The horizontal alignment shall use the tangent or circular curve of a large radius, and the circular curve with superelevation should not be used.

(2) When the intersection angle between the new highway and the existing lower-class highway is less than 70°, the minor highway shall be partially realigned in a certain range before and after the intersection.

Figure 6 - 19 is the schematic diagram of the realignment of the minor highway at the T-shaped intersection. The intersection angle of the highway can meet the requirements through partial realignment. The turning curve after realignment is connected with the original minor highway through the approach

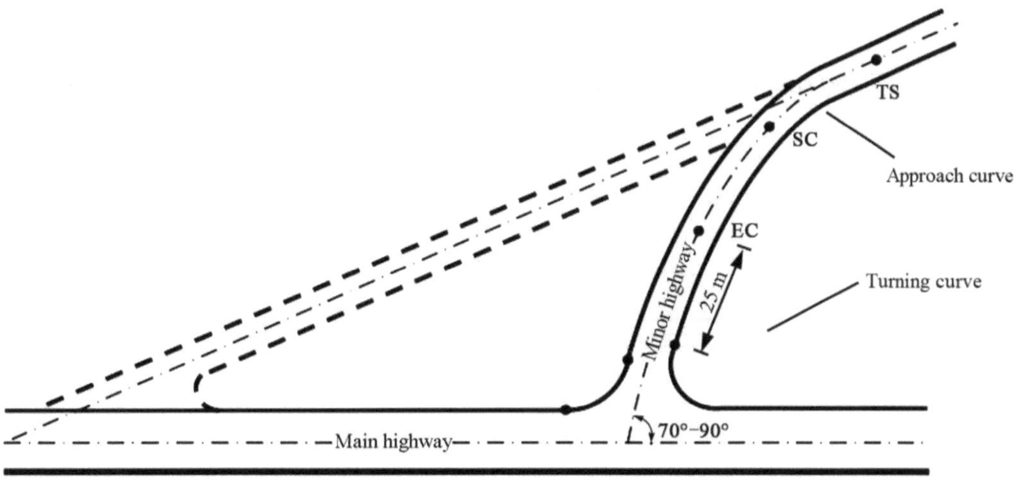


**Figure 6 - 19 Schematic diagram of realignment of the minor highway at the skewed T-shaped intersection**

curve. A tangent with a length of not less than 25 m shall be reserved between the approach curve and the turning curve at the intersection. When the minor road is a Class Ⅱ highway, the radius of the approach curve shall not be less than 80 m; when the minor road is a Class Ⅲ highway or below, the radius of the approach curve shall not be less than 40 – 50 m.

Figures 6 – 20 and 6 – 21 are the schematic diagrams of the alignment of crossroad. When the intersection point remains unchanged (Figure 6 – 20), two curves shall be supplied in each leg of the minor highway. The curve radius far from the intersection shall not be less than the general minimum radius of this highway, and the spiral shall be used per the requirement; the curve radius close to the intersection shall not be less than 45 m, and the spiral shall be used at the end far from the intersection. When the intersecting point is changed (Figure 6 – 21), the S-shaped curve is only applied on one leg of the minor highway, and the radius requirements are the same as above.

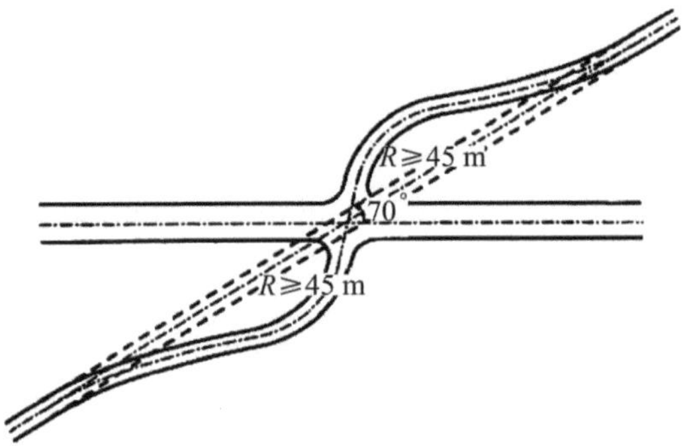


**Figure 6 – 20 Schematic diagram of skewed crossroad realignment (Intersecting point unchanged)**

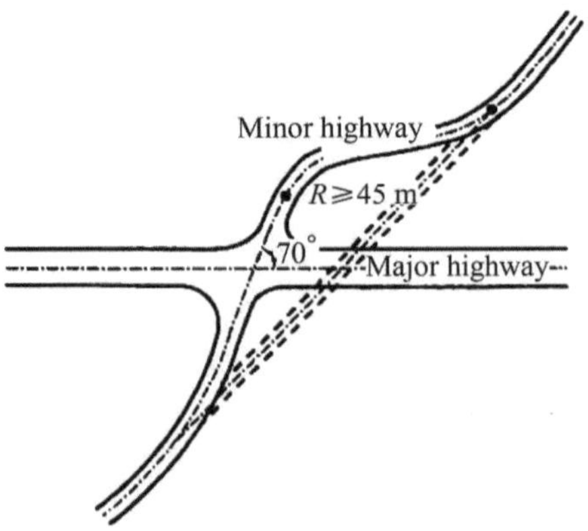


**Figure 6 – 21 Schematic diagram of skewed crossroad realignment (Intersecting point changed)**

If the above-mentioned methods are not applicable to the realignment of the skewed crossroad, the two legs of the minor highway can be changed separately to form two offset T-shaped intersections, as shown in Figure 6 - 22, including the forward offset intersection and reverse offset intersection. The reverse offset intersection is only limited to the case that the proportion of through traffic volume of the minor highway is very small. In the offset intersection, the intersection angle is 90°. And the approach of the minor highway has the same alignment requirement as that of the skewed T-shaped intersection.

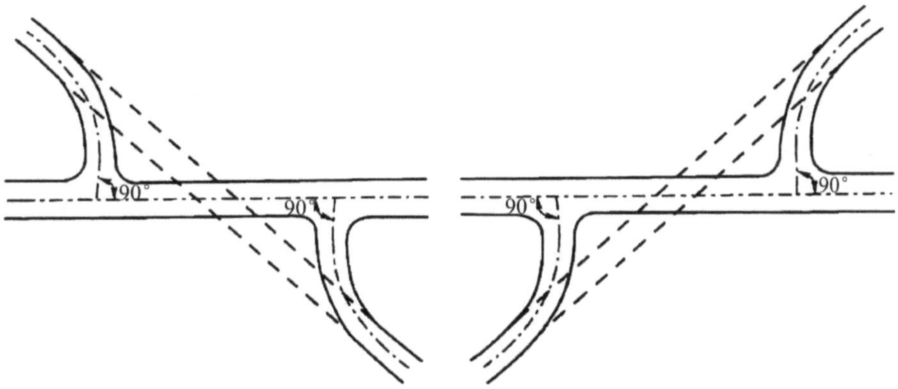


(a) Forward/Sequential offset intersection (b) Reverse offset intersection

**Figure 6 - 22 Schematic diagram of offset intersections**

When the major highway is curved at intersections, and the minor highway is connected to the tangent of the major highway curve, the minor highway should be realigned in the way shown in Figure 6 - 23, so as to introduce the traffic flow into the major highway directly and improve the line of sight at intersections. However, when the curve has a large superelevation rate, the approach is provided with the reverse grade, and the sight distance is limited by grade, it shall be used with caution.

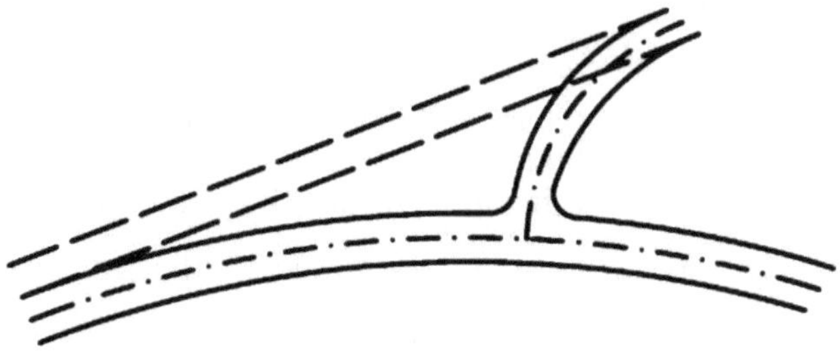

**Figure 6 - 23 Realignment of the curved major highway**

### 6.5.3 Vertical alignment design of intersections

Vertical alignment design of intersections shall meet the following requirements:

(1) Vertical alignments of two intersecting highways should be gentle within the intersection area. And the vertical alignment shall meet the provisions of stopping sight distance.

(2) Grades of the major highway within the intersection area shall be in the range of 0.15%–3%; the approach of the minor highway close to the intersection shall be upgraded in the range of 0.5%–2%.

Driving operation in the intersection area is complex and prone to traffic accidents. Therefore, although the running speed at intersections is lower than that of road sections, it is expected to possess a better vertical alignment at intersections than road sections, so that the driver can observe traffic flows within the intersection area as soon as possible, so as to facilitate speed change or stopping. Figure 6 - 24 is a schematic diagram of the grade of the minor highway approach. The approach of the minor highway close to the intersection shall be upgraded in the range of 0.5%–2%, and the distance from such upgrade section to the margin of the major highway shall not be less than 25 m.

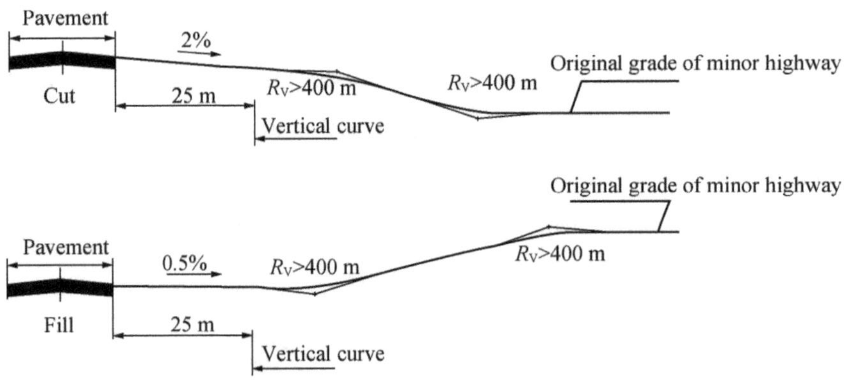


**Figure 6 - 24 Grade of the minor highway approach**

(3) When superelevation is applied on the circular curve of the major highway within the intersection area, the grade of the minor highway shall be subject to the cross slope of the major highway.

As shown in Figure 6 - 25, if the grade direction of the minor highway in a long range before and after the intersection is the same as that of the major highway, it is not necessary to adjust the vertical alignment of the minor highway; if the grade direction is opposite, an S-shaped vertical curve shall be introduced in a certain range of the approach of the minor highway so that the grade of the minor highway complies with the grade of the major highway.

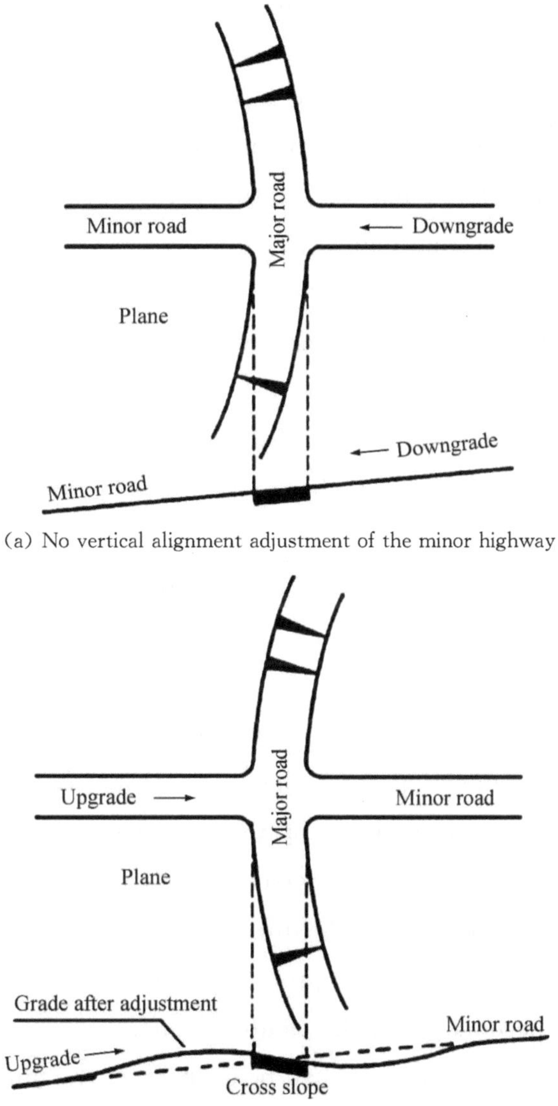


(a) No vertical alignment adjustment of the minor highway

(b) Vertical alignment adjustment of the minor highway

**Figure 6 - 25 Grade of the minor highway for the superelevated major highway**

## 6.5.4 Profile design of intersections

The task of profile design of intersections is to reasonably determine the shape of the common surface between the intersection roads and between the intersection and the surrounding buildings by adjusting the design elevations of the traveled way, sidewalk, and the nearby ground within the intersection area. The profile design needs to meet the requirements of traffic safety, driving comfort, rapid drainage, and architectural art. The essence of profile design is the elevation design of the intersection common plane, which shall meet the following requirements:

(1) The profile of the common part of two intersecting highways and the cross slope of approaches shall be determined according to the functions, classes, horizontal, and vertical alignments, traffic management methods, and other factors of the two intersecting highways.

For the priority intersection, the cross section of the major road shall traverse through the intersection, and the grade of the minor road shall be adjusted to adapt to the cross section of major roads, as shown in Figure 6 - 26; when it is difficult to adjust the profile of the minor road, the cross sections of the two highways shall be adjusted simultaneously, as shown in Figure 6 - 27.

When the functions of two intersecting highways are equal or similar, or when it is a signalized intersection, the two highways shall be properly adjusted, as shown in Figure 6 - 28.

(2) On the separated right-turning lane or the additional right-turning pavement, the elevation and cross slope of each part shall meet the requirements of the elevation of common part of the intersecting roads and adjacent local sections, the superelevation required by turning curves, and pavement drainage and road capacity within the whole intersection area.

(3) Pavement drainage within the intersection area shall be smooth, which is one of the main considerations in the profile design. There shall be no ponding on the pavement anywhere including the painted island.

The grade of at least one road shall be away from the intersection in design, which is conducive to drainage. In the case of special terrain, when the grades of all roads are towards the intersection, rainwater inlets and drainage pipes must be applied in the intersection to ensure drainage.

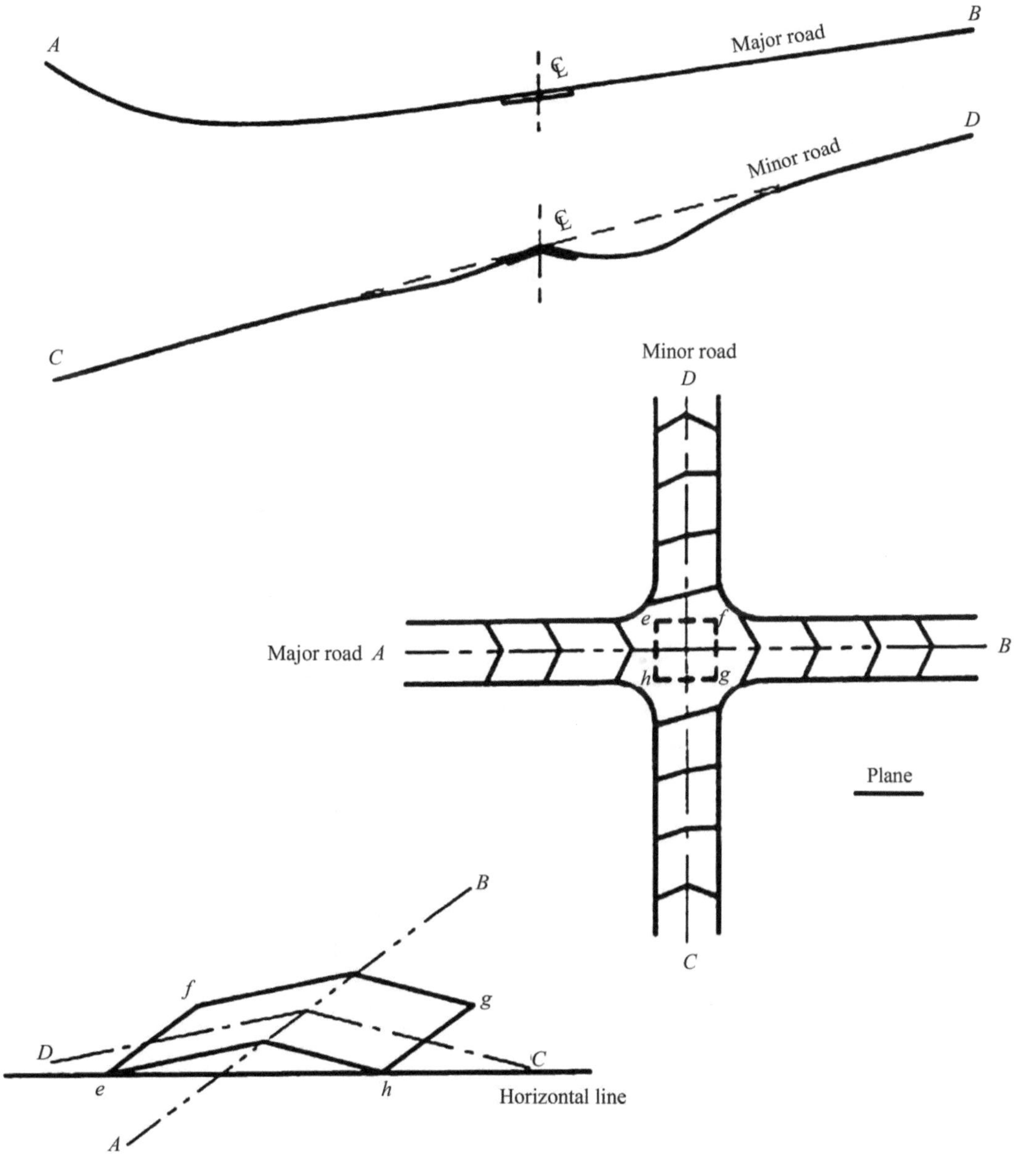


**Figure 6 - 26 Grade adjustment of the minor road for the priority intersection**

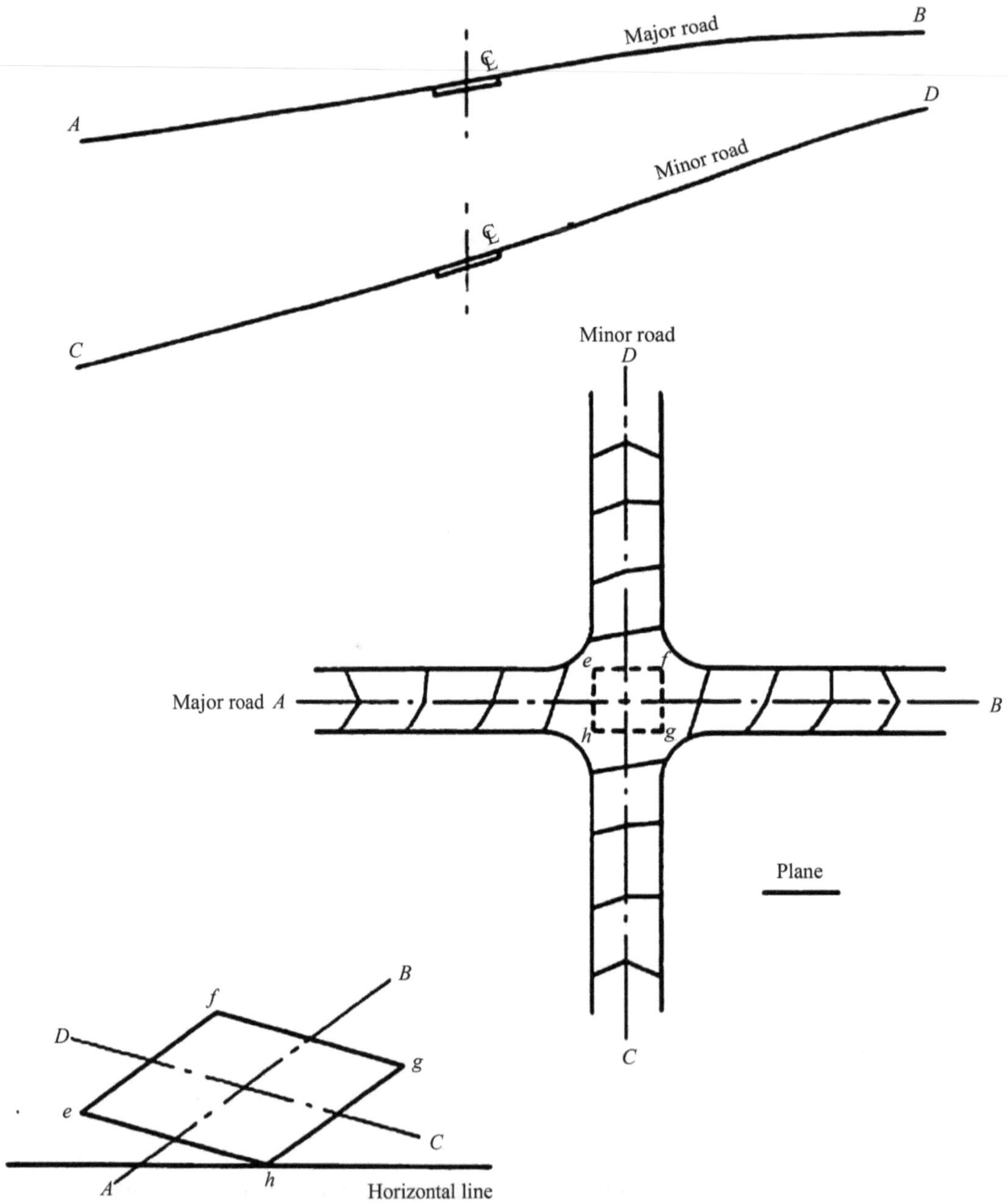


**Figure 6 - 27 Cross slope adjustment of both roads for the priority intersection**

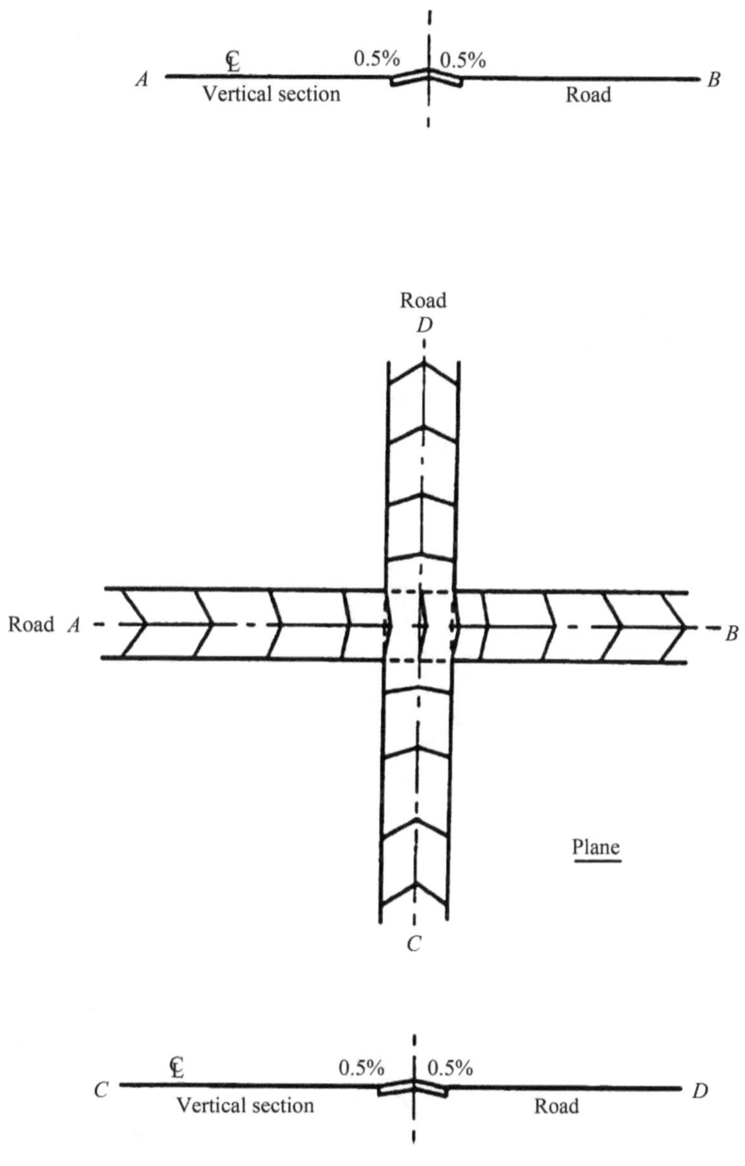


**Figure 6 - 28 Profile design of two intersecting highways of the equal function**

Rainwater inlets, if applied, shall be set in front of the pedestrian crossing or in the low-lying place at the intersection. The rainwater of one road shall not flow through the crosswalk at the intersection or into another road, nor shall it cause ponding at the intersection.

The cross slope within the intersection area shall be gentle, generally not greater than the cross slope of the road section, and the grade should not be greater than 2%, and not greater than 3% under difficult conditions.

At urban intersections with flat terrain, the profile should be designed in the form of an umbrella. The central elevation of intersections shall be slightly raised and inclined to the surrounding, which is conducive to drainage, traffic, aesthetics, and connection.

(4) The design elevation of intersections shall be consistent with the floor elevation of surrounding buildings.

① Forms of intersection profile design

The forms of intersection profile design mainly depend on the class of the intersecting road, traffic volume, cross section, rate and direction of the grade, and surrounding terrain. Taking the cross road as an example, it can be divided into six basic forms according to the terrain and grade direction of the intersecting road, as shown in Figure 6 - 29.

**Figure 6 - 29 Forms of intersection profile design**

(a) All grades away from the intersection on convex terrain

In the design, intersection grades shall be maintained while the cross slopes of roads near the intersection are adjusted appropriately to allow rainwater to flow into street ditches at the four corners of the intersection or out of the roadway for drainage, and it is unnecessary to use a rainwater inlet.

(b) All grades towards the intersection on sag terrain

In this case, surface water concentrates at the intersection, and underground drainage pipes must be used to drain the surface water. In order to avoid rainwater accumulation at the intersection, in addition to raising the elevation of the intersection as much as possible, rainwater inlets shall be set in the low-lying locations at the four corners of the intersection. This form is unfavorable to traffic and drainage and shall be avoided as far as possible. It is desired to strive for the grade of a main road to deviate from the intersection.

(c) Three grades away from and one grade towards the intersection on watershed terrain

In the design, the ridge lines of which the grade points towards the intersection are divided into three directions at the intersection, and the cross sections of the intersecting roads are maintained. Meanwhile, rainwater inlets are applied on both sides of the road with the grade pointing towards the intersection to intercept surface water, so as not to affect traffic.

(d) Three grades towards and one grade away from the intersection on valley terrain

In the design, rainwater inlets are used on both sides of the road with the grade pointing towards the intersection to intercept surface water so as not to affect traffic.

(e) Two grades towards and two grades away from the intersection on slope terrain

In the design, the grades of the intersecting roads are maintained. According to the natural terrain, the cross slopes of the two roads are gradually inclined to the grades of the intersecting roads before reaching the intersection, and a unidirectional inclined slope is formed at the intersection. Rainwater inlets are applied on both sides of the road with the grade pointing towards the intersection.

(f) Two opposing grades towards and the other two grades away from the intersection on saddle terrain

In the design, the grades and cross slopes of the intersecting roads can be adjusted appropriately at the intersection according to the natural terrain.

In addition to these six basic forms, there are also special forms. For example, if the intersection is on the leveling terrain, the design elevation of the intersection shall be slightly raised.

The above are typical profile design forms of intersections. For other types of intersections, the vertical design principles are the same. The performance of profile design is closely related to the combination of grade direction of intersection roads. If an ideal profile design of intersections is expected, the requirements of profile design of intersections need to be considered in the design of road vertical alignment to provide favorable conditions.

② Methods and steps of profile design of intersections

(a) Data collection

i. Survey data: The topographic map with a scale of 1 : 500 or 1 : 200 is generally adopted, on which the grid is plotted with the centerline of the intersecting road as the coordinate baseline. The grid size depends on the specific situation, and is generally 5 m × 5 m or 10 m × 10 m. The grid line is parallel to the road centerline. If the road is skewed, ground elevations of the grid are measured in the direction convenient for construction laying-out.

ii. Traffic data: Traffic volume and traffic composition.

iii. Drainage data: Location and size of drainage facilities.

iv. Road data: Road class, width, grade, cross slope, intersection control elevation and floor elevation of surrounding buildings.

(b) Plot intersection plan

It includes the road centerline, widths of the traveled way and sidewalk, the curb radius, etc.

(c) Determining the intersection design area

The design area is generally 5 – 10 m beyond the tangent points of the curb radius, which is convenient for the transitions of elevation and cross slopes from the intersection to the road section.

(d) Determining the form of profile design

Determine the form of profile design according to the class, grade direction, and terrain of the intersecting road, and decide the contour distance $h$ (generally 0.02 – 0.1 m) of adjacent contour lines according to the elevation difference and precision requirements of the intersection.

(e) Determining the ridge line of roads and control elevation

The ridge line is the one connecting the crown points of roads and is generally the centerline of roads. And the intersection of ridge lines is the control elevation. However, sometimes it is necessary to adjust the ridge line to cater for vehicle operation and road elegance. The control elevation of the intersection is established according to the grade of the intersection road, terrain around intersection, and height of buildings.

As shown in Figure 6 - 30, the X-shaped intersection is divided into four parts by the centerline of the intersecting road within the intersection area. In the profile design of intersections, the design method of each part is the same. Herein, the $A_1OA_2B_2EB_1$ part is employed as an example. In Figure 6 - 30, $OA_1$ and $OA_2$ are the ridge lines, and the elevation of point $O$ is the control elevation.

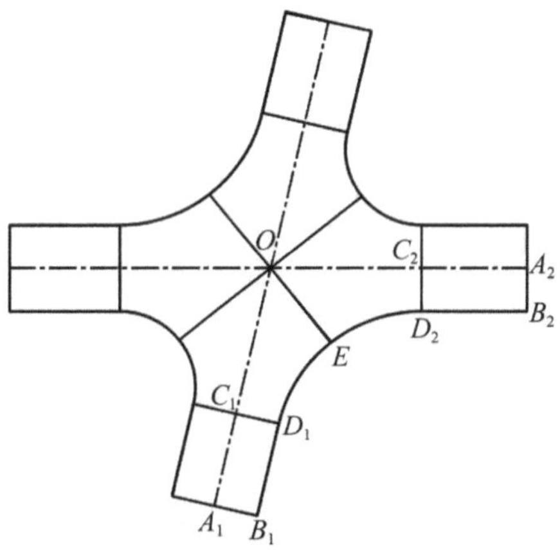


**Figure 6 - 30 Ridge line and feature section of the intersection**

(f) Calculating design elevation of feature points

The feature sections refer to some special sections in the process of intersection profile design, such as the start section of the intersection design area, the section of tangent points of the corner, and the diagonal of the intersection. In Figure 6 - 30, $A_1B_1$, $C_1D_1$, $OE$ etc. are the feature sections. The points $A_1$, $B_1$, $C_1$, $D_1$, etc. on the feature sections are the so-called feature points. The design elevation of feature points may be derived from the control elevation, grade, and cross slope of the intersecting roads. In Figure 6 - 30, if the control elevation of the center point $O$ is $h$, the elevations of $C_1$ and $D_1$ are:

$$h_{C1}=h_O-OC_1\times i_1 \tag{6-3}$$

$$h_{D1}=h_{C1}-\frac{B}{2}\times i_{g1} \tag{6-4}$$

where: $i_1$—grade of road sections where the ridge line $OA_1$ is located;

$B$—traveled way width of road sections where the ridge line $OA_1$ is located;

$i_{g1}$—cross slope of road sections where the ridge line $OA_1$ is located.

The elevations of other feature points can be calculated in a similar way.

Elevations of feature points on the feature sections roughly reflect the profile of intersections. For large and complex intersections, elevations of feature points alone may not meet the precision requirement of intersection design, which need to be densified.

(g) Determining the elevation calculation line network and sketching design contour map

The densification of the elevation calculation line network can adopt the circle center method, equal allocation method, and parallel line method, as shown in Figures 6-31, 6-32, and 6-33.

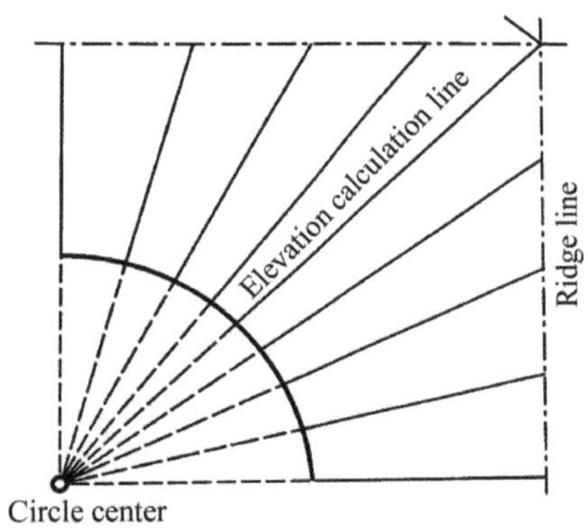


**Figure 6-31 Circle center method**

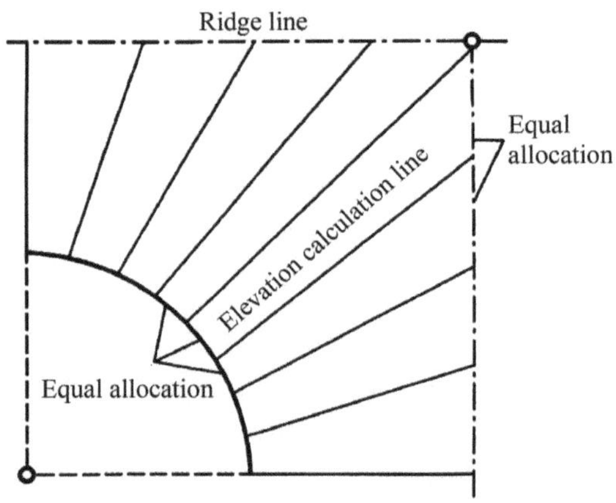


**Figure 6-32 Equal allocation method**

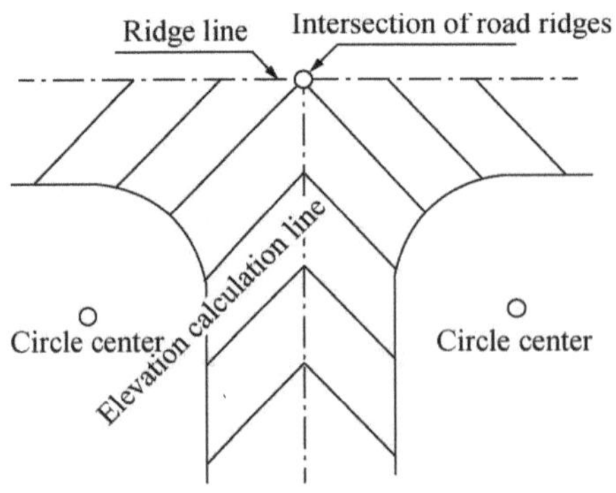


**Figure 6-33 Parallel line method**

The circle center method is to divide the ridge line into several parts which link with the center of the corner curve. These links form an elevation calculation network.

The equal allocation method is to divide the ridge line into several parts, and divide the curb curve into the same parts correspondingly. These corresponding points are connected to form the elevation calculation line network.

The parallel line method is to connect the intersection point of the ridge line and the circle center of each curb curve to form a tangent and then divides points on the ridge line according to the construction precision. The parallel lines of the tangent passing through these points cross with the margin of the traveled way to form the elevation calculation line network.

When the elevation line network is established, the elevations of the elevation line network could be calculated according to the requirements of the crown slope and contour distance. And the preliminary design contour map is obtained by connecting these contour points. The grade of each direction is checked and adjusted until driving and drainage demands are assured.

(h) Calculating the design elevation of each point

According to the design contour map, the design elevation on each node of the grid is calculated by interpolation. The difference between the design elevation and the original ground elevation is the construction height.

## 6.5.5 Intersection sight distance

Approach sight distance corresponding to the design speed shall be provided for each leg and turning lane of intersections, as shown in Figure 6 - 34.

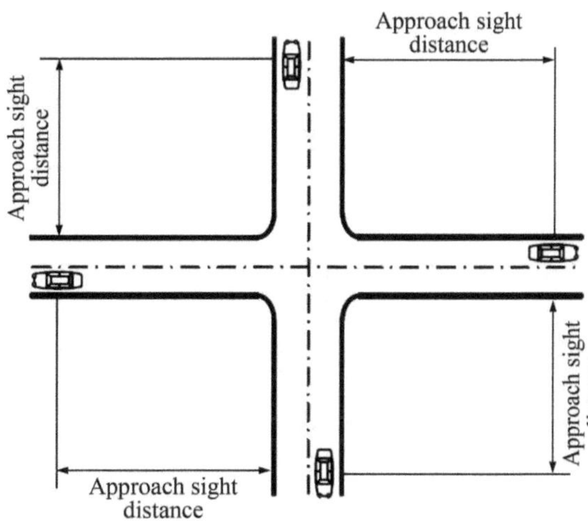


**Figure 6 - 34 Approach sight distance**

Approach sight distance is the one required for drivers to stop after observing the stop markings on roads. Therefore, approach sight distance is the same as stopping sight distance. However, the criteria for deciding approach sight distance are different: the height of the driver's eye is 1.2 m while the height of the object is 0 m. Therefore, the radius of the crest vertical curve required to ensure the approach sight distance shall be larger than that of stopping sight distance. Approach sight distance and the minimum radius of the crest vertical curve corresponding to the design speed shall conform to the provisions in Table 6 - 4.

**Table 6 - 4 Approach sight distance and corresponding minimum radius of the crest vertical curve**

| Design speed/km • $h^{-1}$ | 100 | 80 | 60 | 40 | 30 | 20 |
|---|---|---|---|---|---|---|
| Approach sight distance/m | 160 | 110 | 75 | 40 | 30 | 20 |
| Minimum radius of the crest vertical curve/m | 10,700 | 5,100 | 2,400 | 700 | 400 | 200 |

*Note*: This table refers to the **Specification (2017)**.

Each quadrant of an intersection shall contain a triangular area free of obstructions that might block an approaching driver's view of potentially conflicting vehicles. The length of the legs of this triangular area, along both intersecting roads, shall allow the driver to see any potentially conflicting vehicles in sufficient time to slow down or stop before a collision occurs inside the intersection, as shown in Figure 6 - 35. These specified areas are known as clear sight triangles, which shall be determined under the most adverse circumstances.

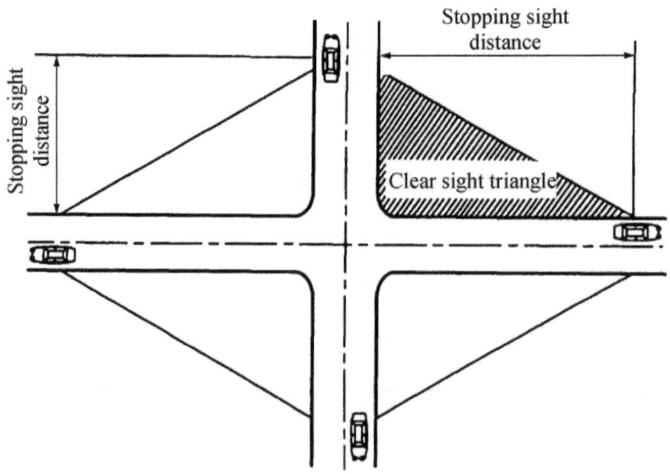


**Figure 6 - 35 Clear sight triangle**

When the clear sight triangle composed of stopping sight distance could not be assured, a clear sight triangle composed of safe-crossing stopping sight distance of the major highway and 5 – 7 m from the minor highway to the centerline of the major highway side lane shall be assured, as shown in Figure 6 – 36. The safe-crossing stopping sight distance shall conform to the provisions in Table 6 – 5.

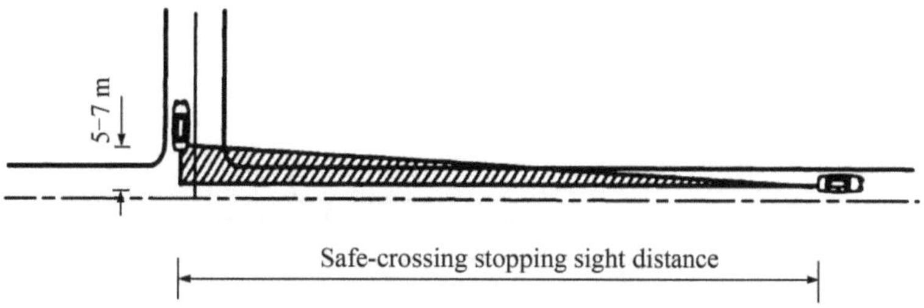


**Figure 6 – 36 Clear sight triangle of safe-crossing stopping sight distance**

**Table 6 – 5 Safe-crossing stopping sight distance**

| Design speed/km • h$^{-1}$ | 100 | 80 | 60 | 40 | 30 | 20 |
|---|---|---|---|---|---|---|
| Stopping sight distance/m | 160 | 110 | 75 | 40 | 30 | 20 |
| Safe-crossing stopping sight distance/m | 250 | 175 | 115 | 70 | 55 | 35 |

*Note*: This table refers to the **Specification (2017)**.

## 6.5.6 Turning roadway design

The alignment and roadway width of the turning curve at intersections are determined according to the turning paths of design vehicles. The turning paths are different for various design vehicles, and the paths of the same design vehicle also vary when turning at different speeds. Therefore, in the design of turning curves, the design vehicle and design speed should be determined first.

**1) Design vehicles for the turning roadway**

Through the analysis of the paths of five design vehicles, it is found that although the total length of the articulated vehicle is the longest, the minimum turning path radius of the truck is larger than that of the articulated vehicle and other design vehicles. On the other hand, when the articulated vehicle turns, the steering clearance required by vehicle dimension is larger than that of trucks and other design vehicles. Through comprehensive analysis and demonstration, in the turning roadway design at intersections, the path of trucks is still used as the design control (inside edge radius of the turning curve). If necessary, the widening and turning clearance of the roadway shall be inspected according to the path of the articulated vehicle and other design vehicles. Figure 6 - 37 shows the minimum turning radius of trucks.

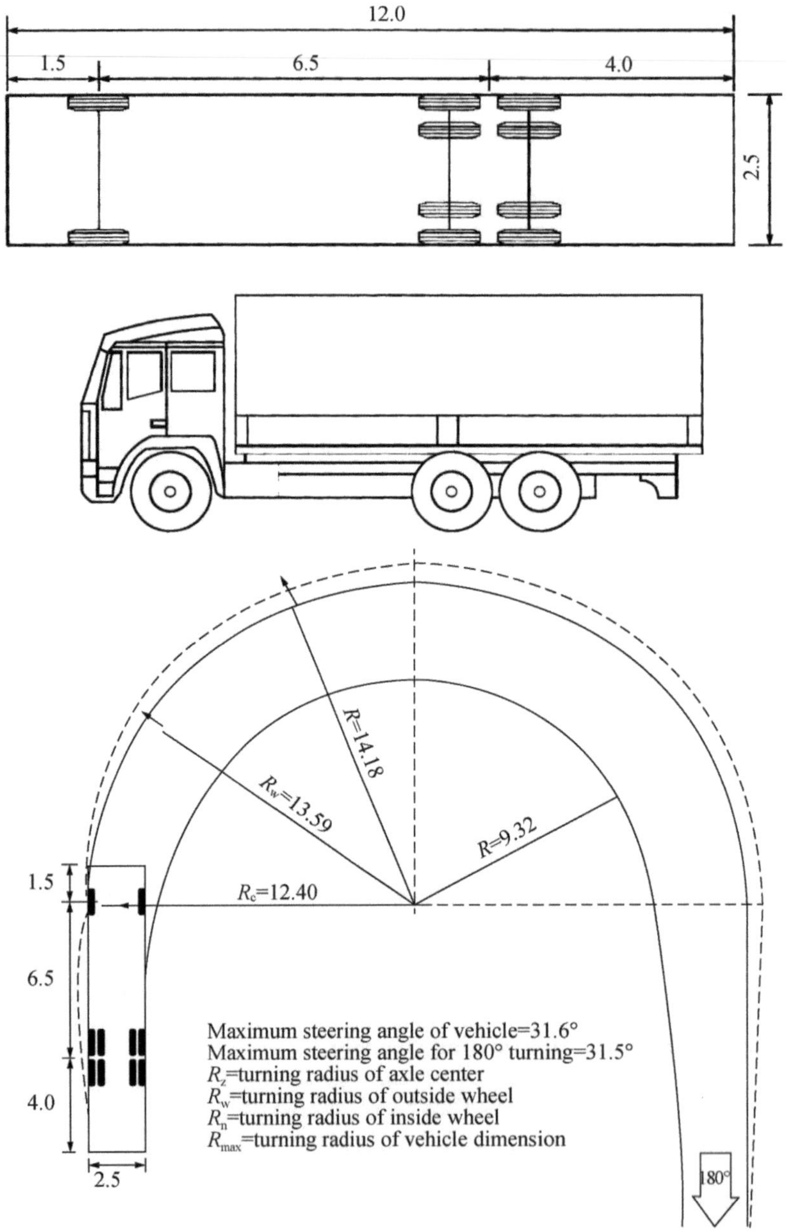


**Figure 6 - 37 Minimum turning radius of trucks (m)**

**2) Design speed of the turning roadway**

Left-turning is sometimes on standby, so it is not necessary to use a high design speed, generally 5 – 15 km/h. The design speed of 5 km/h can be adopted for highways with a small proportion of large vehicles or under limited conditions, but the minimum radius of the left-turning inside edge curve shall not be less than 12.5 m.

When separate right-turning lanes are introduced, the turning design speed should not be greater than 40 km/h. When the design speed of major highways is less than or equal to 60 km/h, the right-turning design speed should not be less than 50%. When the highway class is low and the traffic volume is small, the exclusive right-turning lane is not mandatory.

**3) Alignment and radius of the turning roadway**

According to the turning path, the inside edge of the road is a very complicated curve which could not be expressed by a mathematical model. In practice, it is not necessary to be very precise, so there are two simplified models of roads inside edge curves. Many countries use the three-centered compound curve; some countries adopt the alignment of a circular arc with two ends connected with the spiral of specific parameters. Researches show that the fitting of a three-centered compound curve is satisfactory. The provisions on alignment of the turning roadway in the **Specification (2017)** are as follows:

(1) For the right-turning lane of channelized intersections, the inside road edge shall adopt the three-centered compound curve; for the left-turning inside road edge, a single circular curve is used to regulate the boundary of the divisional island end.

(2) When the design is based on the articulated vehicle, the compound curve conforming to the turning path can be adopted at the road edge.

(3) A circular curve with a radius of 15 m can be used for the turning road edge of a non-channelized intersection.

The minimum circular curve radii of the inside road edge shall be determined according to the turning speed specified in Table 6 – 6.

**Table 6 - 6 Minimum circular curve radii of the inside road edge**

| Turning speed/km · h$^{-1}$ | ≤15 | 20 | 25 | 30 | 40 | 50 | 60 | 70 |
|---|---|---|---|---|---|---|---|---|
| Minimum radius/m | 15 | 20(15) | 25(20) | 30 | 45 | 60 | 75 | 90 |
| Minimum superelevation/% | 2 | 2 | 2 | 2 | 3 | 4 | 5 | 6 |
| Maximum superelevation/% | General value: 6, limit value: 8 | | | | | | | |

*Note*: This table refers to the **Specification (2017)**.

## 6.5.7 Auxiliary lane design

In order to improve the safety of vehicles turning at intersections and ensure the capacity of intersections, auxiliary lanes are applied within the intersection areas.

**1) Design of the right-turning auxiliary lane**

For non-channelized intersections or simple channelized intersections without separated exclusive right-turning lanes, when the major highway has a high speed and large traffic volume, even if the right-turning traffic volume is small, the deceleration of right-turning vehicles will affect the speed of through vehicles and lead to traffic accidents. In this case, the addition of a deceleration divergence lane can avoid the disorder of traffic flow, as shown in Figure 6 - 38.

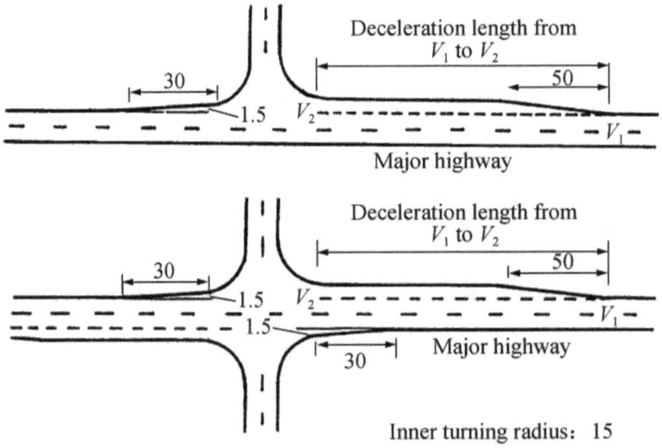


**Figure 6 - 38 Deceleration divergence lane (m)**

In the **Specification (2017)**, it stipulates that when the design speed of the major highway is greater than or equal to 60 km/h, the deceleration divergence lane and acceleration convergence lane shall be used on the major highway.

When two Class Ⅰ highways cross or a Class Ⅰ highway crosses a Class Ⅱ highway with a large traffic volume, the right-turning lane separated by channelization shall be supplied for right-turning vehicles.

For intersections of the Class Ⅰ highway and Class Ⅱ highway, the right-turning lane shall be used when one of the following conditions is met:

(1) The skew angle is close to 70° in the acute quadrant.

(2) Traffic volume is large, and right-turning traffic will cause unreasonable traffic delay.

(3) Large vehicles account for a large proportion of right-turning traffic.

(4) The right-turning speed is greater than 30 km/h.

(5) The right-turning traffic volume at the intersection of the interchange connection line is large.

The channelized right-turning auxiliary lane consists of the separated right-turning lane and speed-change lanes at both ends, as shown in Figure 6 - 39.

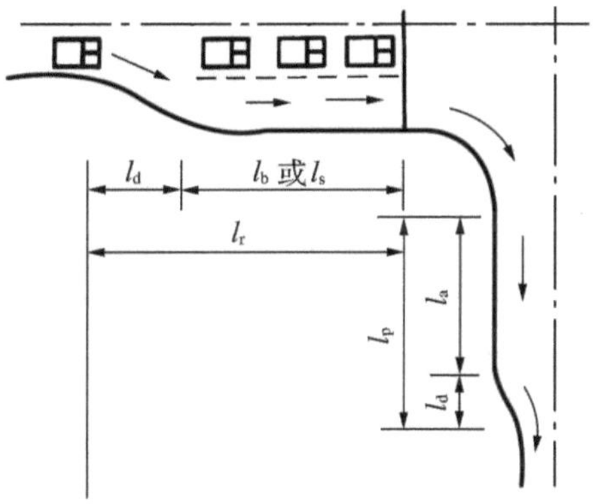


($l_d$—length of transition section; $l_b$—length required for the deceleration; $l_a$—length required for acceleration; $l_s$—queue length; $l_r$—length of the right-turning lane; $l_p$—length of exit acceleration lane)

**Figure 6 - 39 Length of the right-turning lane with equal width**

With the introduction of the right-turning lane at the entrance of intersections, an acceleration lane shall be provided at the exit of the transverse crossing road in order not to affect the through traffic flow on it. The length of the right-turning lane or left-turning lane at the entrance shall meet the length required for the deceleration of right-turning or left-turning vehicles, and shall also ensure that the turning vehicles are not affected by the length of queue fleet. The acceleration lane at the exit shall ensure the length required for acceleration.

The length of the speed-change lane shall consider the type of intersecting highways, design speed, speed-change conditions, etc. and be determined according to Table 6 - 7. The length of the speed-change lane in Table 6 - 7 does not include the length of the transition section. The transition section of the speed-change lane shall also be designed according to the following requirements.

**Table 6 - 7 Length of the speed-change lane**

| Category | Design speed/ $km \cdot h^{-1}$ | Length required for deceleration: $l_b$/m ($a=-2.5\ m/s^2$) | | | Length required for acceleration: $l_a$/m ($a=1.0\ m/s^2$) | | |
|---|---|---|---|---|---|---|---|
| | | To stop | To 20 km/h | To 40 km/h | From stop | From 20 km/h | From 40 km/h |
| Major highway | 00 | 100 | 95 | 70 | 250 | 230 | 190 |
| | 80 | 60 | 50 | 32 | 140 | 120 | 80 |
| | 60 | 40 | 30 | 20 | 100 | 80 | 40 |
| | 40 | 20 | 10 | — | 40 | 20 | — |
| Minor highway | 80 | 45 | 40 | 25 | 90 | 80 | 50 |
| | 60 | 30 | 20 | 10 | 60 | 55 | 25 |
| | 40 | 15 | 10 | — | 25 | 15 | — |
| | 30 | 10 | — | — | 10 | — | — |

(1) When the speed-change lane is of equal width, the length of the transition section listed in Table 6 - 8 shall be added.

**Table 6 - 8 Length of transition section**

| Design speed/$km \cdot h^{-1}$ | 100 | 80 | 60 | 40 |
|---|---|---|---|---|
| Minimum length of the transition section/m | 60 | 50 | 40 | 30 |

(2) When the speed-change lane is of unequal width which changes gradually, its length shall not be less than the calculated value of lane-change process given the side shift rate of 1.0 m/s for deceleration or 0.6 m/s for acceleration.

(3) When the design speed of highways is greater than or equal to 80 km/h and the through traffic volume is large, the right-turning speed-change lane shall be of equal width with the transition section; in other cases, the tapered speed-change lane should be adopted.

(4) When the capacity of the through lane is surplus, or it is difficult to add

an acceleration lane with proper length due to limited conditions, a shorter tapered acceleration lane can be used.

**2) Design of the left-turning auxiliary lane**

The left-turning lane is an exclusive lane opened on the left side of the through lane for left-turning vehicles to diverge, decelerate and wait for left-turning. It is composed of the transition section, deceleration section, and waiting section.

The design of the left-turning lane shall meet the following requirements:

(1) Except for those with a small left-turning traffic volume which pose no obstruction or delay to the through traffic, the left-turning lane shall be used within the intersection area of a four-lane highway.

(2) Left-turning lanes shall be used for the Class Ⅱ highway subject to one of the following conditions:

① Intersections crossing the connecting line of interchanges of the freeway or Class Ⅰ highway.

② Intersections with many non-motor vehicles and without the slow-speed lane.

③ Left-turning traffic causes traffic jams or accidents.

(3) Left-turning lanes shall be composed of the transition section, deceleration section, and waiting section. The length of the left-turning waiting section shall not be less than 30 m. When the traffic volume of left-turning is very small, the waiting section can be ignored.

When there is a wide median (generally not less than 4.5 m), the median width of a certain length at the intersection entrance shall be compressed for the addition of the left-turning lane, as shown in Figure 6 - 40(a).

When there is a narrow median (less than 4.5 m), if the median width is not enough, the one-way or two-way lane marking at the intersection entrance can be offset to the outside to gain space for the left-turning lane. With the offset of lane marking, either the sidewalk and lane separators at both sides or the entrance lane can be compressed according to the specific conditions while the total road width is maintained, as shown in Figure 6 - 40(b).

There are two ways to add left-turning lanes when there is no median at the intersection. One is to widen one side or both sides of the entrance road, increase the total width of the entrance road, and locate the left-turning lane near the center of the entrance road, as shown in Figure 6 - 40(c); the other is not to widen the entrance road but to occupy the opposite lane near the centerline as the left-turning lane.

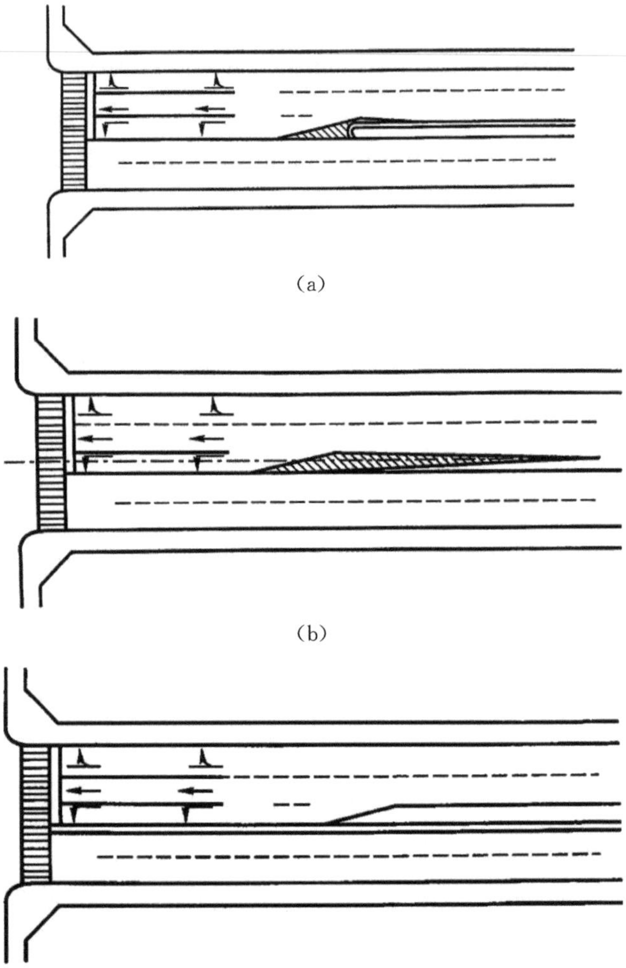


**Figure 6 - 40 Widening left-turning lane**

### 6.5.8 Traffic island design

Traffic islands can be divided into the directional island, refuge island, central island, and divisional island according to their functions in organizing channelized traffic, as shown in Figure 6 - 41. The directional island refers to the traffic island that divides the traffic flow of different types or in the opposite way. The refuge island refers to the traffic island for pedestrians to stay temporarily across the road. The central island refers to the traffic island located in the center of a roundabout, which separates the opposite traffic flow or organizes the left-turning traffic. The divisional island refers to the traffic island that separates traffic flow in the same way.

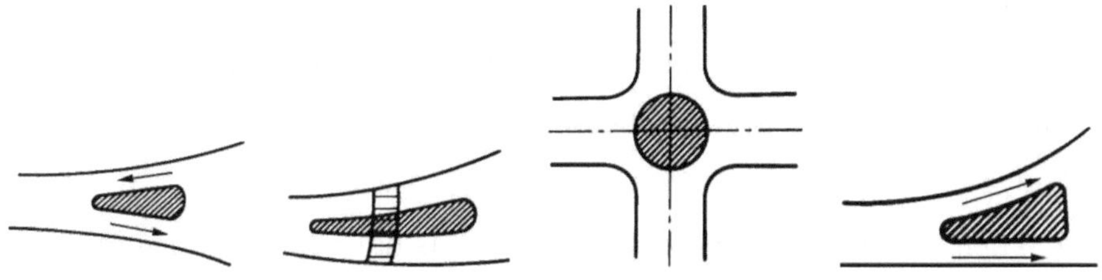


**Figure 6 - 41 Traffic islands**

Applications of traffic islands at the channelized intersections shall conform to the following provisions:

(1) The directional island shall be used when the right-turning lane is needed.

(2) At the signalized intersection, when there are two left-turning lanes, a directional island should be introduced between the left-turning lane and the through lane in the same way.

(3) The divisional island shall be introduced between the left-turning lane and the opposite through lane.

(4) At the T-shaped intersection, a divisional island shall be introduced between the two left-turning paths on the approach of the minor highway.

(5) When it is necessary to provide a pedestrian crossing place between the opposite lanes or to set signs and signal poles, a divisional island shall be employed.

When the design speed is more than 60 km/h, if there are many pedestrians crossing the intersection and the crossing distance is long, a refuge island shall be employed.

According to the structural types, traffic islands can be divided into three types: the solid island surrounded by curbs higher than the surrounding pavement surface, the painted island on pavement surface and the shallow-dish island without curbs. The solid island makes forcible separation of traffic flow so that the separation is absolute. However, when a two-lane highway adopts the solid island, it is easy to cause traffic jams in case of accidents and vehicle malfunction, especially when there is no paved shoulder and wide earth shoulder. Therefore, in the **Specification (2017)**, it stipulates the following provisions on the selection of traffic islands:

(1) When there are more than two lanes separated by traffic islands, when one lane is widened to avoid vehicle malfunction, or when signs and signal poles need to be placed on the island, the solid island surrounded by curbs shall be adopted.

(2) When the area of islands is small, or it is not necessary or not suitable to make a forcible separation, a painted island on the pavement surface should be used.

(3) When the area of islands is large, a shallow-dish island surrounded by the traveled way with a marginal strip width of not less than 0.5 m should be used.

The edge alignment of the traffic island depends on the edge alignment of the adjacent lane. The edge alignment of the traffic island next to the through lane shall be offset by different values according to the curb structure. The edge of island end facing the traffic flow shall be offset and smooth. Figure 6 - 42, Figure 6 - 43 and Figure 6 - 44 show the commonly used edge shape and the relevant dimensions of traffic islands.

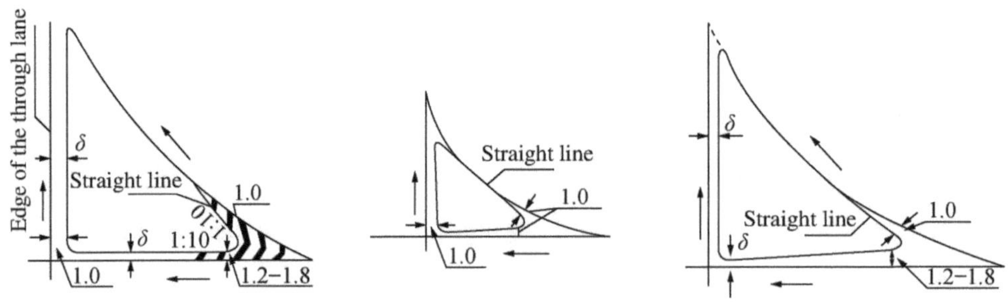


**Figure 6 - 42 Divisional island at corners (m)**

Table 6 - 9 shows the arc radius of the island. Table 6 - 10 shows the offset distance of curbs.

**Table 6 - 9 Arc radius of the island end**

| Island end shape and direction of traffic flow | 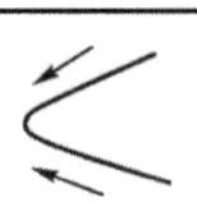 |  | 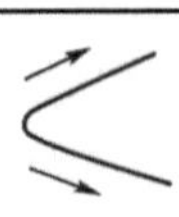 | 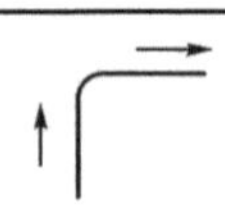 |
|---|---|---|---|---|
| Radius/m | 0.3 | 0.6 | 0.6 | 1.0 |

**Table 6 - 10 Offset distance of curbs**

| Curb style | $\delta$/m |
|---|---|
| bar style | 0.6 |
| semi passable style | 0.3 |
| passable style | 0 |

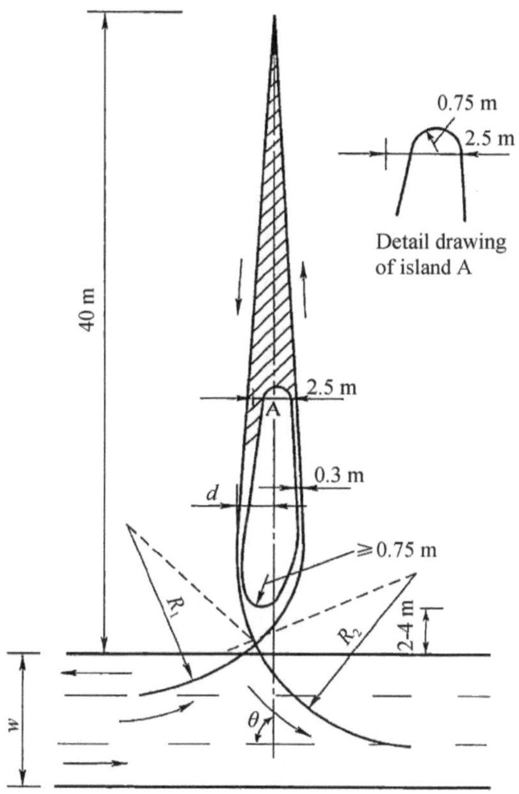


| $\theta$/(°) | 70 | 80 | 90 | 100 | 110 |
|---|---|---|---|---|---|
| $d$/m | 1.5 | 2.0 | 2.5 | 2.0 | 1.5 |

| $W$/m | ≤10 | 11 | ≥11 |
|---|---|---|---|
| $R_1$/m* | 12 | 14 | 20 |

*Note*: $R_2$ is generally equal to $R_1$, but sometimes it needs to be changed to ensure a distance of 2 - 4 m from the island end to the lane edge of the major highway and a width of 2 - 5 m for the island.

* When the articulated truck is used as the design control, $R_1$ and $R_2$ shall not be less than 15 m.

**Figure 6 - 43 Divisional island on the minor highway at the T-shaped intersection**

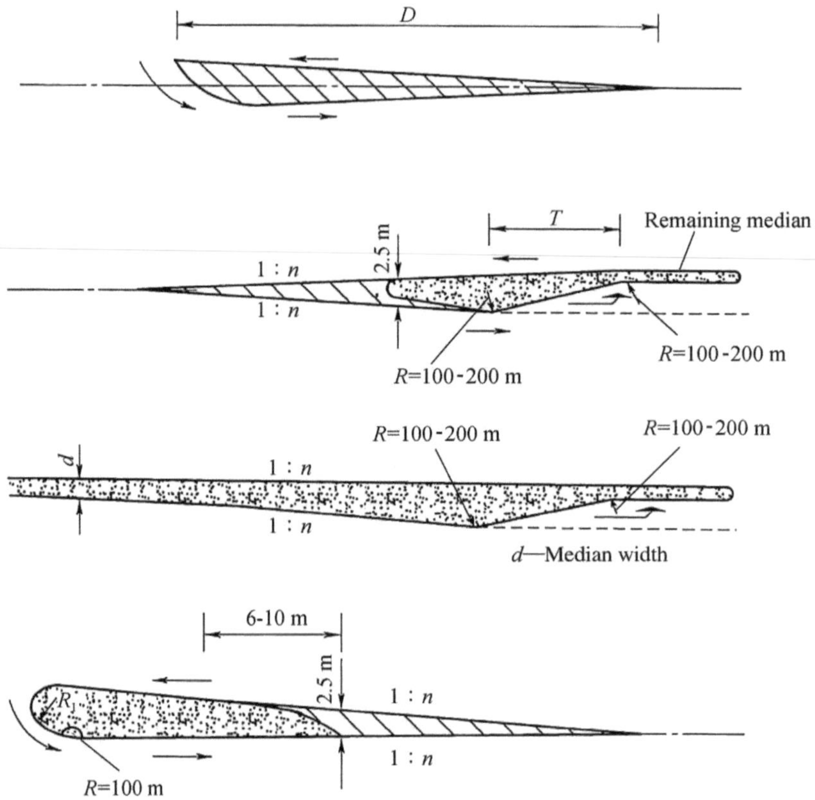


| Speed/(km · h⁻¹) | 45 | 50 | 60 | 80 |
|---|---|---|---|---|
| Gradient parameter $n$ | 15 | 20 | 25 | 30 |
| $D$/m | 40 | 50 | 60 | 80 |
| $T$/m* | 40 | 45 | 55 | 70 |

*Note*: * When the left-turning lane is of right-bend alignment, it shall be properly shortened.

**Figure 6 - 44 Divisional island on the through lane**

## 6.6 Roundabout design

The roundabout is an intersection where a central island is applied in the center of the intersection, and channelized traffic is organized by the rotary lane so that all vehicles at the roundabout travel anticlockwise until exiting. Figure 6 - 45 shows the compositions of the roundabout.

All kinds of vehicles entering the roundabout, whether turning left or right or going through, do not need to stop, and can travel continuously in one way simultaneously, which reduces the delay of vehicles at the intersection; the way

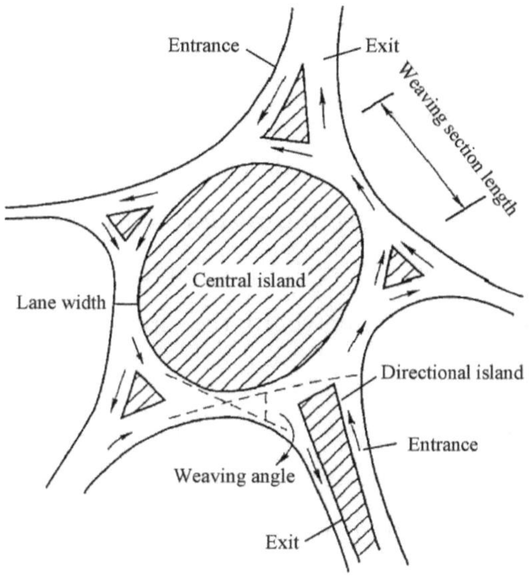


**Figure 6 - 45 Composition of the roundabout**

of traffic flow traveling on the rotary lane is the same, and traffic flow crosses at a small weaving angle so that conflict points are avoided to reduce traffic accidents; for the multi-leg intersection and malformed intersection, it is more effective to organize channelized traffic with a roundabout; greening of the central island can be aesthetic.

However, the roundabout covers a large area, which is difficult to construct in the old city, it increases the detour distance of vehicles, especially for left-turning vehicles, and the cost of the roundabout is generally higher than other intersections.

The application requirements of the roundabout are as follows:

(1) The traffic flow of each intersecting road is uniform, the way of traffic flow is stable, and there are many turning vehicles, especially at multi-leg malformed intersections.

(2) The peak hourly traffic volume of intersections is less than 2000 pcu/h.

(3) Few non-motor vehicles and pedestrians.

In the roundabout design, the key contents include the shape and radius of the central island, layout and width of the rotary lane, length and angle of the weaving section, radii of roundabout entrance and exit curves, sight distance, etc.

### 6.6.1 Central island design

**1) Shape of the central island**

The shape of the central island is determined according to the characteristics of traffic flow, the class of intersecting roads, and terrain. In principle, it shall be ensured that vehicles can successfully complete the weaving operation at a certain speed and that it is conducive to the operation of vehicles from major roads.

Round is the general shape of the central island, and ellipse should be used when the major and minor highways intersect, and the compound curve can be used when the intersection angle is not equal. In addition, the central island with other regular or irregular geometries can be used in combination with terrain, features, and the intersection angle.

**2) Radius of the central island**

The radius of the central island shall first meet the requirements of design speed, and then check whether the distance between adjacent entrances/exits meets the requirements of vehicle weaving according to the number and width of the intersecting roads. Calculations of the radius of the central island are as follows:

(1) According to the requirement of design speed

When a circular central island is used, the radius $R$ of the central island can be calculated according to Formula 6 - 5:

$$R=\frac{V^2}{126(\mu \pm i_h)}-\frac{B}{2} \tag{6-5}$$

Where: $R$—radius of the central island, m;

$V$—design speed of the rotary lane, km/h, being 0.5 - 0.7 times of the design speed of road sections;

$M$—sideway force coefficient, being 0.1 - 0.2;

$i_h$—cross slope of the rotary lane, being 1.5%- 2.0%;

$B$—lane width next to the central island, m.

(2) According to the requirement of weaving section length

Weaving is a process in which two streams of traffic flow converge and then diverge. Two vehicles entering and exiting the roundabout weave when driving on the rotary lanes, and the distance traveled by changing the lane position once is called the weaving length. Weaving length depends on the running speed of vehicles on the rotary lanes. When there is enough distance between adjacent entrances/exits, vehicles entering and exiting the roundabout weave at the appropriate time and drive continuously on the rotary lanes. Such a distance is called the weaving section length, and its value can roughly be the arc length between the intersection points of the extension line of the outside edge of vehicle lanes of adjacent roads and the centerline of rotary lanes, as shown in Figure 6 - 46.

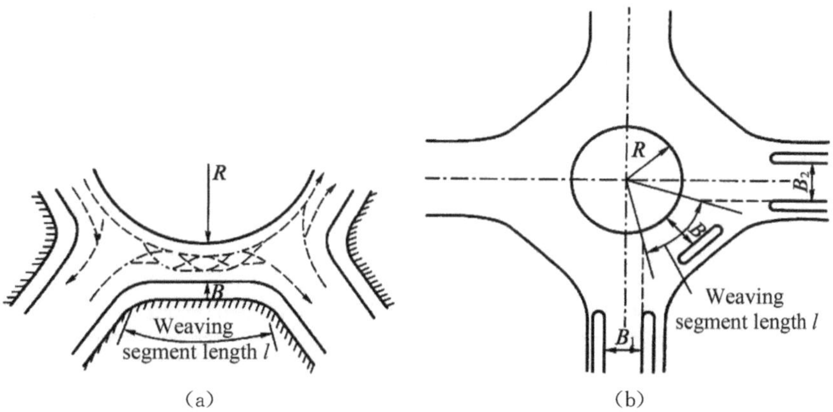


**Figure 6 - 46 Weaving section length**

The radius of the central island shall meet the requirement of the minimum weaving section length between adjacent entrances/exits. Otherwise, vehicles weaving on the rotary lanes have to stop and wait, which is not in line with the traffic characteristics of continuous flow at the roundabout. Table 6 - 11 shows the minimum weaving section length at different speeds.

**Table 6 - 11 Minimum weaving section length**

| | | | | | | | |
|---|---|---|---|---|---|---|---|
| Design speed of rotary lanes/km · $h^{-1}$ | 50 | 45 | 40 | 35 | 30 | 25 | 20 |
| Minimum weaving section length/m | 60 | 50 | 45 | 40 | 35 | 30 | 25 |

*Note*: This table refers to the **Specification (2017)**.

The radius of the central island required by the weaving section length is deduced approximately according to the circumference formed by the weaving section length:

$$R=\frac{n(l+B_p)}{2\pi}-\frac{B}{2} \tag{6-6}$$

Where: $R$—radius of the central island, m;

$n$—number of intersecting roads;

$l$—weaving section length between adjacent entrances/exits, m;

$B_p$—average width of intersecting roads, m;

$B$—width of rotary lanes, m.

As indicated in the formula, in order to ensure the minimum weaving section length, the more the number of intersecting roads, the larger the radius of the central island, which will increase the land occupation of intersections and the detour distance of vehicles on the rotary lanes. Therefore, the number of intersecting roads at the roundabout should not be more than five.

The radius of the central island of the roundabout shall be the larger one as required by design speed and by weaving section length. Table 6 - 12 shows the minimum radius of the central island of the conventional roundabout recommended according to practical experience, which can be used for reference in design.

**Table 6 - 12 Minimum radius of the central island**

| Design speed of rotary lanes/km • $h^{-1}$ | 40 | 35 | 30 | 25 | 20 |
|---|---|---|---|---|---|
| Minimum radius of the central island/m | 60 | 50 | 35 | 25 | 20 |

## 6.6.2 Rotary lane design

### 1) Width of rotary lanes

Rotary lanes are the one-way vehicle lanes around the central island. Its width depends on the traffic volume and traffic organization of the intersecting roads. One lane near the central island is used for detour, the outmost lane is used for right-turning, and one or two lanes in the middle are used for weaving. Therefore, three to four lanes are adopted for rotary lanes, and three lanes are appropriate generally. When the weaving section is long, four lanes can be adopted for rotary lanes. If the vehicle lane of the intersecting road is narrow, two lanes can also be adopted. Practice has demonstrated that too many lanes are not only difficult to function, but also easy to produce traffic chaos.

If three lanes are adopted, the width of each lane is 3.5 - 3.75 m, and the widening value of the single lane on curves is used. When the radius of the central island is 20 - 40 m, the width of the traveled way is generally 15 - 16 m.

In order to ensure traffic safety and reduce mutual interference, non-motor vehicles and vehicles could be separated by a separator (separation pier) or marking. The width of the non-motor vehicle lane shall be determined according to the specific situation, generally not less than the maximum width of the non-motor vehicle lane of the intersecting roads, nor more than 8 m.

**2) Weaving angle**

A weaving angle is the average intersection angle of the vehicle trajectories entering and exiting the rotary lanes. It is represented by the intersection angle of two tangents. One is 1.5 m away from the outside edge of the right-turning vehicle lane and the other is 1.5 m away from the edge of the central island, as shown in Figure 6 - 47.

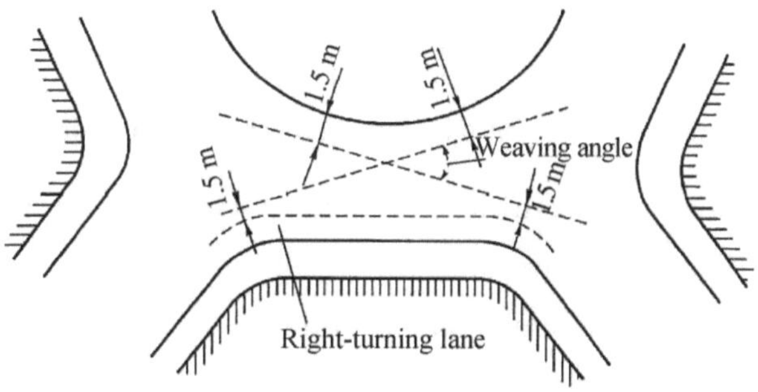


**Figure 6 - 47 Weaving angle**

A weaving angle depends on the width of rotary lanes and the weaving section length. The narrower the rotary lane width is and the longer the weaving section is, the smaller the weaving angle is and the safer the traffic is. However, if the weaving section is longer, the radius of the central island will increase and the land occupation will also increase. According to the experience, the weaving angle should be between 20° and 30°. Generally, when the weaving section length is guaranteed, the weaving angle can meet the requirements.

**3) Alignment of the outside edge of rotary lanes and radii of the roundabout entrance and exit**

In order to meet the traffic needs and save the investment, it is not suitable to use a reverse curve for the horizontal alignment of the outside edge of rotary lanes, such as the solid line shown in Figure 6 - 48. About 20% of the road surface (the area between the solid line and the dashed line in Figure 6 - 48) outside the rotary lanes with such alignment undertakes no vehicles, which is neither reasonable nor economical. The outside edge of rotary lanes should be a tangent with fillet or a compound curve, as shown in the dotted line in Figure 6 - 48.

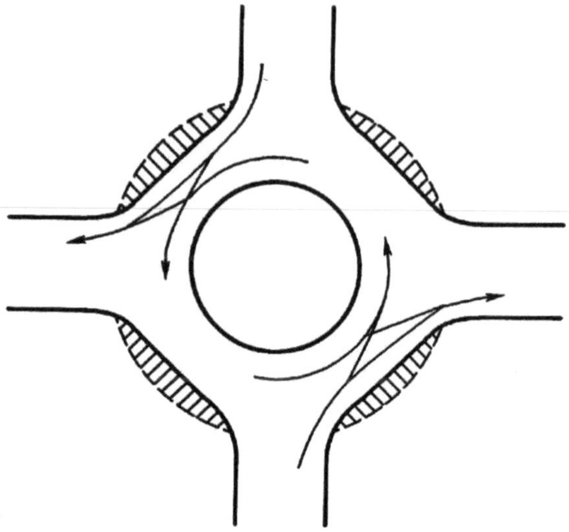

**Figure 6 - 48 Outside edge alignment of rotary lanes**

The radii of roundabout entrance and exit curves depend on the design speed of rotary lanes. In order to sure uniform speed on the rotary lanes, the speed of vehicles entering the roundabout shall be limited. The radius of the exit curve could be greater than that of the entrance curve, so that the vehicles may exit the roundabout quickly. Meanwhile, the curve radius at each entrance shall not differ too much to ensure that the speed difference of vehicles entering the roundabout is small, and the vehicles at the roundabout may move at almost the same speed. The radius of the roundabout entrance curve is often close to or less than the radius of the central island.

**4) Cross section of rotary lanes**

The cross section of rotary lanes has a great influence on driving stability and pavement drainage, which depends on the selection of the ridge line. Generally, the ridge line of the rotary lane cross section is established in the middle of the weaving lanes. If there is a separator between vehicle and non-motor vehicle lanes, the ridge line can also be placed on the separator. The ridge line of rotary lanes is connected with the ridge line of the intersecting roads directly or through the triangular island between the entrance and exit, as shown in Figure 6 - 49. The dotted line in Figure 6 - 49 is the ridge line, and the arrow points to the drainage direction. Rainwater inlets shall be placed around the central island to ensure no ponding on the rotary lanes. In addition, the cross slopes of the entrance and exit should be gentle.

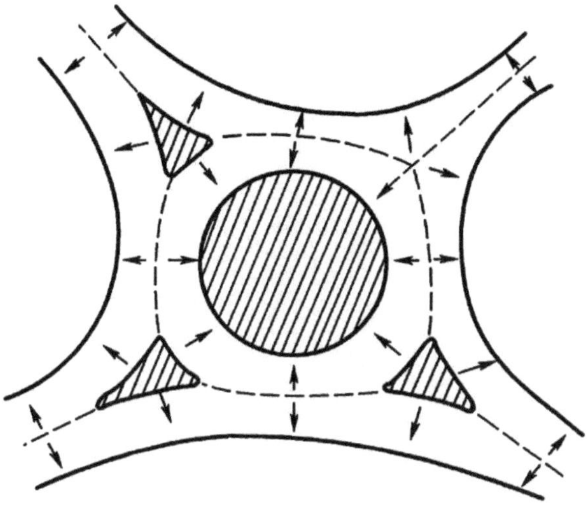

**Figure 6 - 49 Ridge line of the roundabout**

**Questions:**

1. Please describe the main contents of intersection design.
2. What is channelized traffic and what is its function?
3. Please expound on the functions of traffic islands such as the directional island, divisional island, central island, and refuge island.
4. How to set the crosswalk at a crossroad?
5. What is a sight triangle and how to plot it?
6. What is the roundabout? Please comment on its advantages, disadvantages, and practicability.
7. Figure 1 shows an orthogonal crossroad in a city. The width of *AB* street is 25 m (two-way four-lane, 5.5 m non-motor vehicle lane on both sides) with a design speed 40 km/h. The width of *CD* street is 10 m, the street is divided in the middle with a mixture of vehicles and non-motor vehicles on both sides, and the design speed is 30 km/h. The design grades are $I_{AB}=1\%$ and $I_{CD}=2\%$, as shown in the figure.
   (1) Please determine the curb radius of each leg of the intersection.
   (2) Please plot the sight triangle of each entrance and the red line of the intersection area.
   (3) If the elevation of the intersection center Point *O* is 18.0 m, please conduct the profile design within the intersection area.

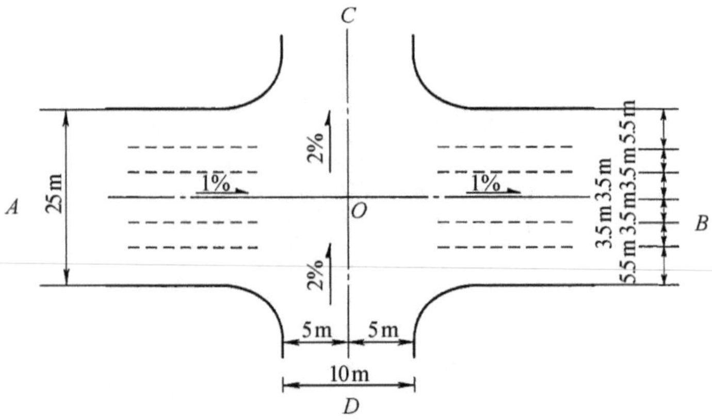


**Figure 1 Diagram of the intersection**

# Chapter 7 Grade Separation and Interchange Design

【**Objectives**: 1. Understand and master the function and classification of interchanges; 2. Master the basic control elements and overall design principles of interchanges; 3. Be familiar with the basic design workflow of interchanges; 4. Preliminarily grasp the design basis, design standard and key points of alignment design of interchanges; 5. Perform comprehensive comparison and selection of interchanges.】

## 7.1 Introduction

### 7.1.1 Definition of the grade separation

The ability to accommodate highway volumes of traffic safely and efficiently through intersections depends largely on the arrangements provided for handling intersecting traffic. The greatest efficiency, safety and capacity are attained when the intersecting traveled ways are grade separated. Grade separations are usually constructed at the junctures between freeways, urban arterial roads, or expressways.

### 7.1.2 Function and type of the grade separation

**1) Function of the grade separation**

The grade separation is a key node and an important part of the freeway and urban expressway networks. The main function of a grade separation is to reduce or eliminate the conflict points that exist when roads join or cross by establishing a graded intersection. The grade separation ensures continuous traffic flow for the intersecting roads and improves road safety. However, compared with the at-grade intersection, the grade separation is of complex forms and difficult construction, and there are more stringent requirements on space occupation and

© The Author(s), under exclusive license to Springer Nature Singapore Pte Ltd. 2025
B. Yu and J. Cheng, *Road Survey and Design*,
https://doi.org/10.1007/978-981-97-4836-5_7

geological conditions for the grade separation. Meanwhile, the grade separation is generally expensive and difficult to rebuild. Therefore, in the decision-making and type selection of the grade separation, it shall be comprehensively determined according to its functional role in the road system, combined with the intersecting road grades, external environmental factors, economic factors and other access control requirements, so as to ensure its matching with the needs of regional traffic and economic development.

**2) Classification of the grade separation**

Grade separations can be divided into different types according to different methods or rules. Figure 7 - 1 summarizes the main classifications of grade separations.

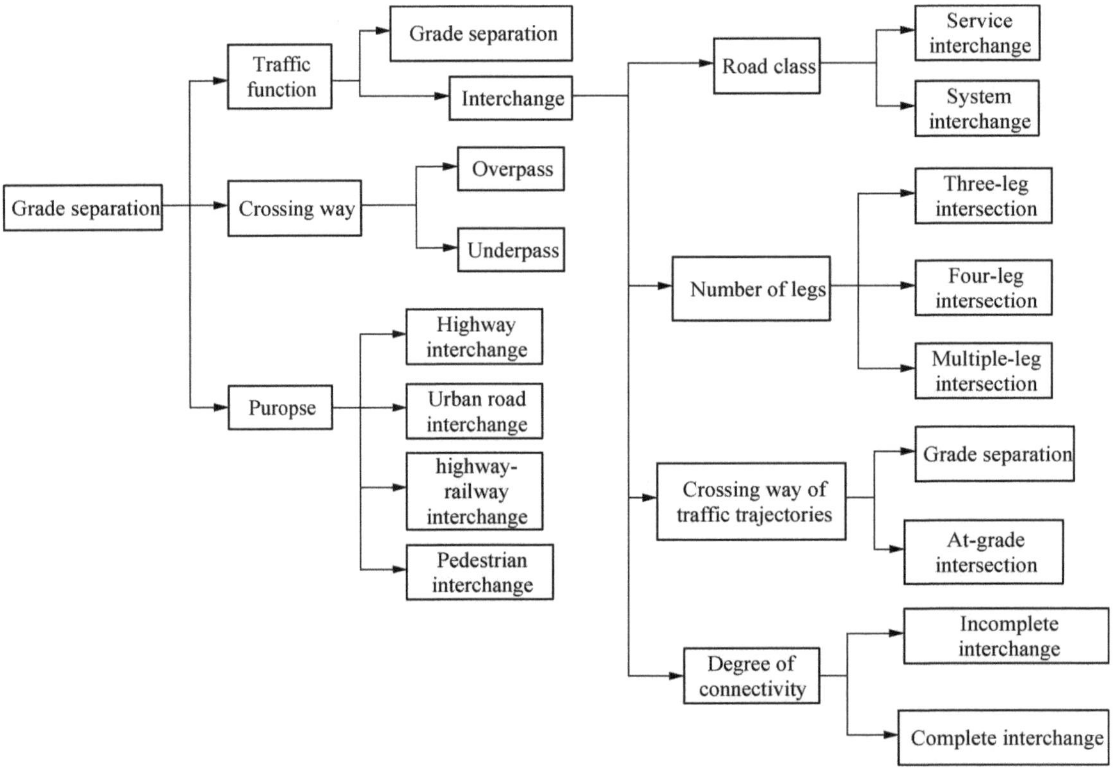


**Figure 7 - 1 Classification of grade separations and interchanges**

According to different classification methods or rules, grade separations can be divided into different types, as shown in Figure 7 - 1.

The grade separation is divided into the grade separation without ramps and interchanges according to the traffic function. As shown in Figure 7 - 2, the grade separation without a ramp has only one crossover structure (crossover bridge or channel), and the upper and lower roads are not connected. Such a grade separation has a simple structure, small area, and low cost. However, vehicles on the intersecting roads cannot turn and the traffic function is limited. The selection of the appropriate type of the grade separation and interchange, along with its design, is influenced by many factors, such as the highway classification, character and composition of traffic, design speed, and degree of access control. In addition to these controls, signing needs, economics, terrain, and right-of-way are of great importance in designing facilities with adequate capacity to safely accommodate traffic demands.

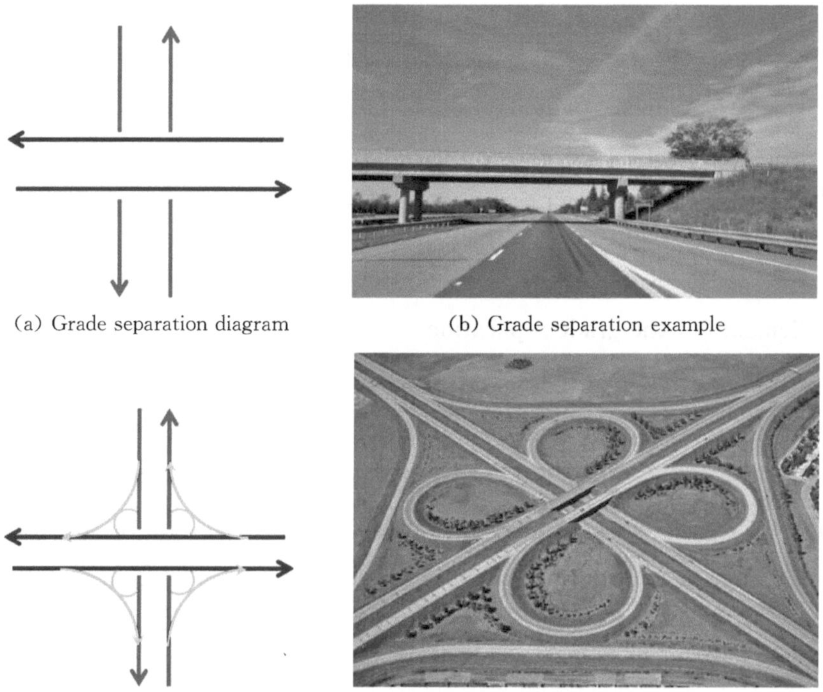

(a) Grade separation diagram (b) Grade separation example

(c) Interchange diagram (d) Interchange example

**Figure 7 - 2 Grade separation and interchange**

An interchange is a system of interconnecting roadways in conjunction with one or more graded separations that provides for the movement of traffic between two or more roadways or highways on different levels. An interchange allows vehicles to turn and completely or partially eliminates the conflict points, which eliminates traffic flow interference in all directions and improves driving safety and traffic capacity.

Compared with the grade separation without the ramp, the interchange is of complex configuration, occupies more land, and costs more. As shown in Figure

7 - 1, interchanges can be further divided into different types according to the class of intersecting roads, the number of intersecting roads, the degree of traffic transfer, and the degree of connectivity. Interchanges can also be divided according to the configuration. The commonly used interchanges under the conditions of different classes of intersecting roads and different numbers of intersecting roads are listed in Table 7 - 1.

**Table 7 - 1 Interchange classification**

| Intersecting road class | Number of roads | Geometry |
|---|---|---|
| Service interchange | Three | Trumpet, cotyledon |
| | Four | Trumpet, cloverleaf, diamond, rotary, combined type |
| | Multiple | Rotary, combined type |
| System interchange | Three | Y-shape, T-shape |
| | Four | Direct, cloverleaf, turbine |
| | Multiple | Combined type |

**3) Elements of the grade separation**

Compared with the grade separation without the ramp, interchanges are more complex in terms of design and construction, which can be applied to a wider range of scenarios in engineering. And interchanges are also the most frequently used types of grade separations. Therefore, the following sections will focus on elements, configuration and design of the interchange and other related contents.

Elements of the interchange and their definitions are listed in Table 7 - 2. Some basic elements are also shown in Figure 7 - 3. In addition to the main elements, an interchange project also includes greenings, drainage, lighting, traffic facilities, etc. within the interchange range.

**Table 7 - 2 Elements of the interchange**

| Name | Definition |
|---|---|
| Interchange range | Sections where the main line and the intersected line are affected by the geometric configuration of the interchange |
| Main line | Roads with the highest class or the largest through traffic volume of the interchange |

**Continued Table 7 - 2**

| Name | Definition |
|---|---|
| Intersected line/Minor line | Roads other than the main line; generally, the main line and the intersected line are also collectively referred to as the cross lines in interchange design, which mainly refers to the through sections of the intersecting roads within the interchange range |
| Ramp | Connecting roads between the intersecting roads, an important part of the interchange |
| Exit ramp | The ramp for vehicles to exit the main line |
| Entrance ramp | The ramp for vehicles to enter the main line |
| Directional ramp | The ramp where vehicles exit and enter in a consistent turning direction: for left-turning it means both exiting and entering are on the left side; for right-turning it means both exiting and entering are on the right side |
| Semi-directional ramp | The ramp where vehicles exit and enter in inconsistent or partially consistent turning direction: for left-turning it means exiting left and entering right, or exiting right and entering left |
| Connection | Passages that connect the ramp with the main line, with the intersected line and with the ramp, including the entrance and exit, speed-change lane, auxiliary lane, collector-distributor lane, nose, etc. |
| Entrance and exit | Referring to the exit and entrance of the ramp from and into the cross lines; the exit is for vehicles to move from the cross lines to the ramp, and the entrance is for vehicles to move from the ramp into the cross lines |
| Acceleration lane | An additional lane for vehicles to accelerate from a low-speed lane to a high-speed lane |
| Deceleration lane | An additional lane for vehicles to decelerate from a high-speed lane to a low-speed lane. Acceleration lanes and deceleration lanes are collectively referred to as the speed-change lane |
| Auxiliary lane | An additional lane provided outside the cross lines near the divergence and convergence spots of two-lane ramps at the interchange in order to balance the lane numbers of ramps and the cross lines and maintain the basic lane number of the cross lines |
| Collector-distributor lane | An auxiliary lane provided outside the main line and separated from the main line in order to isolate the weaving area and reduce the number of entrances and exits of the main line |
| Gore nose | At the connections of divergence or convergence, the circular terminal formed by the intersection of adjacent road edges |

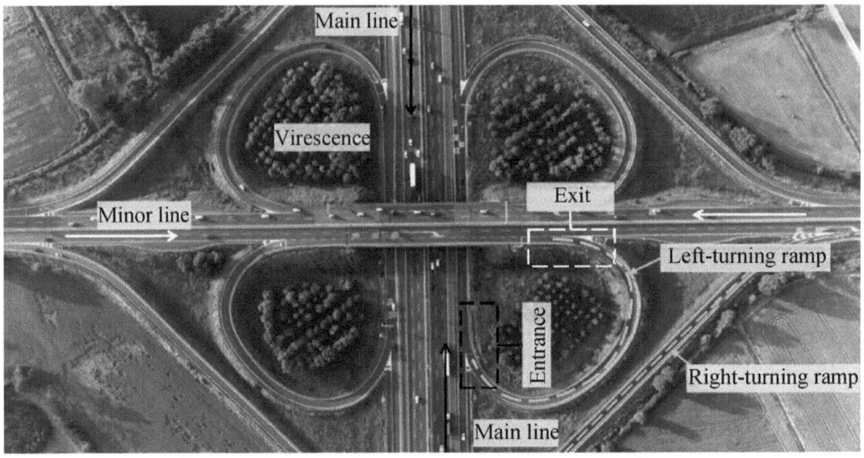


**Figure 7 - 3 Basic elements of the interchange**

## 7.2 Selection of the interchange configuration

The selection of the interchange configuration is to determine that it meets the requirements of design traffic volume, design speed, and normal operation of vehicles according to its functional role in the road network, ensures traffic efficiency and driving safety, and coordinates with the surrounding environment. A reasonable interchange configuration has a positive effect on promoting its transportation function, strengthening the connection with the surrounding road networks, and benefiting regional transportation development. On the contrary, if the interchange configuration does not match the needs, it will limit its function and fail to serve the smooth passage of the intersecting roads, resulting in a waste of investment and an increase in long-term management costs.

### 7.2.1 Factor influencing interchange configuration selection

Selection of the interchange configuration generally requires comprehensive consideration of the following aspects, including nature, environment, traffic, road, and others, as shown in Table 7 - 3. Various aspects shall be integrated when choosing the interchange configuration.

**Table 7 - 3 Consideration of interchange configuration selection**

| Aspects | Main contents |
|---|---|
| Natural condition | It mainly includes the interchange design scope and the topographical, geological, hydrological, and climate conditions of the area where it is located |
| Environmental condition | It mainly includes the neighboring land planning, current status of surrounding buildings, land use, and the existence of cultural relics and historic sites |
| Traffic condition | It mainly considers the traffic development, volume and composition of the traffic, development and planning of traffic network, and design traffic capacity |
| Road condition | It mainly includes the basic conditions of the intersecting roads (nature, function, class, etc.), design speeds of the intersecting roads, and road-land range |
| Other condition | Charging system, investment, and relevant requirements of the administrative department |

## 7.2.2 Principles of interchange configuration selection

The selection of the interchange configuration shall abide by the following principles:

(1) According to the road network layout and planning, the interchange configuration on one road shall be unified as far as possible, and the location and form of the entrance and exit shall be consistent.

(2) The interchange configuration shall consider the class, nature, task, and traffic volume of the intersecting roads to ensure safe and smooth driving and continuous traffic flow. The directions with large traffic volumes and high design speeds require high alignment parameters, short routes, and the gentle grade; when the vehicle composition is complex, the needs of individual traffic characteristics shall also be considered. On urban roads, if vehicles and non-motor vehicles have a large traffic flow and are separated, a three-layer or four-layer interchange can be used.

(3) The interchange configuration shall be adapted to the natural and environmental conditions of the location, and take full account of regional planning, topography, geological conditions, possible land range, cultural relics protection, distributions of surrounding buildings and facilities, etc. Based on the extensive analysis and research on the premise of meeting traffic requirements,

the interchange configuration shall make rational use of terrain and geological conditions, reduce land acquisition, and demolition, realize economical operation, coordinate with the surrounding environment, and present an aesthetic, novel, and reasonable structure.

(4) The interchange configuration shall fully consider the integration of short-term and long-term needs. It shall not only meet the requirements of short-term transportation to reduce investment, but also consider the needs of long-term development and the possibility of reconstruction so that the short-term project could be updated in the future.

(5) The interchange configuration shall consider the incorporation of the toll gate. If it is a tolled interchange, the quadrant of the connecting road shall be determined according to the turning traffic volume, and the specific location of the connecting road shall be decided according to the length requirements of the speed-change lane. The configuration of a three-way intersection at both terminals of connecting roads shall be selected according to the function, class, and site constraints of the intersecting roads.

(6) The interchange configuration shall consider project construction and take into account the appearance and investment of the project, and be conducive to construction, maintenance, and drainage. New techniques, technologies and structures are advocated to improve project quality, shorten the construction period, and reduce cost.

(7) The interchange configuration shall be decided together with the layout of ramps with clear priority. When designing the horizontal and vertical alignments of the intersecting roads, the layout of horizontal alignment and the control of vertical elevation of ramps shall be regarded. The relationship between the major road and the minor road shall be properly handled so that the requirements of the major road are met first. The interchange configuration shall be compatible with the cross lines' shape, structure, overall layout and environment. When the high-speed roads intersect with the other roads, the high-speed roads remain unchanged or less changed while the other roads are either raised or lowered; and the grade separation of urban roads shall warrant that non-motor vehicle lanes remain unchanged or less changed, which is conducive to the passage of pedestrians and bicycles.

(8) The interchange configuration shall be integrated with the function. The

configuration varies with the terrain, features and environmental conditions where the interchange is located. It is usually to determine the function first and then the configuration and unites the configuration selection and function.

## 7.2.3 Commonly used interchange configurations

This section introduces the commonly used interchange configurations in engineering, and summarizes the advantages and disadvantages of some interchange configurations.

**1) Incomplete interchange**

An incomplete interchange is the one with conflict points on at least one plane between the traffic trajectories of the intersecting roads. It is generally used for the intersection of the major road and minor road, and can also be used when the traffic volume in a certain direction is small or is constructed in stages, or when a ramp cannot be installed in a certain direction due to topography, features, and road network planning. The representative types include partial cloverleaf interchange and diamond interchange. Table 7 - 4 summarizes the advantages and disadvantages of them.

**Table 7 - 4 Incomplete interchange**

| Interchange configuration | Description | Advantage | Disadvantage |
|---|---|---|---|
| Partial cloverleaf interchange | Partial left-turning traffic has no left-turning ramp, and the interchange is of incomplete cloverleaf shape; it is suitable for the intersection of the main road and minor road (Figure 7 - 4). When used, the impact of turning vehicles on the traffic on the main line shall be mitigated as much as possible; when necessary, channelized traffic or signal control can be used on the minor road | The through traffic on the main line is smooth and fast; single exit direction facilitates vehicle operation and simplifies traffic signs on the main line; only one overpass structure is needed so the land area and engineering cost are relatively small; it is convenient to construct in stages and can be expanded into the cloverleaf interchange in the future | The at-grade intersection exists on the minor road, affecting traffic capacity and safety; there is the possibility of stopping and wrong-way entrance |

**Continued Table 7 - 4**

| Interchange configuration | Description | Advantage | Disadvantage |
|---|---|---|---|
| Diamond interchange | There are four one-way ramps leading to the intersecting roads and there is an at-grade intersection on the minor road; it is applicable to situations where the main urban road and minor urban road intersect and the land use is constrained, which is less used on highways (Figure 7 - 5). When used, the at-grade intersection is set on the minor road; the main road shall over-cross or under-cross according to terrain and drainage conditions, and the underpass is generally appropriate. On the minor road, traffic can be organized by channelization or traffic signals | The through traffic on the main line is smooth and fast; the left-turning vehicles have a shorter detour distance; there is a high-standard entrance and exit on the main line, and traffic signs are simple; when the main line underpasses, the grade of the ramp is apt to the deceleration of exiting and the acceleration of entering; the form is simple, only one overpass structure is needed, and the land area and cost are small | An at-grade intersection exists at the junction of the minor line and ramp, affecting traffic capacity and safety |

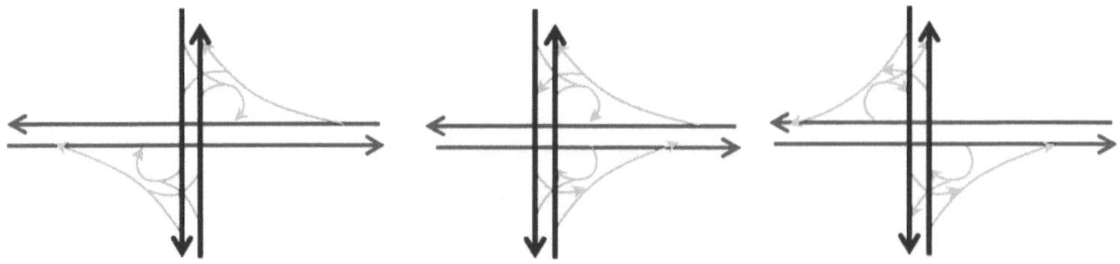

**Figure 7 - 4 Partial cloverleaf interchange**

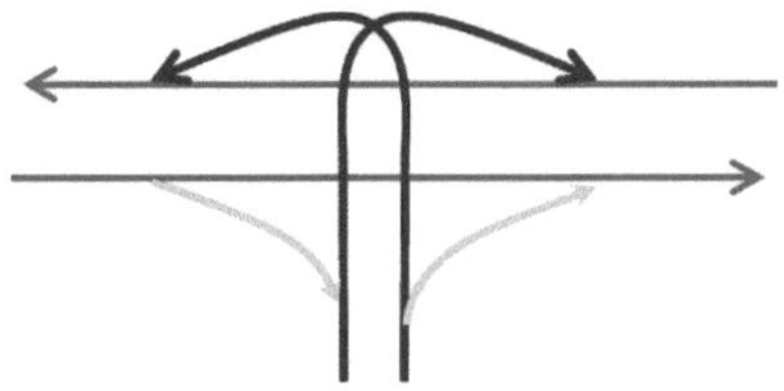

(a) Three-way diamond interchange

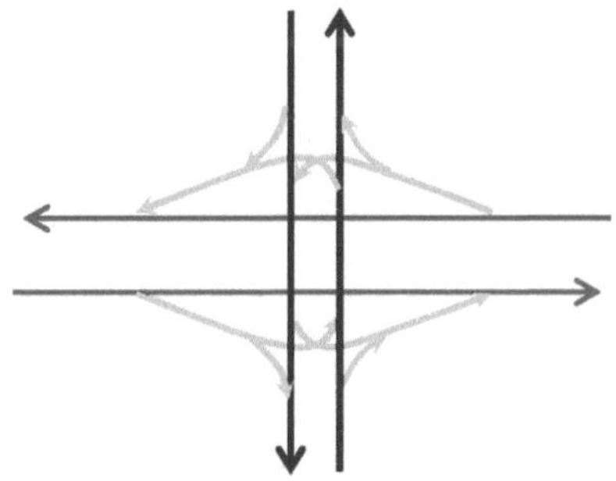

(b) Four-way diamond interchange

**Figure 7 - 5 Diamond interchange**

**2) Complete interchange**

The intersection where the traffic trajectories of the intersecting roads are all separated in space is called a complete interchange (interchange in short). The number of ramps at the interchange is equal to the number of turning directions, and each turning direction has a dedicated ramp with no conflict points, which enjoys the merit of safe, fast driving and large traffic capacity. It is applicable to intersections between high-speed roads or between high-speed roads and other roads with large traffic volumes. The representative forms of interchanges include the trumpet interchange, cloverleaf interchange, cotyledon interchange, Y-shaped interchange, X-shaped interchange, turbine interchange, combined interchange, etc. Advantages and disadvantages of each interchange are summarized in Table 7 - 5.

**Table 7 - 5 Complete interchange**

| Interchange configuration | Description | Advantage | Disadvantage |
|---|---|---|---|
| Trumpet interchange | It consists of a left-turning loop ramp (about 270°) and a semi-directional left-turning ramp, which is a representative type of three-way interchange (Figure 7 - 6). It is type A if the vehicle drives into the cross lines (or the main lines) through the left-turning loop ramp and type B if the vehicle drives out | One overpass structure, less investment; no conflict points and weaving, large traffic capacity, safe driving; elegant appearance, easy-to-identify driving direction; except for the loop ramp, other ramps provide higher-speed directional or semi-directional operation for turning vehicles | The loop ramp has low alignment parameters and a low operation speed; it is mainly suitable for the three-way intersection in which the high-speed road and normal road intersect |

**Continued Table 7 - 5**

| Interchange configuration | Description | Advantage | Disadvantage |
|---|---|---|---|
| Cloverleaf interchange | An interchange consists of four symmetrical loop ramps for left-turning vehicles (Figure 7 - 2). It is applicable to toll-free grade separations between high-speed roads or on the urban peripheral ring roads. In order to eliminate weaving on the main line, to avoid double exits, to simplify traffic signs and improve traffic capacity and driving safety, a collector-distributor lane is frequently employed outside the main line (Figure 7 - 7), so that the entrance and weaving section are arranged in the collector-distributor lane | Ramps are independent of each other; there is no conflict point, and the traffic operates continuously and naturally; An overpass structure is used, which can be constructed in phase | It covers a large area; the left-turning vehicles have a long detour distance; loop ramps, adapt to lower speeds; there is a weaving section up and down the overpass structure, which limits the traffic capacity |
| Cotyledon interchange | It uses two loop ramps for left-tuning vehicles (Figure 7 - 8). It is mostly used as the pre-project of the cloverleaf interchange. When used, it is appropriate to underpass the main line | One overpass structure is used, which has low cost, symmetrical ramps and elegant appearance | Traffic environment is not as good as the trumpet interchange; there is a weaving section on the cross lines, and the left-turning vehicles make a long detour distance |

**Continued Table 7 - 5**

| Interchange configuration | Description | Advantage | Disadvantage |
|---|---|---|---|
| Y-shaped interchange | It uses the directional or semi-directional ramp for left-tuning vehicles (Figure 7 - 9).<br>It is suitable for the three-way interchange with heavy traffics in all directions | It can provide high-speed directional or semi-directional operation for turning vehicles, with large traffic capacity; it is free of weaving and conflict points;<br>the driving direction is clear, the path is short, and the operation is smooth;<br>the occupied width outside the cross lines is small | There are more overpass structures and the cost is high |
| X-shaped interchange | It consists of four semi-directional ramps for left-turning vehicles (Figure 7 - 10).<br>It is mostly used for the system interchange between high-speed roads, with large left-turning traffic volume, high-speed requirements, and large traffic capacity | Turning vehicles in all directions move clearly, freely and smoothly;<br>it has a single exit or entrance to facilitate vehicle operation and simplify traffic signs;<br>it is free of weaving and conflict points and is apt to high speed and large capacity | The overpass structure is long and with more layers;<br>the cost is high and the area is large |
| Turbine interchange | It consists of four semi-directional ramps for left-turning vehicles (Figure 7 - 11).<br>It is suitable for the system interchange with low turning speed between high-speed roads | The ramp grade is gentle and adapts to higher speeds;<br>vehicles enter and exit the cross lines safely and smoothly;<br>it is free of weaving and conflict points and has greater traffic capacity | Left-turning vehicles have a longer detour distance and higher operation cost;<br>five two-story overpass structures are needed;<br>the cost is relatively high and the area is large |

Continued Table 7 - 5

| Interchange configuration | Description | Advantage | Disadvantage |
|---|---|---|---|
| Combined interchange | According to traffic volume and in association with terrain and feature constraints, two or more different types of left-turning ramps are used in one interchange (Figure 7 - 12). It is suitable for the system interchange with small left-turning traffic on one to two ramps | The combinations are diverse; the ramp layout shall be compatible with traffic volume, make full use of terrain and features, and adjust subject to local conditions | It is complicated, the cost is relatively high, and the area is large |

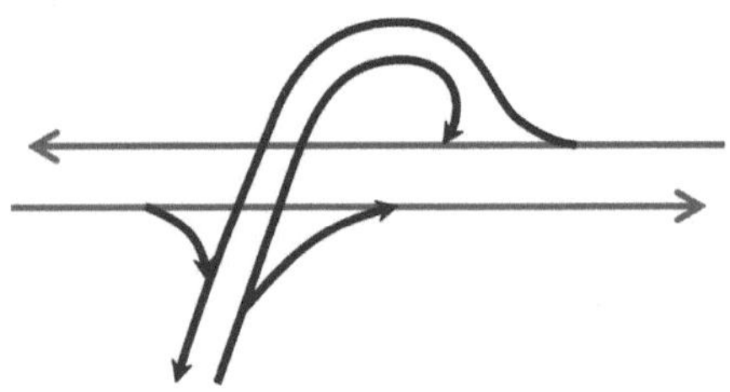

(a) Type A

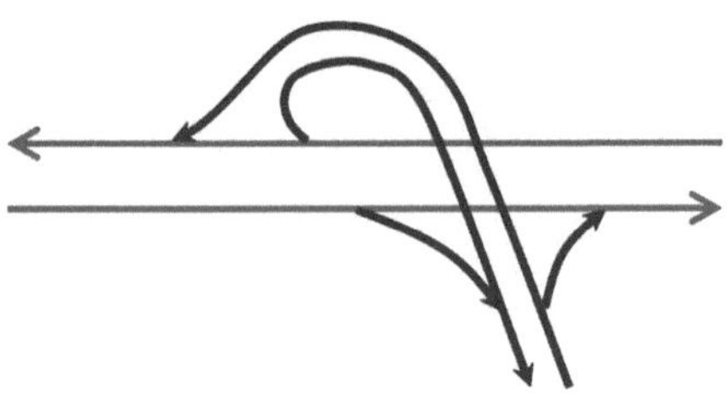

(b) Type B

**Figure 7 - 6 Trumpet interchange**

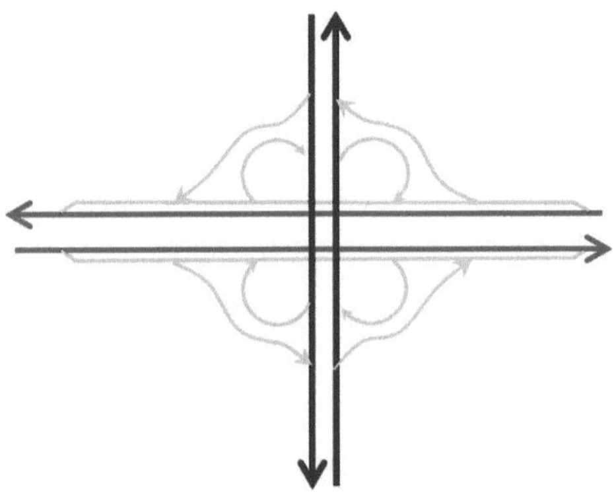

**Figure 7 - 7 Cloverleaf interchange with the collector-distributor lane on the main line**

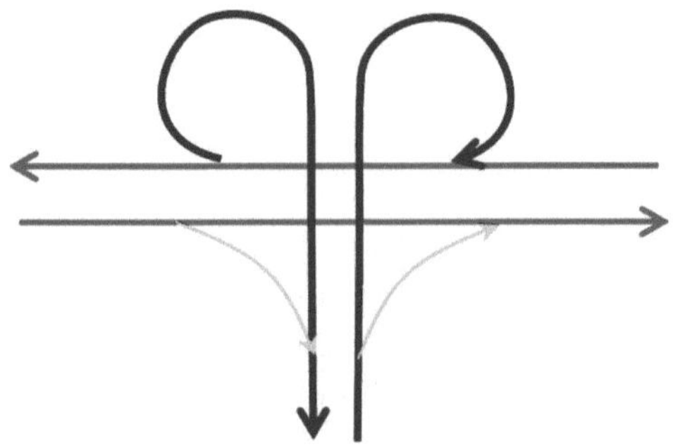

**Figure 7 - 8 Cotyledon interchange**

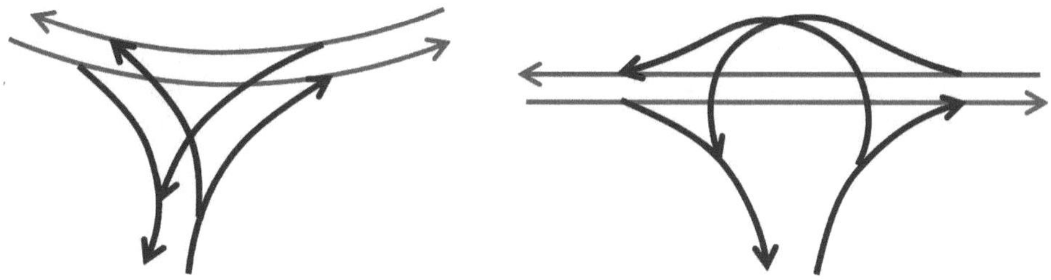

(a) Directional Y-shaped interchange (b) Semi-directional Y-shaped interchange

**Figure 7 - 9 Y-shaped interchange**

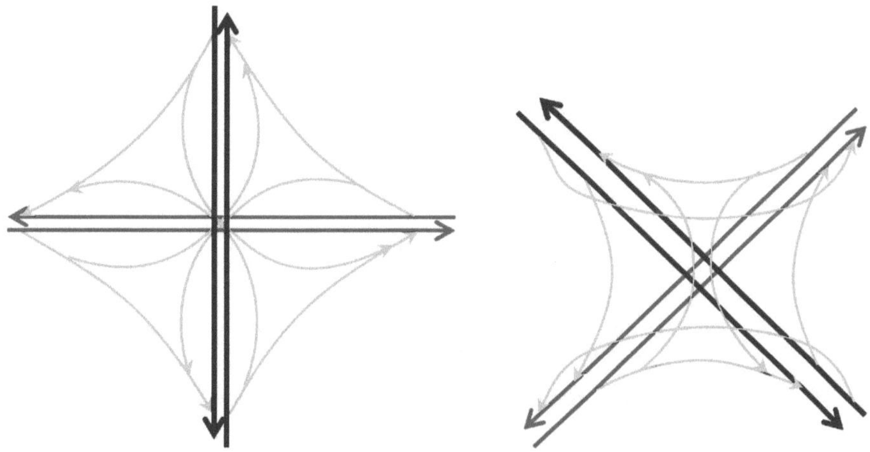

(a) Diagonal close arrangement of opposite left-turning ramps (b) Diagonal opening arrangement of opposite left-turning ramps

**Figure 7 - 10 X-shaped interchange**

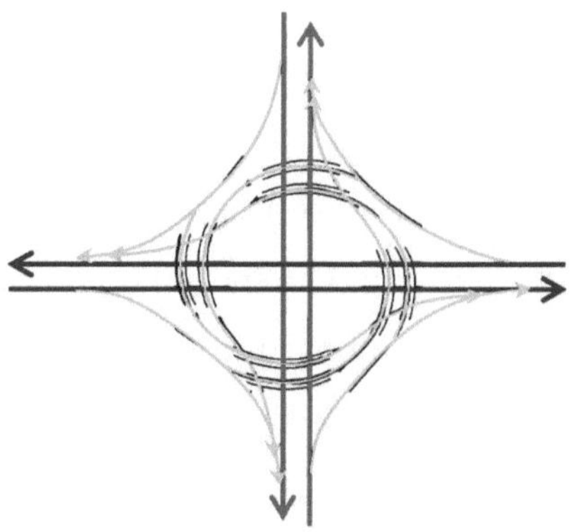

**Figure 7 - 11 Turbine interchange**

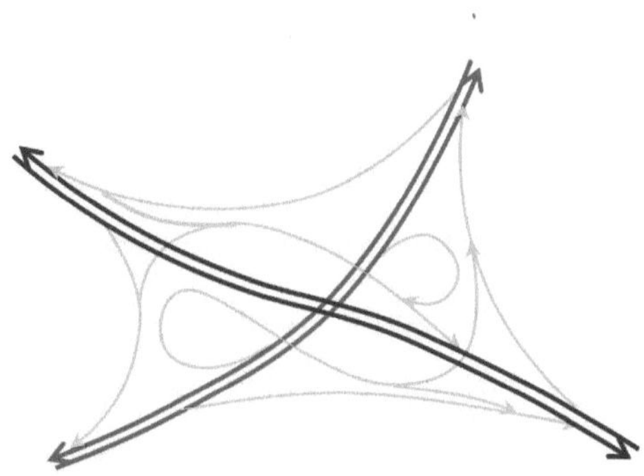

**Figure 7 - 12 Combined interchange**

## 7.2.4 Process of interchange configuration selection

When the intersection location is determined, the selection of the specific interchange configuration is carried out, following the process shown in Figure 7 - 13.

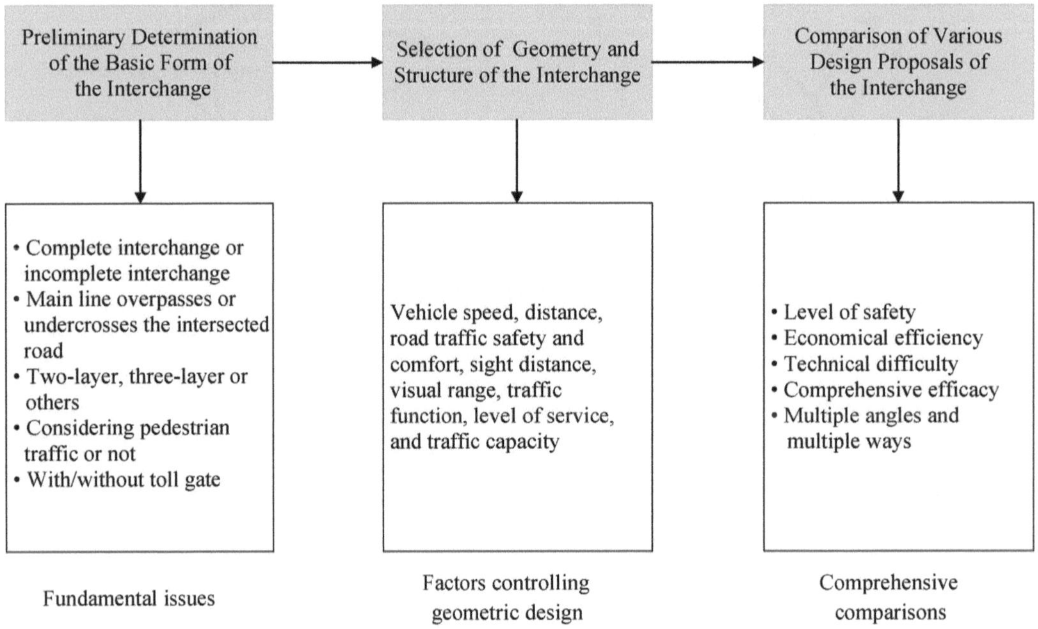


**Figure 7 - 13 Process of interchange configuration selection**

**1) Determination of the basic interchange configuration**

In the preliminary selection of the interchange configuration, it shall first answer the questions listed in Figure 7 - 13, and then further select the commonly used interchange configurations. The interchange configuration shall be determined according to the traffic volume in each direction, in association with the consideration of terrain, ground features, and traffic conditions. The following provisions shall be followed:

(1) When the through and turning traffic volumes are relatively large, the design speeds of intersecting highways are high, and a higher speed is required for the collector-distributor lane, directional or semi-directional interchanges can be used. The turbine or combined interchange can also be used.

(2) When the freeway intersects with the Class Ⅰ highway and there is no toll station, the combined interchange can be used. When the traffic volume is high, directional and semi-directional ramps can be used; when the left-turning traffic volumes in some directions are small, loop ramps can be used.

(3) When two Class Ⅰ highways intersect, the cloverleaf interchange, rotary interchange or combined interchange should be used.

(4) When the freeway intersects with the Class Ⅰ highway or Class Ⅱ

highway of heavy traffic, and there is a toll station, a double-trumpet interchange should be used.

(5) When the freeway intersects with the Class Ⅱ highway of low traffic volume, the partial cloverleaf interchange should be used of which the single-trumpet intersection exists on the crossed road. When there is no toll station on the ramp, the diamond interchange should be used.

(6) When the Class Ⅰ highway intersects with Class Ⅱ, Class Ⅲ and Class Ⅳ highways, the diamond interchange and partial cloverleaf interchange should be adopted when the interchange is applied due to traffic transfer.

**2) Selection of geometry and structure of interchanges**

The geometry and structure of interchanges have a great influence on the speed, length, driving safety and comfort, sight distance, line of sight, traffic function, level of service, and traffic capacity. With the determination of the interchange configuration, it is necessary to arrange the layouts of the overall structure and ramps of interchanges, such as the layout of overpass structure, the location of entrance and exit, the quadrant of ramp layout, and cross section of ramps (integrated or separated), the ramp's horizontal alignment, profile and cross section, and the layout of speed-change lanes.

**3) Comparison and selection of interchange schemes**

After the determination of the interchange configuration, geometry, and structure, multiple interchange schemes are proposed. The reasonable configuration and appropriate scale shall be selected through the comparisons of safety, technology, economy, and benefit of multiple schemes. The expected scheme shall meet the requirements of traffic function, be suitable for site conditions, and have small quantities, aesthetic appearance, and low investment. For complex large-scale interchanges, perspective drawings or three-dimensional animations may also be performed for scheme inspection and comparison.

## 7.3 Interchange design

### 7.3.1 Steps of interchange design

Interchange design shall be conducted with the collection of necessary design data through field survey and investigation, with the relevant data listed in Table 7 - 6. Interchange design includes a series of processes from planning, feasibility study, scheme design to technical design. Among them, the scheme design and technical design are generally carried out in the order shown in Figure 7 - 14.

**Table 7 - 6 Elements of the interchange**

| Data type | Contents |
|---|---|
| Natural data | Topographic map (scale 1 : 500—1 : 2000) including interchange range, and the building lines, types, story heights, and various ground and underground poles and pipelines and other features of buildings need to be marked in detail; survey and collect the land properties and development planning, hydrology, geology, soil, climate, plants, wildlife and mineral resources of the project area and construction site |
| Traffic data | Road network structure and planning of the area where the interchange is located, the comprehensive transportation system and planning, the distribution of traffic volume, its composition, etc. At each design stage, the node traffic distribution map shall be provided to clarify the traffic volume, composition, and origination in each direction of the node |
| Road data | Classes, horizontal and vertical alignments, cross sectioning of the intersecting roads; intersection angle, coordinates and elevation of control; types and thickness of pavements; determining the clearance height, design load, design speed, horizontal and vertical design indicators, etc. When intersecting with railways, the number of railway tracks, spacing, rail top elevation, train passing times, and track breaking time, requirements of clearance width and height shall also be investigated |
| Drainage data | The current status and planning of the drainage system in the area where the interchange is located, including the location, depth, and size of various pipes and canals |
| Document data | Interchange design assignments, approvals, the specific requirements, opinions, and other relevant documents from the administrative department and local governments; relevant technical standards, specifications, and other documents |
| Other data | Borrow, spoil and material sources; construction company, construction season, construction period, transportation organization, energy supply, environmental protection, etc. |

Propose preliminary schemes: The preliminary scheme is to search various possible interchange schemes on the topographic map according to the road, traffic and natural conditions at interchanges. The scheme shall meet the basic requirements of interchange design, topographical conditions, planning, and relevant regulations of the interchange location.

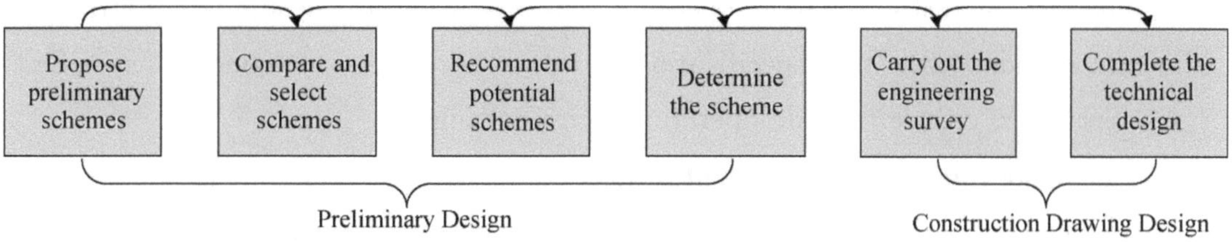


**Figure 7 - 14 Process of interchange scheme design**

Compare and select schemes: when making analysis and comparison of the preliminary schemes, we shall take into account whether the alignment is suitable, whether the radius can meet requirements, whether the floors can be overpassed, and whether the demolition is reasonable. It is general to choose 2 - 4 schemes.

Recommend potential schemes: It is necessary to complete the geometric design, bridge span layout plan and rough engineering calculation of each scheme, and make a comparison table of each scheme. Generally, 1 - 2 schemes are recommended after comprehensive comparison in terms of the coordination of the interchange configuration and turning traffic volume, ramp geometric design index and safety, form and length of the structure, land occupation and demolition, construction difficulty, project cost, and maintenance and operation conditions, etc.

Determine the scheme: After consulting the opinions of relevant parties and comprehensively weighing the relationship between investment and the scheme, between short-term and long-term developments, and between local and overall situations, the scheme is finally determined.

Carry out the engineering survey: Carry out field laying-out and detailed survey of the adopted scheme, and further collect all the information required for technical design.

Complete the technical design: Complete all construction drawing designs and project budgets.

## 7.3.2 Principles of interchange design

**1) Meet specification requirements**

Relevant design indicators shall meet the requirements of "Guidelines for

Design of Highway Grade-separated Intersections" (JTG/T D21—2014, "Guidelines" hereafter). Under special conditions, when the design is not strictly in accordance with the specifications, it is necessary to conduct specific discussions to ensure the safe operation of interchanges.

**2) Make full use of the original project**

Make full use of the original project and land to save project investment, reduce new land use, and shorten the construction window.

**3) Meet the demand of long-term traffic growth**

Reconstruction and expansion of interchanges shall be conducted based on the predicted long-term traffic volume, which shall be rationally determined in combination with the analysis of the road network, including the upgrade of interchange ramp standards, expansion of toll stations, and traffic divergence measures of the road network.

**4) Reduce interference to traffic during construction**

During the reconstruction and expansion of interchanges, it should maintain the existing traffic entering and exiting the freeway or shorten the interruption time as much as possible. When the traffic have to be interrupted due to ramp reconstruction, the timing of traffic interruption shall be selected reasonably (such as at night) to minimize the impact on the existing traffic.

**5) Compare and select multiple interchange schemes**

The new interchange shall, under the premise of meeting the requirements of transportation transfer functions and technical indicators, pay attention to obtaining compact layout of interchanges, conduct multi-scheme comparison, and achieve the goal of "minimum land use, scale, and cost".

### 7.3.3 Control elements on interchange design

The main control elements of interchange design include the design vehicle, design speed, sight distance, traffic volume, level of service and boundary line. Each control element is the basis for the interchange design. Table 7 - 7 lists the requirements of each control element. The specific requirements for some design control elements in the ramp design and connection design will be detailed in subsections 7. 3. 3 and 7. 3. 4.

**Table 7 - 7 Information on interchange design**

| Control elements | Requirements |
|---|---|
| Design vehicle | Passenger cars, buses, articulated buses, trucks, and articulated trucks (trailers) are used as design vehicles. In general, the passenger car is used as the standard vehicle to calculate traffic volume; when there are a large number of containers, large equipment, and national defense transportations, the largest vehicle shall be used to verify key geometric parameters (radius, widening, sight distance, etc.) of ramps |
| Design speed | The design speed of normal sections is generally used within the interchange range. Under special circumstances, such as when the turning radius could not meet the standard due to constraint, the design speed can be appropriately reduced, but the difference between the design speeds of adjacent sections shall be no more than 20 km/h theoretically;<br>When the design is finished, it is recommended to check the operation speed of ramp sections. For adjacent sections with weak operation speed continuity, alignment needs to be refined.<br>For the design speeds of the ramp and ramp connections, please refer to subsections 7. 3. 3 and 7. 3. 4 |
| Sight distance | Within the specified visual range, the sight of drivers within the interchange range shall not be blocked or impaired by obstacles; sight distance inspection is generally required for the following road sections:<br>• When there are bridge piers, guardrails, cutting slopes, and plants inside circular curves, and the radius of the circular curve is small, it is necessary to check the stopping sight distance of the innermost lane on the curve, and the identification sight distance for sections in front of the divergence nose;<br>• When the separator is provided with guardrails, anti-glare boards, plants, and other visual barriers, and the radius of the circular curve is small, it is necessary to check the stopping sight distance of lanes near the separator on the outside curve.<br>For the sight distances of the ramp and ramp connections, please refer to subsections 7. 3. 3 and 7. 3. 4 |
| Traffic volume | In engineering feasibility study, annual average daily traffic (AADT) is used in interchange design, and AADT is the forecasted traffic volume in the 20th year after the completion of interchanges |
| Level of service | The levels of service of the intersecting roads, ramp, diverging-converging area, weaving section, and collector-distributor lane within the interchange range are divided into six levels; the design level of service of road sections shall be selected according to the corresponding function and class; in theory, the design level of service of the ramp, divergence-convergence zone, weaving section, and collector-distributor lane can be one level lower than that of the main line, but shall not be lower than the fourth level |

**Continued Table 7 - 7**

| Control elements | Requirements |
| --- | --- |
| Boundary line | The boundary line of road construction in interchange design shall conform to the relevant provisions in the **Standard (2014)** |

## 7.3.4 Ramp design

As an important part of interchanges, the ramp is the connection segment responsible for the smooth passing of vehicles turning on the intersecting roads. Reasonable ramp design is beneficial to accomplish the traffic function of interchange and ensure the safety and smoothness of driving. Ramps shall be designed in accordance with the relevant design basis and the actual environment of interchanges.

### 1) Ramp classification

Ramp is divided into two classifications, as shown in Table 7 - 8.

**Table 7 - 8 Ramp classification**

| Classification | Type | Characterization | Diagram |
| --- | --- | --- | --- |
| Classification by the function of the ramp and its relationship with the intersecting roads | Right-turning ramp | Vehicles on the right-turning ramp exit on the right side of the cross lines and enter on the right side of the intersecting road. Generally, there is no overpass structure; according to the form of the interchange and land restriction, the right-turning ramp can be designed using four types: single (or compound) curve, reverse curve, parallel line, or oblique line. In special cases, the right-turning ramp can also be realized by continuous left-turning of about 360°. The right-turning ramp is a directional ramp with a right exit and right entrance, which is simple in form, straightforward, and smooth in operation | 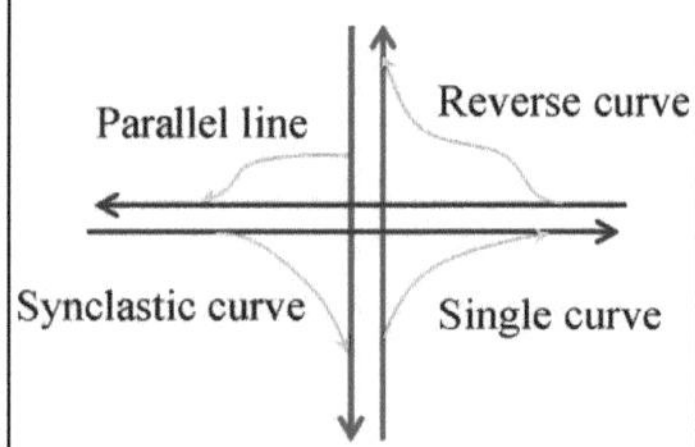 |

**Continued Table 7 - 8**

| Classification | Type | Characterization | Diagram |
|---|---|---|---|
| Classification by the function of the ramp and its relationship with the intersecting roads | Left-turning ramp | Vehicles on the left-turning ramp must turn 90°- 270° to cross the opposite lane. In addition to the loop ramp, at least one overpass structure is required for the ramp. According to the relationship between the ramp and intersecting road, the left-turning ramp can be classified into three types: directional ramp, semi-directional ramp, and loop ramp | Refer to Table 7 - 9 |
| Classification by the ramp cross section and lane type | One-way one-lane ramp (type I cross section) | A commonly used form of ramps. Regardless of the right-turning or left-turning ramp, it is applicable when the turning traffic volume is relatively small and does not exceed the design capacity of one-lane ramp | 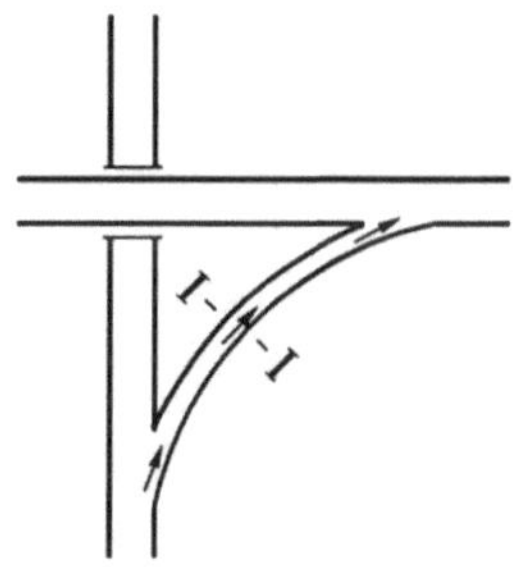 |
| | One-way two-lane ramp (type II cross section) | There are two lanes at the segments between the ramp entrance and exit, but only one lane at the entrance and exit. The two lanes can be separated by lineation, and no emergency parking strip is needed on the right side. It is mainly applicable when the turning traffic volume does not exceed the design capacity of the one-lane ramp and the passing demand is considered | 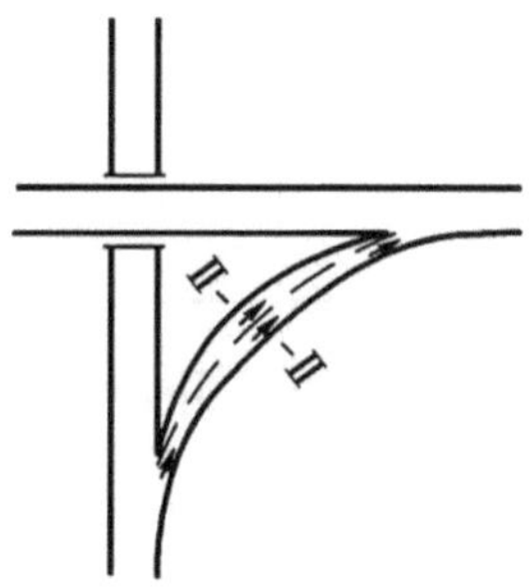 |

**Continued Table 7 - 8**

| Classification | Type | Characterization | Diagram |
| --- | --- | --- | --- |
| | One-way two-lane ramp (type Ⅲ cross section) | There are two lanes on the ramp (including entrance and exit). The two lanes can be separated by lineation, and the emergency parking strip is set on the right side. It is applicable when the turning traffic volume exceeds the design capacity of the one-lane ramp, and the demands of passing and emergency parking are considered | 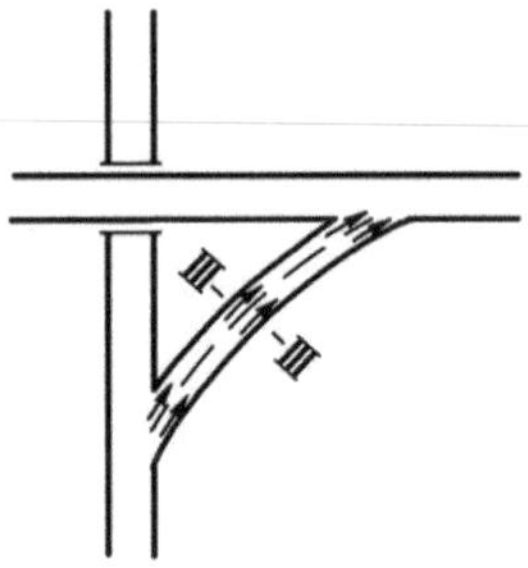 |
| | Two-way two-lane ramp (type Ⅳ cross section) | Generally, the medial separator is used between the opposing lanes, which is suitable when the turning traffic volume reaches the design capacity and the land use is abundant. If land is limited, it can also be separated by lineation which is only applicable when the turning traffic volume is less than the design capacity of the one-lane ramp. According to the distribution of two-way traffic volume, the two-way three-lane or two-way four-lane ramp, can also be applied | 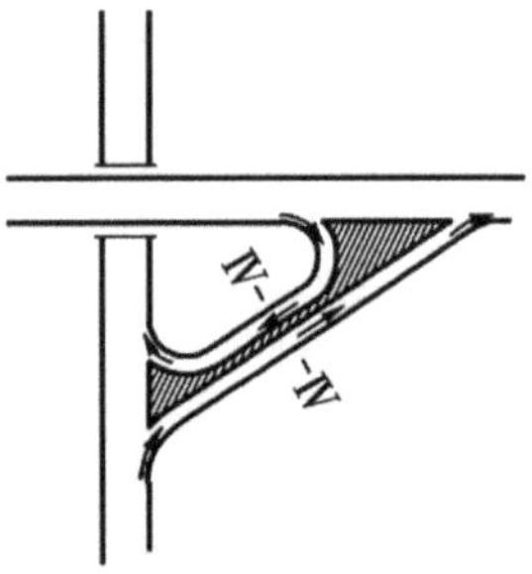 |

The classification of left-turning ramps is more complicated than that of right-turning ramps. As described in Table 7 - 9, left-turning ramps can be divided into three types: directional, semi-directional, and loop ramps. The three types of left-turning ramps and their diagrams are listed in Table 7 - 9.

**Table 7 - 9 Classification of left-turning ramps**

| Type | Characterization | Diagram |
| --- | --- | --- |
| Directional ramp | The left-turning vehicle exits directly from the left side of the traveled lane, which turns left about 90° and enters on the left side of the intersecting road.<br>Advantages: ramp length is the shortest, which can reduce operation costs; there is no reverse circuitous operation, which is suitable for high vehicle speeds and has a large capacity.<br>Disadvantages: there are more overpass structures, so two one-way overpass structures with two-story or one one-way overpass structure with three-story are required; there must be sufficient spacing between the two-way traffic on the intersecting road to over-cross (or cross) the opposite lane. Due to the principle of left-out and left-in on the directional ramp, it is not compatible with the right-hand driving rules in China and is not recommended normally | 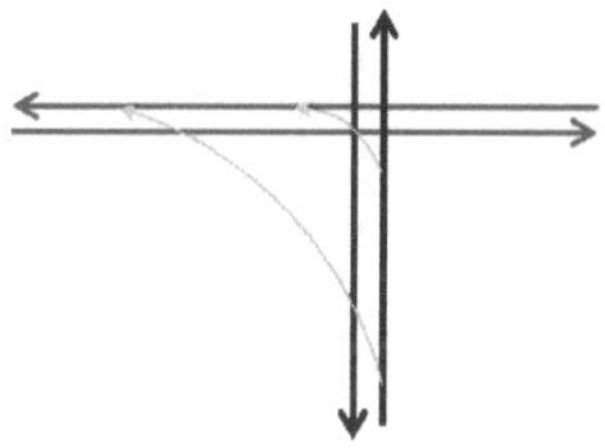 |
| Semi-directional ramp | ***Left-exit and right-entrance***: the left-turning vehicle exits on the left side of the traveled lane, turns left on the ramp, and enters on the right side of the traveled lane. Compared with the directional left-turning ramp, the right-entrance improves the shortcomings of the left-entrance so that vehicles are safe and convenient to enter. But it still suffers from the concern of left-exit. Vehicles on the ramp detour slightly; there must be sufficient spacing between the two-way lanes to exit. All of the three types shown in the diagram can be used, which shall be selected according to the terrain and ground features | 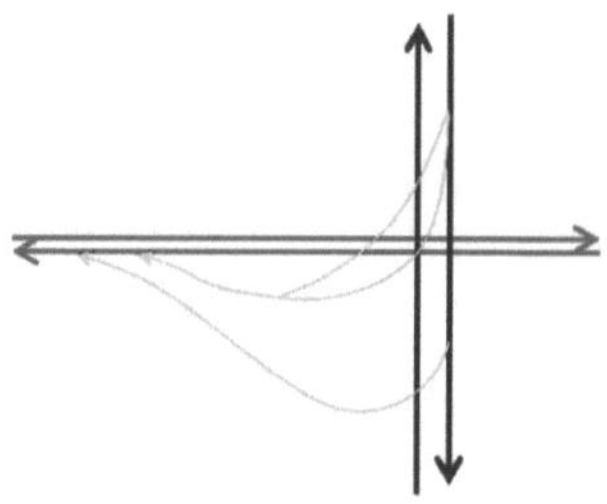 |

**Continued Table 7 - 9**

| Type | Characterization | Diagram |
|---|---|---|
| Semi-directional ramp | ***Right-exit and left-entrance***: the left-turning vehicle exits from the right side of the traveled lane, turns left on the ramp, and enters on the left side of the traveled lane. The shortcomings of left-exit are improved so that vehicles are safe and convenient to exit, but it still suffers from the concern of left-entrance. There must be sufficient spacing between the two-way lanes to enter | 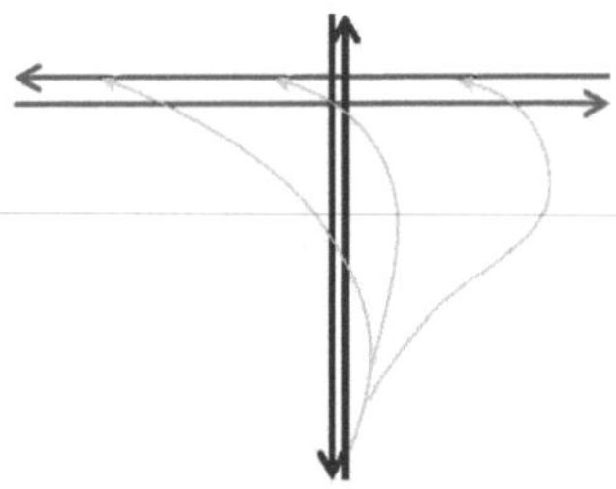 |
| | ***Right-exit and right-entrance***: the left-turning vehicles all exit and enter on the right side of the traveled lane, and turn left on the ramp to change direction. Right-exit and right-entrance is the most commonly used form of the left-turning ramp. It completely eliminates the shortcomings of left-exit and left-entrance, which is safe and convenient to drive. However, it suffers from longer operation distances and more overpass structures. The five types in the diagram shall be used depending on the terrain, features, and alignment | 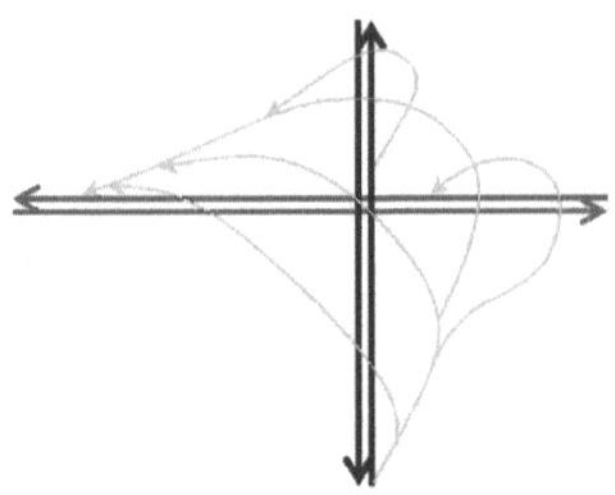 |
| Loop ramp | The left-turning vehicle travels on the ramp of the cross lines and turns right for about 270° to realize left-turning, and enters on the right side of the intersected road. The loop ramp is characterized by right-exit and right-entrance, which is safe to drive. There is no need to build an overpass structure on the ramp, and the cost is the lowest. However, the ramp has a low alignment index, low speed, small capacity, covers a large area, and suffers a long detour distance for left-turning movements | 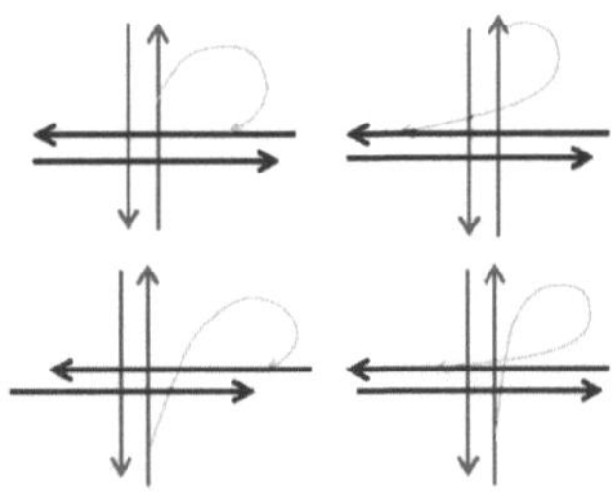 |

**2) Ramp connection method**

The connection between the ramp and the main line between ramps and between the main lines, should adopt the connection methods shown in Table 7 - 10, and shall conform to the following provisions:

(1) The weaving of ramps shall be accommodated by introducing a weaving section outside the through lanes of the freeway, and the ramp shall not weave with the through lanes of the freeway.

(2) Weave shall not happen between the through lanes of the freeway.

(3) The connection on the left side of the freeway is not recommended.

**Table 7 - 10 Appropriate ramp connections**

| Connection method | | Divergence | Convergence | Weave |
|---|---|---|---|---|
| Connections between the ramp and main line | | Right divergence | Right convergence | Right weave |
| | |  |  | 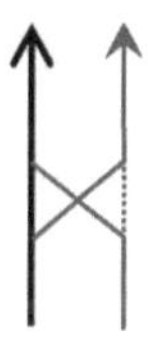 |
| Interconnections | Ramp | Ramp divergence | Ramp convergence | Ramp weave |
| | | 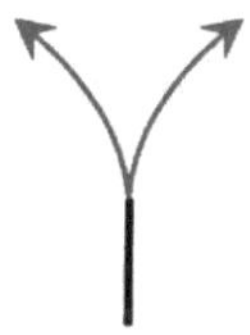 |  | 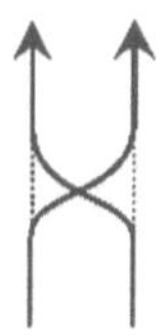 |
| | Main line | Main line divergence | Main line convergence | Main line weave |
| | |  |  | — |

When there are two or more ramps connected to the main line, it belongs to a combination of multiple traffic flows. According to geometric principles, there are 20 combinations of multiple traffic flows. However, only 4 of them are the safest in light of design consistency, and the connection methods recommended in Table 7 - 11 should be used. Continuous exits are likely to induce drivers' hesitation in judgement of exits, and it is easy to cause wrong-way operation. Continuous entrances generate a significant impact on the downstream through traffic flows. When conditions are limited, continuous entrances and exits can be used, but the minimum distance between accesses shall conform to the provisions.

**Table 7 - 11 Connection methods of continuous divergence and convergence**

| Connection method | Continuous divergence | Continuous convergence | Divergence and convergence | Convergence and divergence |
|---|---|---|---|---|
| Recommended connection methods | 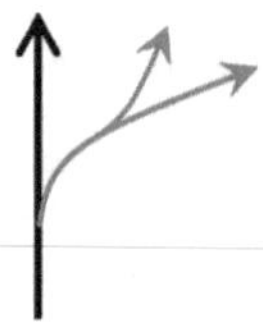 |  |  |  |
| Connection methods under restriction |  |  | — | — |

**3) Control elements of ramp design**

Table 7 - 7 lists the main control elements considered in interchange design. For ramp design, the main control elements include the design speed, design traffic volume, level of service and sight distance. The design vehicles of the ramp are generally consistent with that of the main line, and are not elaborated herein.

(1) Design speed

The design speed of the ramp shall be in harmony with the interchange configuration, the form of the ramp, and the traffic volume on the ramp. Restricted by terrain and economics, the design speed of the ramp shall generally be lower than that of the cross lines. Meanwhile, the design speed differences between the main line, the intersected line and the ramp shall not be too large to prevent excessive acceleration and deceleration from causing great adverse impacts on driving comfort and safety. The design speeds of interchange ramps for highways and urban roads are shown in Table 7 - 12.

**Table 7 - 12 Design speed of interchange ramps for highways and urban roads (km/h)**

| Types | Ramp form | Directional ramp/km • $\text{h}^{-1}$ | Semi-directional ramp/km • $\text{h}^{-1}$ | Loop ramp/km • $\text{h}^{-1}$ |
|---|---|---|---|---|
| Highway | System interchange | 80, 70, 60, 50 | 80, 70, 60, 50, 40 | 40 |
| | Service interchange | 60, 50, 40 | 60, 50, 40 | 40, 35, 30 |
| Urban road | Type A, Type B, Type C | 80, 70, 60, 50, 40, 35, 30, 25, 20 | | |

According to the ramp type, the design speed can be decided according to Table 7 - 13. The design speed shall also be able to satisfy the requirements of optimal vehicle speed, adapt to the needs of continuous deceleration or pre-acceleration of vehicles, and facilitate traffic organization of ramps. With the completion of ramp alignment design according to the design speed, the continuity of operating speed shall be verified on road sections with large changes in the alignment index. If the operating speed continuity of adjacent road sections or the consistency of design speed and operating speed fails to pass the verification, the horizontal and vertical alignments or superelevation of ramps shall be adjusted.

**Table 7 - 13 Design speed of the basic segment of ramps**

| Types | | Directional ramp | | Semi-directional ramp | | Loop ramp | |
|---|---|---|---|---|---|---|---|
| | | Standard | Non-standard | Inward turning | Outward turning | Standard | Non-standard |
| System interchange | Design speed/ $km \cdot h^{-1}$ | 60 - 80 | 50 - 60 | 60 - 80 | 40 - 60 | 40 | 40 |
| | Ramp alignment | 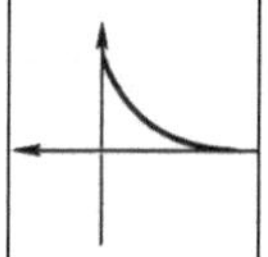 |  | 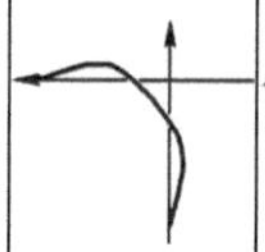 |  |  | 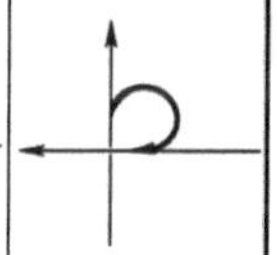 |
| Service interchange | Design speed/ $km \cdot h^{-1}$ | 40 - 60 | 30 - 40 | — | 40 - 60 | 30 - 40 | 30 - 40 |
| | Ramp alignment |  |  | — |  | 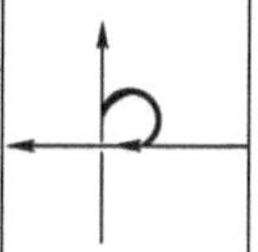 | 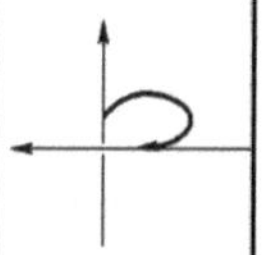 |

*Note*: This table refers to the **Guidelines (2014)**.

(2) Sight distance

Sight distance inspections of ramp design mainly include the stopping sight distance of the basic segment of ramps, the decision sight distance in front of the divergence nose, and the sight triangle in front of the convergence terminal. The parameters used in the sight distance inspections shall be determined according to the types of vehicles and sight distances, and are generally calculated according to the following parameters:

- Stopping sight distance: eye height 1.2 m, object height 0.1 m

- Stopping sight distance of trucks: eye height 2.0 m, object height 0.1 m
- Decision sight distance: eye height 1.2 m, object height 0 m

The one-way ramp and two-way separated ramp shall meet the requirements of stopping sight distance, which shall not be less than that specified in Table 7 - 14. When there is a possibility of vehicles meeting on the two-way non-separated two-lane ramp, the meeting sight distance shall be used which shall not be less than twice the stopping sight distance. Stopping sight distance can be adopted in restricted sections with lane separation measures. In sections where the traffic composition is dominated by large vehicles or there is an impact on the sight distance of trucks, the intersecting roads and ramps at the interchange shall be inspected according to the sight distance of trucks.

**Table 7 - 14 Stopping sight distance of ramps**

| Design speed of ramp/km • $h^{-1}$ | | 80 | 70 | 60 | 50 | 40 | 35 | 30 | 25 | 20 |
|---|---|---|---|---|---|---|---|---|---|---|
| Highway/m | General area | 110 | 95 | 75 | 65 | 40 | 35 | 30 | — | — |
| | Snowy and icy area | 135 | 120 | 100 | 70 | 45 | 35 | 30 | — | — |
| Urban road/m | | 110 | 90 | 70 | 55 | 40 | 35 | 30 | 25 | 20 |

*Note*: This table refers to the **Guidelines (2014)**.

The road section where the driver is likely to misjudge information or make wrong decisions before the divergence terminal shall meet the requirements of decision sight distance. When conditions permit, the decision sight distance of the highway interchange should adopt the value specified in Table 7 - 15, and the high limit value in the table should be adopted for road sections with complex information. When it is restricted, the decision sight distance shall not be less than 1.25 times the stopping sight distance of the main line.

**Table 7 - 15 Decision sight distance of ramps at the interchange**

| Design speed of the main line/km • $h^{-1}$ | 120 | 100 | 80 | 60 |
|---|---|---|---|---|
| Decision sight distance/m | 350 - 460 | 290 - 380 | 230 - 300 | 170 - 240 |

*Note*: This table partially refers to the **Guidelines (2014)**.

In front of the convergence terminal, a sight triangle is established by 100 m from the main line to the convergence nose and 60 m from the ramp to the convergence nose (Figure 7 - 15), which shall be clear of obstacles that hinder the inter-visibility between the main line and ramp.

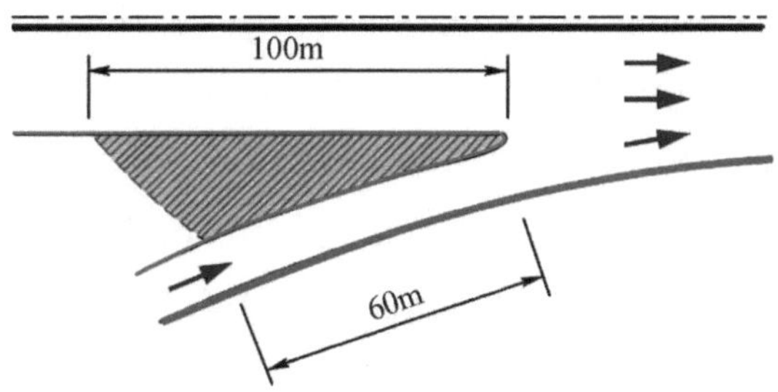


**Figure 7 - 15 Sight triangle before the convergence nose**

(3) Design traffic volume

The design traffic volume is the basis for determining the ramp type, design speed, lane number, geometry, incomplete or complete interchanges, and for whether to build in stages. In interchange design, the design hourly traffic volume is adopted which conforms to the following provisions: the design hourly traffic volume uses the 30th hourly traffic volume of the year, or the most economical and reasonable hourly traffic volume between the 20th and 40th hours according to the function of interchanges and the variation characteristics of local hourly traffic volume. The design hourly traffic volume is converted according to formula (7 - 1). Distribution diagram of node traffic volumes in the interchange shall also be provided to clarify the design hourly traffic volume of each direction and each road section.

$$DDHV = AADT \cdot K \cdot D \tag{7-1}$$

where: $DDHV$—design hourly volume, pcu/h;

$AADT$—annual average daily traffic, pcu/d;

$K$—coefficient of design hourly traffic volume, which is determined according to the functions of the intersecting road, traffic volume, regional climate and topography, and other conditions;

$D$—uneven coefficient of traffic volume direction, which is determined according to the observation data of local traffic volume. In case of a lack of data, it can be selected in the range of 50%- 60%.

(4) Traffic capacity and level of service

The traffic capacity of the ramp depends on the capacity of the basic segment of the ramp and the capacity of the exit and entrance, of which the smaller of the three is adopted. Generally, the capacity of the exit and entrance is very small compared with the capacity of the basic segment of the ramp. Therefore, the traffic capacity of the ramp is mainly controlled by the capacity of the exit or entrance, and is also affected by the capacity of the main line, lane number, and design traffic volume. When the design level of service adopts the fourth level, the design capacities of one-lane and two-lane in the basic segment of the ramp may refer to Table 7 - 16.

**Table 7 - 16 Design traffic capacity of the basic segment of the ramp**

| Design speed of the ramp/km • $h^{-1}$ | | 80 | 70 | 60 | 50 | 40 | 35 | 30 |
|---|---|---|---|---|---|---|---|---|
| Design traffic capacity/pcu • $h^{-1}$ | One-lane | 1,500 | 1,400 | 1,300 | 1,200 | 1,000 | 900 | 800 |
| | Two-lane | 2,900 | 2,600 | 2,300 | 2,000 | 1,700 | 1,500 | 1,300 |

*Note*: This table refers to the **Guidelines (2014)**.

**4) Cross section design of the ramp**

Cross section composition and lane number of the ramp shall be determined prior to the horizontal and vertical alignment design of the ramp. Cross section composition shall meet the needs of vehicle operation, safety, management, maintenance, and emergency rescue, and shall fully consider factors such as the interchange configuration and environmental impact. The lane number of the ramp and the standard cross section shall be determined according to the design hourly traffic volume, traffic composition, design speed, design level of service, and whether passing needs are met.

(1) Cross section composition of the ramp

Cross section of the ramp is generally composed of the traveled lane, marginal strip, paved shoulder, and soil subgrade. The two-lane separated ramp shall also include a medial separator. When non-motor vehicles are incorporated into the interchange ramps of urban roads, the side separator and non-motor vehicle lane are mandatory. The width of each part shall conform to the following provisions:

- Lane width should be 3. 5 m when the ramp design speed is less than 70 km/h; lane width shall be 3. 75 m otherwise. Specific lane width shall be comprehensively determined in consideration of the ramp design speed, vehicle type, and restriction.
- Marginal strip width shall be 0. 50 m.
- When an emergency parking strip is provided on a one-way two-lane ramp,

the width of the left paved shoulder is 0.75 m; for the other ramps, it shall be 1.00 m.

- The width of the right paved shoulder for emergency parking should be 3.0 m; it shall not be less than 1.5 m under constraint. The width of the right paved shoulder should be 2.0 m for the two-way two-lane separated ramp under constraint, and it can be 1.0 m when emergency parking is not a necessity.
- Width of the earth shoulder should be 0.75 m, and it can be 0.5 m under constraint.

Width of each element of the cross section of the interchange ramp is shown in Table 7-17.

**Table 7-17 Width of each element of the cross section of the interchange ramp**

| Composition | Lane/m | Marginal strip/m | Earth shoulder/m | Paved shoulder (including marginal strip)/m | | Medial separator/m |
|---|---|---|---|---|---|---|
| | | | | left | right | |
| Highway | 3.5(3.75) | 0.5 | 0.75(0.5) | 1.0(0.75) | 3.0(1.0, 1.5,2.0) | ≥1.0 |
| Urban road | 3.5(3.75, 3.25,3.00) | 0.5(0.25) | 0.75(0.5) | 1.0(0.75) | 2.5(1.0, 0.75,0.5) | ≥1.5 |

(2) Type of the ramp cross section

Ramp cross sections mainly include four types, as shown in Table 7-18.

**Table 7-18 Type of the ramp cross section**

| Cross section type | Diagram/cm |
|---|---|
| Type Ⅰ—one way one-lane ramp | 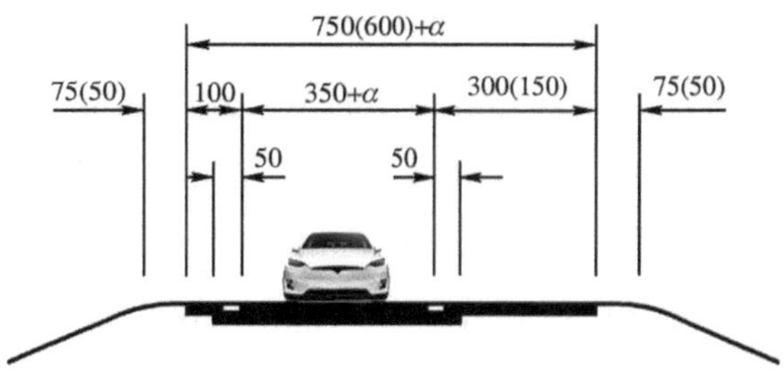 |

**Continued Table 7 - 18**

| Cross section type | Diagram/cm |
|---|---|
| Type Ⅱ —one-way two-lane ramp without the emergency parking strip | 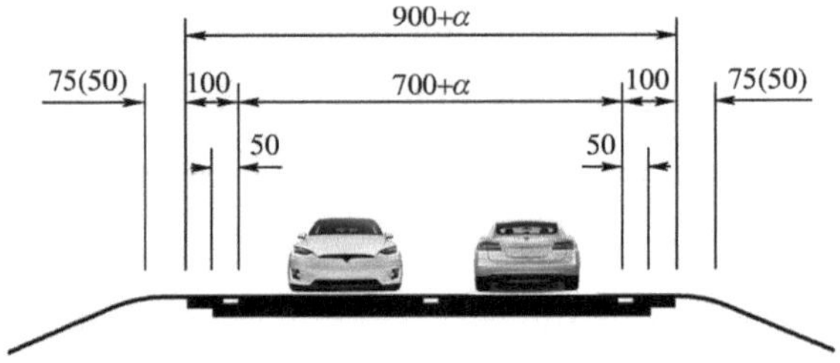  |
| Type Ⅲ —one-way two-lane ramp with the emergency parking strip | 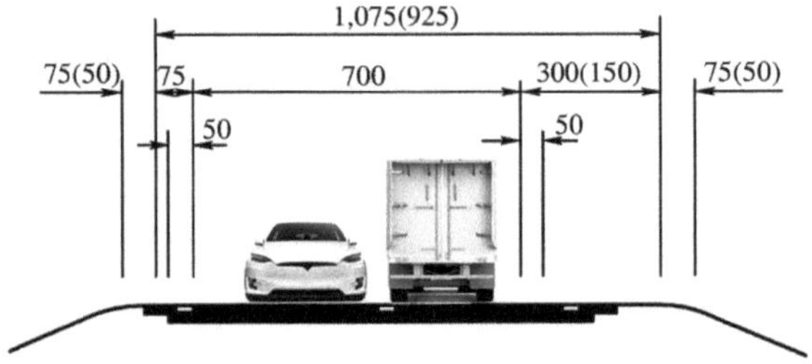  |
| Type Ⅳ —two-way two-lane separated ramp | 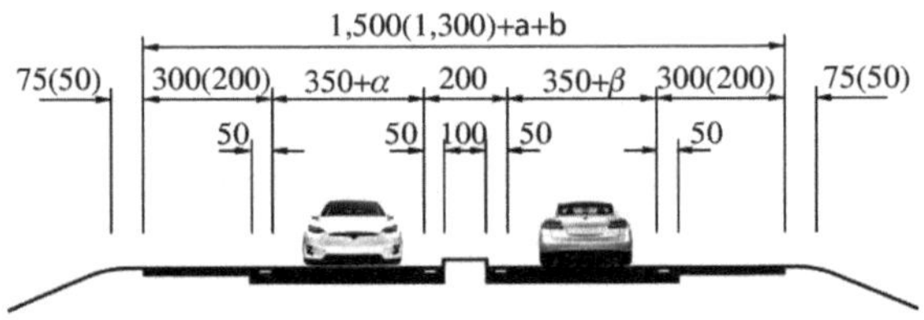  |

*Note*: α and β are the widened values.

The cross section of the one-way ramp and the number of the speed-change lanes shall be selected according to the design speed, design hourly volume and ramp length according to Table 7 - 19. When the design hourly volume of the ramp is less than that of one lanes, but the ramp has two lane, the speed-change lane should adopt one lane. When the design hourly volume of the ramp is greater than or equal to that of one lane, the speed-change lane should adopt two lanes. When the design hourly volume of the main line upstream of the deceleration lane or downstream of the acceleration lane approaches the design capacity of the main line, the traffic capacities of the divergence-convergence zone shall be checked. When it cannot meet the design traffic capacity, the length of the speed-change lane or the lane number shall be increased. If necessary, the cross section of the ramp may be modified.

**Table 7 - 19 Selection of the cross section of the one-way ramp and the number of the speed-change lanes**

| Design speed of the ramp/ km • h$^{-1}$ | 80 | 70 | 60 | 50 | 40 | 35 | 30 | Ramp length/ m | Cross section of the ramp | Number of the speed-change lane |
|---|---|---|---|---|---|---|---|---|---|---|
| Design hourly volume (DDHV) of the ramp/ pcu • h$^{-1}$ | DDHV <400 | DDHV <400 | DDHV <400 | DDHV <400 | DDHV <400 | DDHV <400 | DDHV <400 | ≤500 | Ⅰ | One |
| | | | | | | | | >500 | Ⅱ | One |
| | 400≤ DDHV <1500 | 400≤ DDHV <1500 | 400≤ DDHV <1500 | 400≤ DDHV <1500 | 400≤ DDHV <1500 | 400≤ DDHV <1500 | 400≤ DDHV <1500 | ≤350 | Ⅰ | One |
| | | | | | | | | >350 | Ⅱ | One |
| | 400≤ DDHV <1500 | 400≤ DDHV <1500 | 400≤ DDHV <1500 | 400≤ DDHV <1500 | 400≤ DDHV <1500 | 400≤ DDHV <1500 | 400≤ DDHV <1500 | Unlimited | Ⅱ | Two |
| | 400≤ DDHV <1500 | 400≤ DDHV <1500 | 400≤ DDHV <1500 | 400≤ DDHV <1500 | 400≤ DDHV <1500 | 400≤ DDHV <1500 | 400≤ DDHV <1500 | Unlimited | Ⅲ | Two |

*Note*: This table refers to the **Guidelines (2014)**.

(3) Superelevation and transition of the ramp

Similar to road alignment design, when the ramp radius is smaller than the minimum radius without a superelevation, a superelevation shall be introduced on the circular curve section, and a superelevation transition section shall be introduced between road sections with different cross slopes.

① Superelevation rate

It is not necessary to set a superelevation when the ramp is on the tangent or the radius of the circular curve is greater than the minimum radius without a superelevation. For sections without a superelevation, a one-way cross slope is recommended for one-way ramps; either a two-way or one-way cross slope can be used for two-way ramps according to the length and alignment of ramps.

The maximum superelevation of the ramp should be 6%, and it shall not exceed 6% in snowy-icy areas. In non-snowy-icy areas, when the traffic composition is dominated by passenger cars, the maximum superelevation of the ramp can be increased as appropriate but shall not exceed 8%. The earth shoulder on both sides of the ramp shall be provided with the cross slope inclined to the outside of the subgrade. On both sides of the road section without a superelevation and the road section with a superelevation, the slope of the earth shoulder is recommended to be 3%. On the inside of the curve where it is superelevated, when the superelevated slope is greater than 3%, the cross slope of the earth shoulder should be the same as the superelevation rate; when the superelevation rate is less than 3%, the cross slope of the earth shoulder shall be 3%.

② Superelevation transition

Superelevation transition shall be provided between the tangent and superelevated circular curve or between two circular curves of different superelevation rates on the ramp. The length of superelevation transition shall be determined according to factors such as the design speed of the ramp, cross section type, position of rotation axis, and superelevation gradient. The superelevation transition of the ramp should be placed on the spiral, and the calculation formula for the superelevation transition length is the same as that of the cross lines, which can refer to Chapter Ⅳ. There are three types of superelevation transition, as shown in Table 7 - 20. The Superelevation gradient shall not be greater than the values specified in Table 7 - 21. The Superelevation transition of the ramp shall be smooth and gentle, free of distortion and abrupt changes.

**Table 7 - 20 Superelevation transition methods**

| Transition method | Description | Diagram |
|---|---|---|
| Revolving about the centerline of traveled lanes | Pavement revolves about the center-line of the traveled lanes until the superelevation rate is attained | 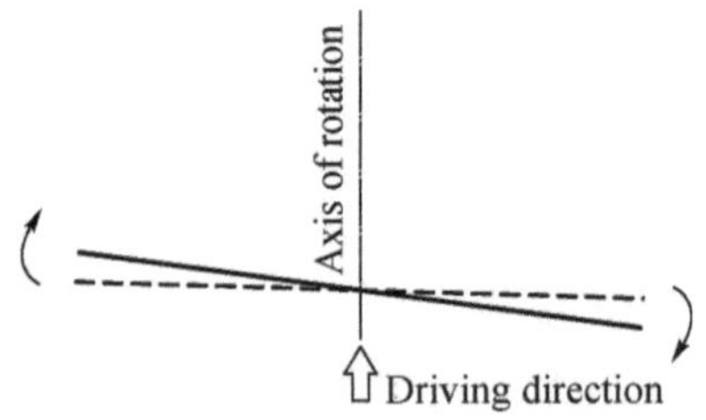 |

**Continued Table 7 - 20**

| Transition method | Description | Diagram |
|---|---|---|
| Revolving about the centerline of traveled lanes | When there is a medial separator, the rotation axes are the centerlines of the traveled lanes on both ways. The roadway on both ways revolves about the rotation axes until the superelevation rate is attained respectively, making two independent one-way superelevations | 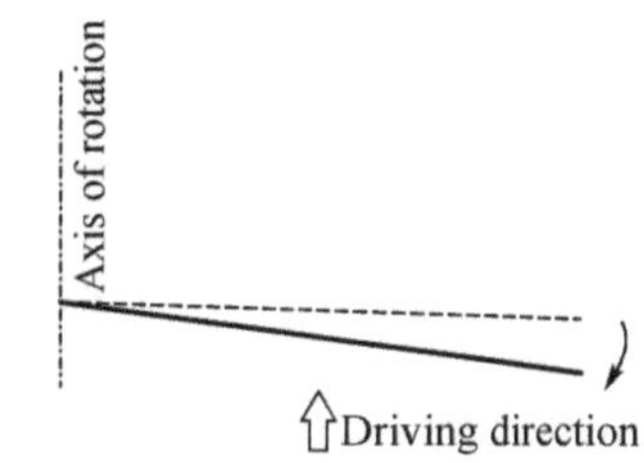  |
| Revolving about the outside-edge of the left marginal strip | With the outside edge line of the left marginal strip as the rotation axis, the roadway revolves about the rotation axis until the superelevation rate is attained. When there is a medial separator, the rotation axes are the two outside edge lines of the medial separator, and the roadway on both ways revolves about the rotation axes respectively, viz. independent one-way superelevation. The medial separator remains horizontal | 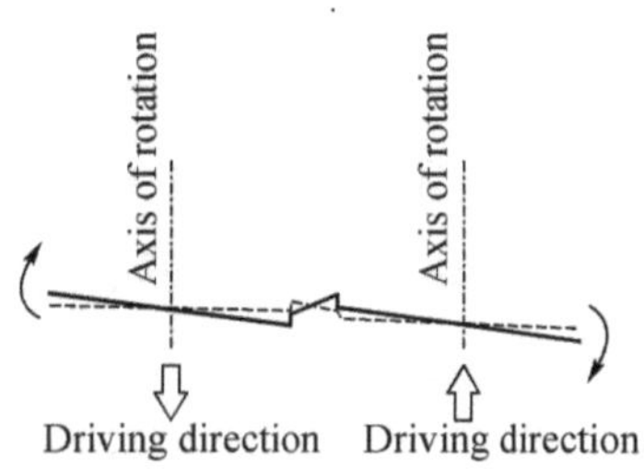  |
| Revolving about the centerline of the median | When the median of the two-way ramp is paved, the centerline of the median can be used as the rotation axis, about which the roadway on both ways revolves until the superelevation rate is attained | 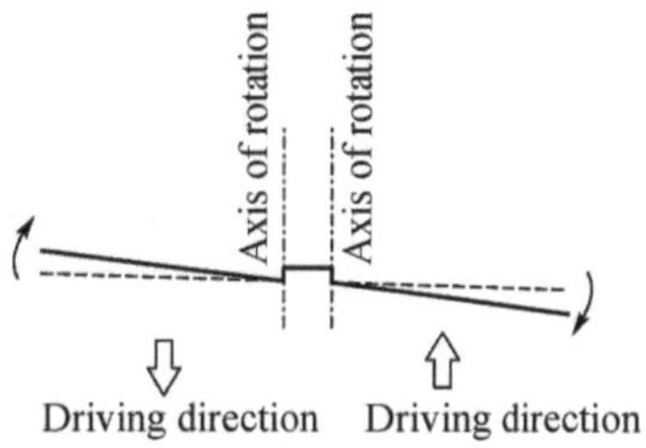  |

**Table 7 - 21 Maximum superelevation gradient of the ramp**

| Rotation axis | About the centerline | | About outside edge lines of the left marginal strip | |
|---|---|---|---|---|
| Cross section of the ramp | One-way one-lane ramp, two-way two-lane separated ramp | One-way two-lane ramp, two-way two-lane unseparated ramp | One-way one-lane ramp, two-way two-lane separated ramp | One-way two-lane ramp, two-way two-lane unseparated ramp |

Continued Table 7 - 21

| Rotation axis | | About the centerline | | About outside edge lines of the left marginal strip | |
|---|---|---|---|---|---|
| Design speed of the ramp/ km • $h^{-1}$ | 80<br>80 | 1/250 | 1/200<br>1/200 | 1/200<br>1/200 | 1/150 |
| | 70 | 1/240 | 1/190 | 1/175 | 1/140 |
| | 60 | 1/225 | 1/175 | 1/150 | 1/125 |
| | 50 | 1/200 | 1/150 | 1/125 | 1/100 |
| | ≤40 | 1/150 | 1/150 | 1/100 | 1/100 |

(4) Widening and transition of the ramp

Similar to the general road design, when the width of the circular curve section on the ramp is unable to meet the field traffic demand, the roadway of the circular curve section shall be widened, and a transition section shall be provided between road sections with different widths. The widening value of the circular curve on the ramp shall be determined according to the radius of the circular curve, ramp cross section, and traffic conditions (Table 7 - 22). Two-way two-lane separated ramps (type Ⅳ) shall be widened separately according to the widening value corresponding to the radius of the circular curve of respective lanes. Widening is generally located on the inside of the circular curve, and widening transition can be carried out in the same way as that of the cross lines. Two-way separated ramps should be widened inside and outside respectively.

**Table 7 - 22 Traffic conditions on the ramp**

| Ramp cross section | Traffic condition | |
|---|---|---|
| | General condition | Special condition |
| One-way one-lane ramp (Type Ⅰ) | When a truck stops on the shoulder, an articulated truck may pass slowly | When a passenger car stops on the shoulder, the articulated truck may pass slowly |
| Two-way two-lane separated ramp (Type Ⅳ) | | |
| One-way two-lane or two-way two-lane unseparated ramp (Type Ⅱ) | Two articulated trucks may pass in parallel at a slow speed or in opposing directions | Articulated trucks and trucks may pass at a slow speed in parallel or in opposing directions |

Adoption of widening value shall also conform to the following provisions:

• When the type Ⅰ ramp and type Ⅳ ramp are connected on curve sections of equal radii, the one-way widening value of the type Ⅳ ramp shall be used.

- When the traffic condition or ramp width changes, the widening value shall be recalculated.
- When the shoulder width of the type Ⅲ ramp is 3.00 m and the radius of the circular curve is greater than 32 m, no widening is required.

**5) Horizontal and vertical alignment design of the ramp**

Horizontal and vertical alignments of the ramp shall be properly combined and coordinated to ensure visual continuity and driving safety. At the same time, horizontal and vertical alignment designs of the ramp shall consider the variation of operating speed of the whole section and adjacent sections, and horizontal and vertical alignment indexes shall adapt to the operating speed and its variation.

(1) Horizontal alignment design

The horizontal alignment index of the interchange ramp shall be determined according to the interchange configuration, ramp design speed, traffic volume, terrain, land conditions, construction cost, etc., and ensure that vehicles can move continuously and safely to achieve engineering and operating economy.

① Radius of the ramp circular curve

The radius of the ramp circular curve depends on the design speed of ramps, and shall consider economy, safety, and comfort. The circular curve of the highway interchange ramp shall not be less than the minimum radius specified in Table 7 - 23. Generally, a radius larger than the general value shall be selected, and the limit value can only be used when restricted by terrain or other special circumstances. Limit values shall not be used in snowy-icy areas.

**Table 7 - 23　Minimum radius of the circular curve of the highway interchange ramp**

| Ramp design speed/km • h$^{-1}$ | | 80 | 70 | 60 | 50 | 40 | 35 | 30 |
|---|---|---|---|---|---|---|---|---|
| Minimum radius of the circular curve/m | General value | 280 | 210 | 150 | 100 | 60 | 40 | 30 |
| | Limit value | 230 | 175 | 120 | 80 | 50 | 35 | 25 |

*Note*: This table refers to the **Guidelines (2014)**.

The minimum radius of the circular curve of the urban road interchange ramp is shown in Table 7 - 24. The minimum radius with the maximum superelevation greater than or equal to 2% should be used, and the minimum radius without the superelevation can be used when feasible.

**Table 7 - 24 Minimum radius and minimum length of the circular curve of the urban road interchange ramp**

| Ramp design speed/km • $h^{-1}$ | | 80 | 70 | 60 | 50 | 40 | 35 | 30 | 25 | 20 |
|---|---|---|---|---|---|---|---|---|---|---|
| Snowy-icy area/m | | — | — | 240 | 150 | 90 | 70 | 50 | 35 | 25 |
| General area/m | Without the superelevation | 420 | 300 | 200 | 130 | 80 | 60 | 45 | 30 | 20 |
| | Maximum superelevation 2% | 315 | 230 | 160 | 105 | 65 | 50 | 35 | 25 | 20 |
| | Maximum superelevation 4% | 280 | 205 | 145 | 95 | 60 | 45 | 35 | 25 | 15 |
| | Maximum superelevation 6% | 255 | 185 | 130 | 90 | 55 | 40 | 30 | 25 | 15 |
| Minimum length of the horizontal curve/m | | 150 | 140 | 120 | 100 | 90 | 80 | 70 | 50 | 40 |

② Transition curve of the ramp

The transition curve of the ramp uses a spiral. When a spiral is adopted at the ramp and its terminal, its parameter and length should not be less than those specified in Table 7 - 25.

**Table 7 - 25 Parameter and length of the ramp spiral**

| Ramp design speed/km • $h^{-1}$ | | 80 | 70 | 60 | 50 | 40 | 35 | 30 | 25 | 20 |
|---|---|---|---|---|---|---|---|---|---|---|
| Highway | Spiral parameter A/m | 140 | 100 | 70 | 50 | 35 | 30 | 20 | — | — |
| | Spiral length/m | 70 | 60 | 50 | 40 | 35 | 30 | 25 | — | — |
| Urban road | Spiral parameter A/m | 135 | 110 | 90 | 70 | 50 | 40 | 35 | 25 | 20 |
| | Spiral length/m | 75 | 70 | 60 | 50 | 45 | 40 | 35 | 25 | 20 |

*Note*: This table refers to the **Guidelines** (**2014**).

The spiral length shall not be less than the one required for superelevation transition, and the parameter $A$ should be less than or equal to 1.5 times the radius of the connected circular curve. The parameters of the two spirals between reverse curves should be equal or approximating. When the parameter difference is large, the ratio of the two parameters should not be greater than 1.5. For compound curves, the ratio of the radii should not be greater than 1.5; otherwise, a spiral shall be used.

③ Minimum radius of the ramp at the divergence nose

At the divergence nose of the exit ramp, because the running speed is higher after separation from the cross lines, it shall have a larger radius, and make the subsequent curvature change adapt to the change of running speed. The minimum radius at the divergence nose of the highway interchange ramp is specified in Table 7 - 26.

**Table 7 - 26 Minimum radius at the divergence nose of the highway interchange ramp**

| Design speed of the main line/km • $h^{-1}$ | | 120 | | 100 | 80 | 60 |
|---|---|---|---|---|---|---|
| Design speed at the divergence nose/km • $h^{-1}$ | | 80 | 70 | 65 | 60 | 55 |
| Minimum radius/m | General value | 450 | 350 | 300 | 250 | 200 |
| | Limit value | 400 | 300 | 250 | 200 | 150 |

*Note*: The table refers to the **Guidelines** (**2014**).

④ Key points of horizontal alignment design of the ramp

(a) Running speed on the ramp changes gradually from high to low and then to high. In the design of the ramp horizontal alignment, the curvature change of the horizontal alignment shall adapt to the state of speed change.

(b) The ramp horizontal alignment index shall be consistent with the design speed of ramps. The radius of the circular curve shall be greater than or equal to the general minimum radius normally. Attention shall be paid to the continuity of the ramp alignment index, and the alignment with an abrupt change of the design index shall not be used as far as possible.

(c) The ramp horizontal alignment shall adapt to the traffic volume. The ramp with a large turning traffic volume has a large traffic capacity and higher requirement of running speed. Higher technical indexes shall be adopted, and the vehicle track shall be as short as possible.

(d) On the premise of satisfying the land use condition and budget requirement, the right-turning ramp and the left-turning directional or semi-directional ramps shall be of the higher horizontal alignment index.

(e) The horizontal alignment index of the exit ramp shall be higher than that of the entrance ramp.

(f) Divergence and convergence spots shall provide the sound horizontal alignment and visual condition.

(g) Ramps shall be reasonably combined to minimize the quantity of demolition and land use on the premise that ramp horizontal alignments meet the traffic conditions, site conditions, and technical indicators.

(h) Horizontal alignment design of the ramp shall be combined with the vertical alignment design. The length of the ramp plane shall meet the requirements of ramp profile design, neither making the grade too steep nor too gentle. Combination design with the vertical section and coordination with the layout of bridge span shall be considered, so that the three-dimensional alignment may induce the driver's line of sight and meet the control elevation requirement of structures.

(1) Design of the vertical alignment

① Maximum grade of the ramp

The ramp is controlled by the elevations of the upper and lower lines of the ramp. In order to overcome the height difference, save land, reduce demolition, and consider the low running speed on the ramp, the grade of ramp is generally larger than that of the cross lines. The maximum grades of highway and urban road ramps are shown in Tables 7 - 27 and 7 - 28.

**Table 7 - 27 Maximum grade of highway interchange ramps**

| Design speed of the ramp/km • $h^{-1}$ | | | 80, 70 | 60, 50 | 40, 35, 30 |
|---|---|---|---|---|---|
| Maximum grade/% | Exit ramp | Upgrade | 3 | 4 | 5 |
| | | Downgrade | 3 | 3 | 4 |
| | Entrance ramp | Upgrade | 3 | 3 | 4 |
| | | Downgrade | 3 | 4 | 5 |

*Note*: The table refers to the **Guidelines (2014)**.

The maximum grade of the ramp may increase by 1% due to topography or land constraints; the upgrade of the exit ramp and the downgrade of the entrance ramp may increase by 2% under special difficulties in non-snowy-icy areas. The resultant grade of the ramp should not be greater than 9%, and should not be greater than 7.5% in snowy-icy areas.

The maximum grade of urban road ramps shall not be greater than that specified in Table 7 - 28. If vehicles and non-motor vehicles are mixed on the ramp, it is necessary to take into account the requirements of the non-motor vehicles. The maximum grade shall conform to the provisions of non-motor vehicle lanes, which should not be greater than 3% generally.

**Table 7 - 28 Maximum grade of urban road interchange ramps**

| Design speed of the ramp/km • $h^{-1}$ | | 80 | 70 | 60 | 50 | ≤40 |
|---|---|---|---|---|---|---|
| Maximum grade/% | General area | 5 | 5.5 | 6 | 7 | 8 |
| | Snow-icy area | 4 | 4 | 4 | 4 | 4 |

② Minimum radius and minimum length of the ramp vertical curve

In the design of the ramp vertical curve, the minimum radius and the minimum length of the vertical curve shall meet the requirements at the same time, which are shown in Table 7 - 29.

**Table 7 - 29 Minimum radius and minimum length of the ramp vertical curve**

| Design speed of the ramp/km • $h^{-1}$ | | | 80 | 70 | 60 | 50 | 40 | 35 | 30 | 25 | 20 |
|---|---|---|---|---|---|---|---|---|---|---|---|
| Minimum radius of the vertical curve/m | Crest | General value | 4,500 (4,500) | 3,500 (3,000) | 2,000 (1,800) | 1,600 (1,200) | 900 (600) | 700 (450) | 500 (400) | — (250) | — (150) |
| | | Limit value | 3,000 (3,000) | 2,000 (2,000) | 1,400 (1,200) | 800 (800) | 450 (400) | 350 (300) | 250 (250) | — (150) | — (100) |
| | Sag | General value | 3,000 (2,700) | 2,000 (2,025) | 1,500 (1,500) | 1,400 (1,050) | 900 (675) | 700 (525) | 400 (375) | — (255) | — (165) |
| | | Limit value | 2,000 (1,800) | 1,500 (1,350) | 1,000 (1,000) | 700 (700) | 450 (450) | 350 (350) | 300 (250) | — (170) | — (110) |
| Minimum length/m | | General value | 100 (105) | 100 (105) | 90 (90) | 70 (75) | 60 (60) | 40 (55) | 35 (45) | 30 (40) | — (30) |
| | | Limit value | 75 (70) | 75 (70) | 60 (60) | 50 (50) | 40 (40) | 35 (35) | 30 (30) | 25 (25) | — (20) |

*Notes*: (1) The values in parentheses are the minimum radius and minimum length of the vertical curve of urban road interchange ramps.

(2) The table refers to the **Guidelines (2014)**.

③ Key points of ramp profile design

(a) For the ramp and its connection with the cross lines, the profile shall be continuous to avoid abrupt changes so that the main line and the ramp divergence point could be connected smoothly.

(b) The grade of the ramp shall conform to the provisions of the maximum grade and the minimum grade.

(c) A gentle grade is desired to ensure driving comfort and safety, and the maximum grade shall be used with caution. In particular, a gentle grade shall be adopted for the acceleration upgrade ramp and deceleration downgrade ramp.

(d) When there is a toll station on the ramp, the grade of the toll station section shall meet the requirements. The grade of the road section adjacent to the toll plaza shall be gentle, and a large downgrade shall not be close to the toll plaza.

(e) In addition to the coordination with grades at the junction of the main line and the intersected line, the ramp profile design is mainly restricted by the control elevation, such as ramps crossing the main line and the intersected line, ramps over-crossing or under-crossing the non-motor vehicle lanes and sidewalks, and mutual-crossing between ramps, etc. During the design, these control elevations shall be coordinated with each other, and the vertical profile shall be raised or lowered reasonably to meet the requirements of building height and clearance height of structures.

(f) The vertical curve with a larger radius shall be used where grades of the ramp and its terminal change to ensure sufficient stopping sight distance. Vertical curves at the diverging and converging points and their vicinity shall also meet the requirements of decision sight distance to observe the road ahead.

### 7.3.5 Connection design

The connections include the speed-change lane, auxiliary lane, collector-distributor lane, etc. The connections between the ramp and cross lines, between the cross lines, or between ramps are very important to ensure the function of interchanges. Connections shall generally meet the following requirements: smooth and safe access, consistent alignment with the cross lines, convenient entrance and exit, good visibility between the cross lines and the ramp, and they shall also meet the requirements of lane balance and lane continuity.

**1) Speed-change lane**

The speed-change lane includes the deceleration lane and acceleration lane. The additional lane required for deceleration when the vehicle enters the ramp from the cross lines is called the deceleration lane; the additional lane required for acceleration when the vehicle enters the cross lines from the ramp is called the acceleration lane.

(1) Type of speed-change lanes

Table 7 - 30 summarizes the characteristics of the two types of speed-change lanes, and Figure 7 - 16 shows the diagrams of acceleration and deceleration lanes.

**Table 7 - 30 Comparison of speed-change lanes**

| Type | Description | Advantage | Disadvantage |
|---|---|---|---|
| Parallel type | An additional lane is added in parallel outside the cross lines. The parallel speed-change lane shall have a taper to connect with the cross lines | Clear lane separation, easy to recognize when driving | Vehicle trajectory is a reverse curve, which is unfavorable for driving |
| Tapered type | It is gradually widened at a certain angle from the cross lines to form an additional lane connected with the ramp | The alignment is smooth and coincides with the traffic trajectory, which is favorable for driving | The start of the speed-change lane is not easy to identify |

When the speed-change lane has a single lane, the deceleration lane should adopt the tapered type and the acceleration lane should adopt the parallel type; when the speed-change lane has a dual-lane, the acceleration and deceleration lanes shall both adopt the tapered type. When the radius of the circular curve of the main line is smaller than the general minimum radius specified within the interchange range and it is difficult to use the tapered type, the speed-change lane outside other curves should be the parallel type.

(2) Cross section of speed-change lane

The cross section of the speed-change lane consists of the left marginal strip (shared with the lane of cross lines), traveled ways, and right shoulder (including the right marginal strip). Widths of each element are shown in Figure 7 - 17. The right shoulder is not compulsory for urban roads but the marginal strip shall retain. The width of speed-change lane should use the width of the ramp lane. The marginal strip shall be used between the speed-change lane and the through lane of the main line, with a recommended width of 0.5 m. The width of the right paved shoulder should be the wider value of the paved shoulders on ramps. Under constraint, the width of the right paved shoulder may be reduced as appropriate, but shall not be less than 1.5 m.

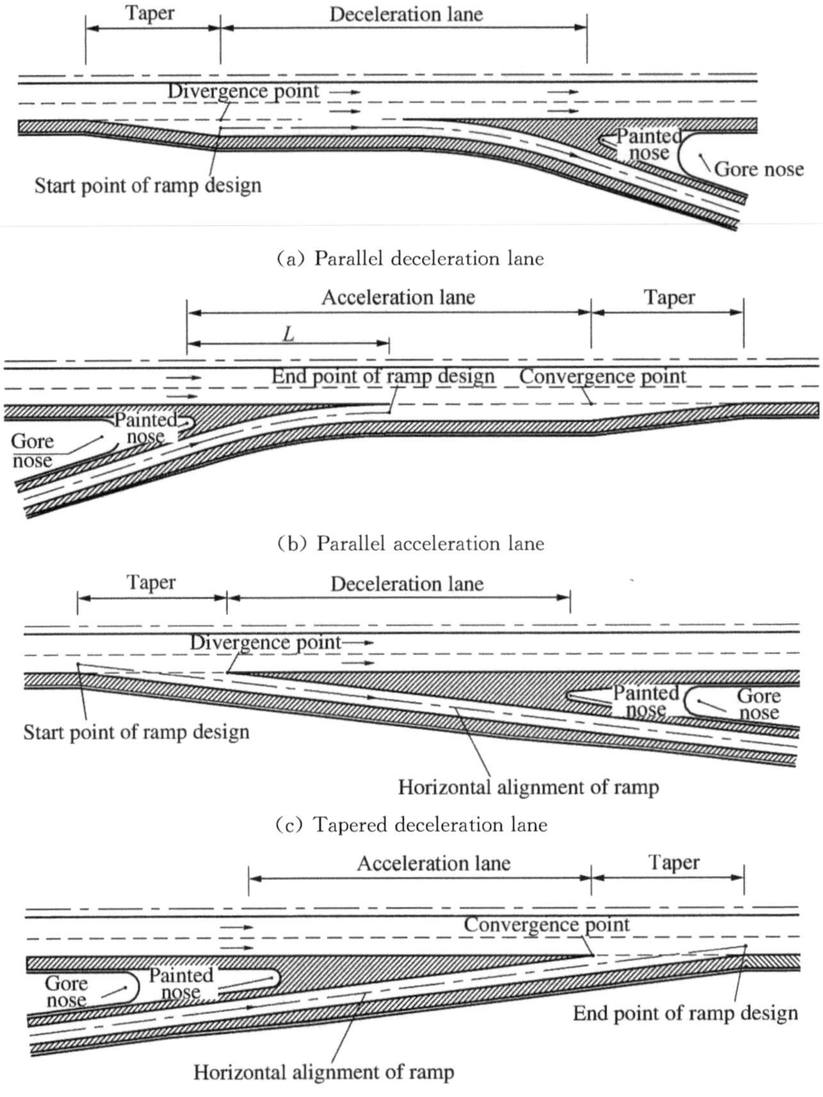


(a) Parallel deceleration lane

(b) Parallel acceleration lane

(c) Tapered deceleration lane

(d) Tapered acceleration lane

**Figure 7 - 16 Type of speed-change lanes**

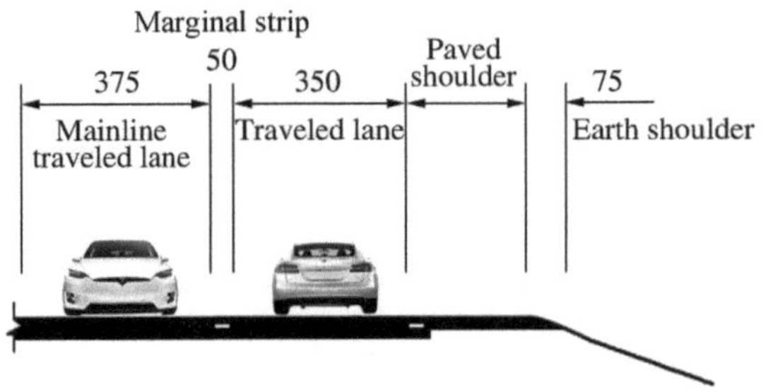


**Figure 7 - 17 Width of the speed-change lane (cm)**

(3) Length of the speed-change lane

The length of the speed-change lane is the sum of the length of the acceleration or deceleration lane and the taper length, where the length of the acceleration or deceleration lane refers to the distance between the spot where the width of the taper reaches one-lane and the divergence or convergence nose.

① Deceleration lane length

According to the research, drivers may adopt two deceleration modes during deceleration: one is to exit at an unchanged deceleration speed, that is, at an even deceleration rate; the other is to exit at two different deceleration speeds. The first stage is to lift or completely release the accelerator pedal and decelerate with the engine and other driving resistances, which is of a small deceleration rate; the second stage is that after the driver sees the alignment and width of the exit ramp, in order to exit the line and enter the ramp safely, the driver decelerates with the main brake at a larger deceleration rate. The length of each deceleration stage of both deceleration modes is calculated as:

$$L=\frac{V_1^2-V_2^2}{26a} \tag{7-2}$$

where: $L$—distance traveled by vehicles in each deceleration stage, m;

$V_1$—speed before deceleration, km/h;

$V_2$—speed after deceleration, km/h;

$a$—average acceleration (deceleration) rate of vehicles, m/s$^2$, being 2 - 3 m/s$^2$ normally.

③ Acceleration lane length

The acceleration lane is an additional lane set in the outermost lane of the main line, which shall meet the need for acceleration of the merging vehicle, and also provide the length for the vehicle to drive on the additional lane while waiting for a gap to merge into the main line (referred to as the waiting length). Therefore, the first part of acceleration lane length is the length required to accelerate the merging vehicle, calculated according to Equation (7 - 3):

$$L_1=\frac{V_1^2-V_2^2}{26a} \tag{7-3}$$

where: $L_1$—distance that the vehicle accelerates, m;

$V_1$— final speed of merging acceleration, km/h;

$V_2$—initial speed at the convergence nose, km/h;

$a$—acceleration rate of merging, m/s², being 0.8 - 1.2 m/s² normally.

The waiting length $L_2$ is related to the gap between vehicles on the outer lane of the main line (the minimum headway that can be used for vehicles to change the lane and merge into the target lane smoothly), the maximum serviced traffic volume, etc. Therefore, the length of the acceleration lane is the sum of the length of the acceleration section and the waiting length.

When the speed-change lane is located in sections with a grade greater than 2%, the length of the deceleration lane on downgrades and the length of the acceleration lane on upgrades shall be corrected according to the average grade of the main line, and the correction coefficient is shown in Table 7 - 31.

**Table 7 - 31 Correction coefficient of the speed-change lane length on grades**

| Average grade of the main line $i$/% | $i \leqslant 2$ | $2 < i \leqslant 3$ | $3 < i \leqslant 4$ | $i > 4$ |
|---|---|---|---|---|
| Correction coefficient of the deceleration lane on downgrades | 1.00 | 1.10 | 1.20 | 1.30 |
| Correction coefficient of the acceleration lane on upgrades | 1.00 | 1.20 | 1.30 | 1.40 |

In addition to the minimum length requirement, the length of the speed-change lane shall also be calculated by taking into account the operating speed of the cross lines, traffic volume, the proportion of large vehicles, etc., which shall be increased if necessary.

③ Taper length

Taper length refers to the transition length from the position where the taper width reaches one lane to the cross lines. Taper length shall ensure the stability and comfort of driving when the vehicle changes lanes.

**2) Basic number of lanes**

In the whole length of the high-speed roads or in the longer section between important nodes, the basic number of lanes must be maintained. And the number of lanes shall be balanced at the convergence or divergence points of the cross lines and ramp or between ramps, which is coordinated through the auxiliary lanes.

(1) Basic number of lanes

The basic number of lanes refers to a certain number of lanes necessary according to the requirements of traffic volume and capacity in a road or a certain section. The basic number of lanes in a fairly long section shall not change. The basic number of lanes shall not be changed in a fairly long section of roads, and the basic number of lanes does not change due to the interchange. The purpose is to prevent the possible traffic bottleneck caused by the construction of the interchange, resulting in the compromise of traffic functions of the interchange.

(2) Principle of lane balance

Traffic volumes on the cross lines at the interchange vary inevitably due to divergence and convergence. In order to adapt to the change in the traffic flow and ensure smooth traffic flow and engineering economy, the number of lanes at the convergence or divergence spots shall be balanced. The principle of lane balance is:

- The number of lanes on the cross lines after the merging of two traffic flows shall not be less than the sum of all lanes on the intersection roads before merging minus one.
- The number of lanes on the cross lines shall not be less than the sum of all lanes on the intersection roads after diverging minus one.
- The number of lanes in one direction on the cross lines shall not be reduced by more than one each time.

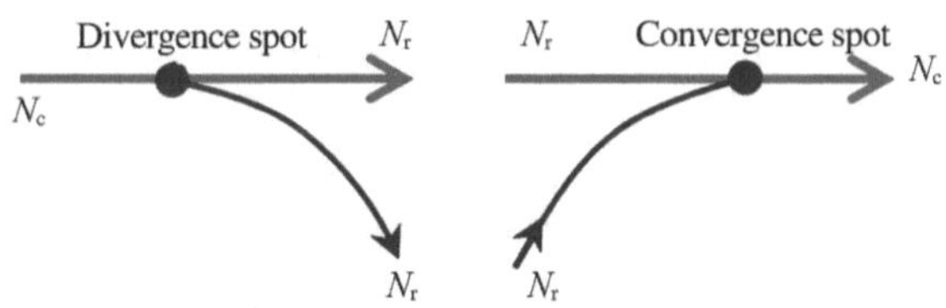


**Figure 7 - 18 Balance of lane numbers at divergence and convergence spots**

As shown in Figure 7 - 18, according to the principle of lane balance, the lane balance shall be calculated by Formula (7 - 4) at the divergence and convergence spots to check whether the number of lanes is balanced.

$$N_C \geqslant N_F + N_E - 1 \tag{7-4}$$

where: $N_C$—number of lanes on the cross line before divergence or after convergence;

$N_F$—number of lanes on the cross line after divergence or before convergence;

$N_E$—number of lanes on the ramp.

**3) Auxiliary lane**

At the divergence and convergence spots, the number of lanes shall be balanced and the basic number of lanes shall be continuous. In case of a contradiction between the two, it can be solved by adding auxiliary lanes on the cross line before the convergence spot or after the divergence spot, as shown in Figure 7 - 19.

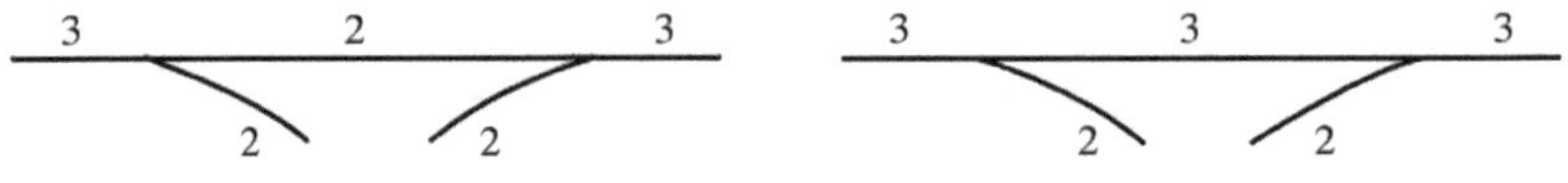


(a)

(a) Lane number balanced but basic number of lanes discontinuous

(b)

(b) Basic number of lanes continuous but lane number unbalanced

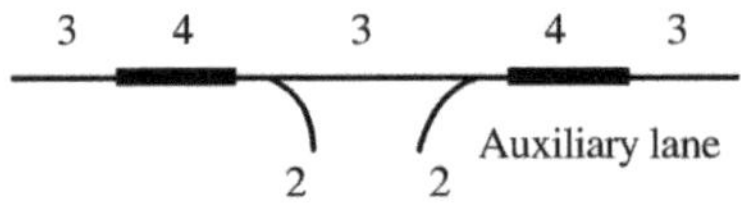


(c)

(c) Lane number balanced and basic number of lanes continuous

The numbers in the figure represent the number of lanes.

**Figure 7 - 19 Auxiliary lane**

In the context that the basic number of lanes is continuous, the general one-lane ramp may meet the requirements of lane number balance; when adopting two-lane ramps, the number of lanes is unbalanced, and auxiliary lanes need to be added. The length of the auxiliary lane is determined according to the maximum traffic volume at the design level of service, the length of the weaving section, the number of lanes, and the lane connection way at the weaving section, and meanwhile considers the length required for vehicle lane change.

The width of the auxiliary lane is the same as that of the cross line, and there is no marginal strip between the auxiliary lane and the cross line. The width of the paved shoulder on the right side of the auxiliary lane is generally the same as that of the cross line; when limited by land area or other conditions, it can be narrowed, but not less than 1.50 m.

**4) Collector-distributor lane**

Within the interchange range, the collector-distributor lane is an additional road arranged outside the main line in parallel and separated from the main line in order to isolate the weaving section and reduce the number of entrances and exits of the main line. When the specified distance between the entrances and exits of the main line could not be warranted or the weaving operation of turning traffic flow interferes with the through traffic flow, the collector-distributor lane separated from the main line shall be used to connect the entrances and exits in series (Figure 7 - 20), to avoid the interference of weaving traffic flow to the through traffic flow on the main line and improve the traffic capacity and driving safety. The minimum length of the weaving section on the collector-distributor lane is determined according to the maximum traffic volume at the design level of service, the length of the weaving section, the number of lanes, and the configuration of the weaving section.

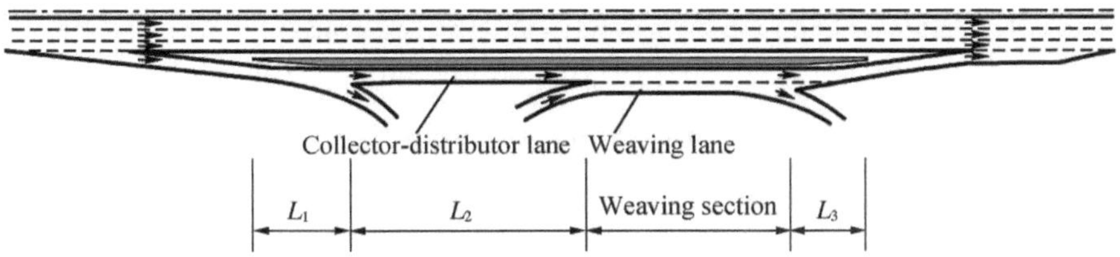


**Figure 7 - 20 Diagram of the collector-distributor lane**

(1) Layout of the collector-distributor lane

① Connection between the collector-distributor lane and the main line shall be designed according to the entrance-exit and conforms to the principle of lane balance. When the one-lane entrance-exit can meet the needs of traffic volume, the one-lane ramp or two-lane ramp of the one-lane entrance-exit may be used.

② Spacing of adjacent entrance-exit on the collector-distributor lane shall conform to the provisions specified in Table 7 - 32 and spacing of the entrance and its adjacent exit shall accommodate weaving operation.

③ Diverging and converging on the collector-distributor lane shall conform to the provisions of divergence and convergence between ramps.

④ When the traffic volume of the weaving section is large, the capacity of the weaving section shall be tested, and if necessary, the grade separation can be adopted to eliminate the weaving section or reduce weaving traffic volume, as shown in Figure 7 - 21.

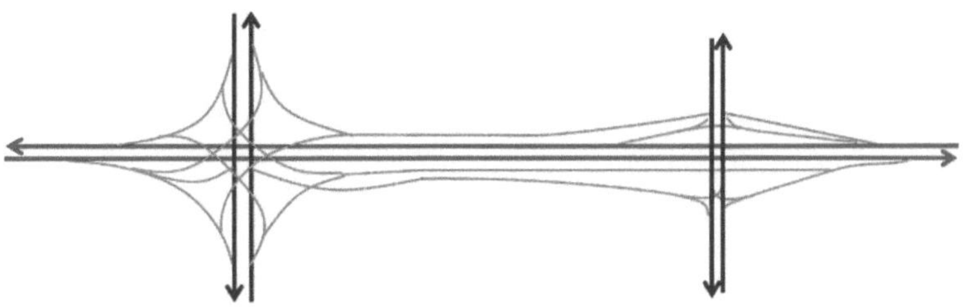

**Figure 7 21 Elimination of the weaving section or reduction of weaving traffic volume**

(2) Cross section of the collector-distributor lane

The collector-distributor lane consists of the traveled way and paved shoulder. A separator (side separator) shall be used between the collector-distributor lane and the main line, which shall be located outside the paved shoulder of the main line, and the width of the paved shoulder of the main line shall not be compressed in general (Figure 7 - 22). A separator may be convex or use a painted island. The collector-distributor lane is generally of two-lane; and when the traffic volume is small, the non-weaving section may use a single lane. The width of the paved shoulder on the right side is generally 2.5 m. When there is a large surplus of capacity for two lanes (the traffic volume is less than or slightly greater than the capacity of the single lane), the width of the paved shoulder may be reduced to 1.0 m. An earth shoulder of 0.75 m shall be provided outside the paved shoulder of the collector-distributor lane.

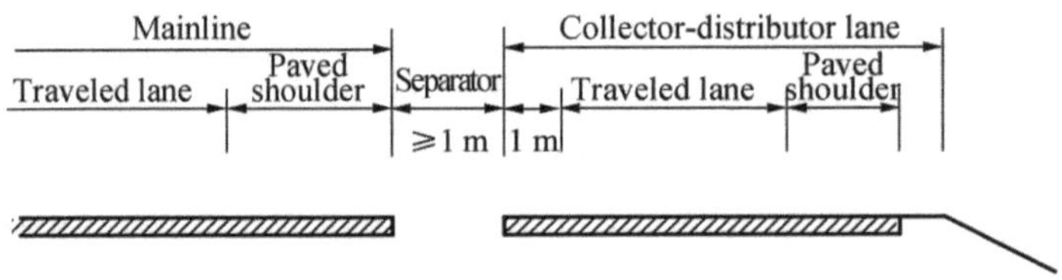


**Figure 7 - 22 Diagram of the collector-distributor lane**

**5) Entrance-exit spacing**

In order to ensure smooth traffic between entrances and exits and reduce the mutual interference between vehicles, the spacing between the noses of adjacent connections of the main line and ramps should not be too small. The minimum distance between the continuous divergence and convergence noses on the main line of freeways and the minimum distance between the adjacent divergence noses and convergence noses on the ramp shall not be less than the values listed in Table 7 - 32 through Table 7 - 34. When the spacing between entrances and exits of the main line cannot be warranted or the weaving of turning traffic interferes with the main line traffic, the collector-distributor lane separated from the main line shall be used to connect the entrances and exits in series. The types of entrances and exits are shown in Figure 7 - 23.

| Adjacent entrances or exits on the main line | Adjacent entrances or exits on the ramp |
|---|---|
| 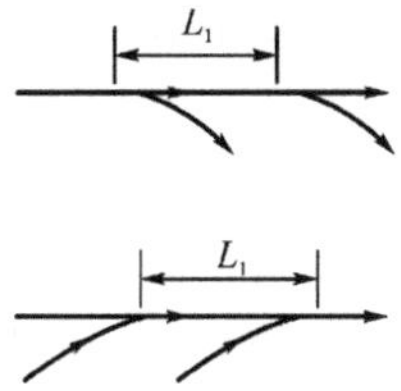  | 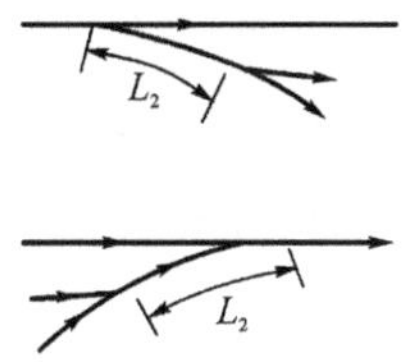  |

**Figure 7 - 23 Types of entrances and exits**

**Table 7 - 32 Minimum distance between continuous divergence and convergence noses on the main line**

| Design speed of the main line/km • h$^{-1}$ | | 120 | 100 | 80 | 60 |
|---|---|---|---|---|---|
| Minimum distance between continuous divergence and convergence noses/m | General value | 400 | 350 | 310 | 270 |
| | Minimum value | 350 | 300 | 260 | 220 |

**Table 7 - 33 Minimum distance between adjacent divergence noses on the ramp**

| Design speed of the main line/km • h$^{-1}$ | 120 | 100 | 80 | 60 |
|---|---|---|---|---|
| Minimum distance between adjacent divergence noses/m | 240 | 210 | 190 | 170 |

**Table 7 - 34 Minimum distance between adjacent convergence noses on the ramp**

| Design speed of the main line/km • h$^{-1}$ | 80 | 70 | 60 | 40 | 40 | 35 | 30 |
|---|---|---|---|---|---|---|---|
| Minimum distance between adjacent convergence noses/m | 210 | 180 | 160 | 140 | 120 | 110 | 100 |

## 7.4 Cases of interchange design

### 7.4.1 Design of Caiyuanba Interchange in Chongqing

**1) Project Overview**

Caiyuanba interchange is located in the center of the main urban area of Chongqing. There are public transportation facilities such as Chongqing Railway Station, Chongqing Coach Station, Zhongshan 3rd Road, Nanqu Road, Binjiang Road, Caiyuan Road, and other urban arterial roads in this area which attract large traffic and pedestrian flows. Within the Caiyuanba interchange range, rail-transit line 3 and interchange ramps are staggered, and the transfer between the rail-transit and vehicles has become a constraint in the design of Caiyuanba interchange. Fates of the underground shopping malls, civil defense facilities, and Bund Mall under Caiyuanba Square are also key factors to be considered in the design of this interchange.

**2) Design challenges**

Caiyuanba interchange is a complex municipal project, and the design challenges can be summarized as follows:

(1) Relationship between the existing ramps of Caiyuanba interchange and the intersecting roads is complex, with a total of six approach and three-storey. It is challenging for the ramps of the new interchange to over-cross or underpass the existing roads. And the interchange faces a large height difference and has up to five-story at most (including the rail transit), which directly affects the horizontal alignment and vertical profile design of ramps.

(2) Bund Mall has a great influence on the selection of the interchange scheme. For various reasons, it is necessary to retain Bund Mall and minimize the impact of the interchange on it. The gap between Bund Mall and the high-rise buildings on the east side is only 18 m, which makes the layout of the ramp very difficult.

(3) The underground shopping malls, civil defense facilities, and passageways under the railway station square are complicated. The above-complicated factors shall be considered in the ramp plane design and bridge span layout, which greatly restricts the ramp alignment design.

(4) The ramps underpass the main line several times and are affected by the bridge span and high-rise buildings, so the ramp alignment design is subject to great restriction.

(5) Rail-transit line 3 intersects with ramps several times, and profiles of the interchange ramp must leave sufficient spacing for the rail traffic.

**3) Scheme selection**

The interchange ramp design is based on the traffic volume survey. Traffic function analysis is based on the traffic volume survey and traffic volume prediction. Each ramp, entrance-exit, and speed-change lane shall undertake necessary traffic functions. Ramp capacity shall not be lower than the design traffic volume. The forecasted traffic volume 20 years after the opening is shown in Figure 7 - 24.

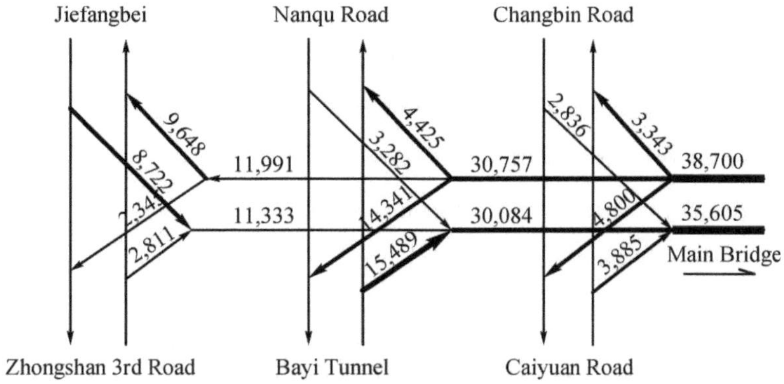


**Figure 7 - 24 Forecasted traffic volume (pcu/d)**

(1) Scheme 1

Scheme 1 is a three-storey semi-directional interchange (Figure 7 - 25). The features of scheme 1 are as follows.

① It has a simple form and clear traffic direction.

② It needs small amount of demolition to occupy the railway station square, but has a great impact on the landscape of the railway station square.

③ It is restricted by the Bayi Tunnel (two-way four-lane), and traffic capacity of the main line cannot satisfy the predicted traffic volume.

④ The existing interchange needs to be modified, which is not conducive to traffic maintenance during construction.

The biggest functional defect is that the scheme does not connect Zhongshan 3rd Road. Therefore, no further comparison will be made.

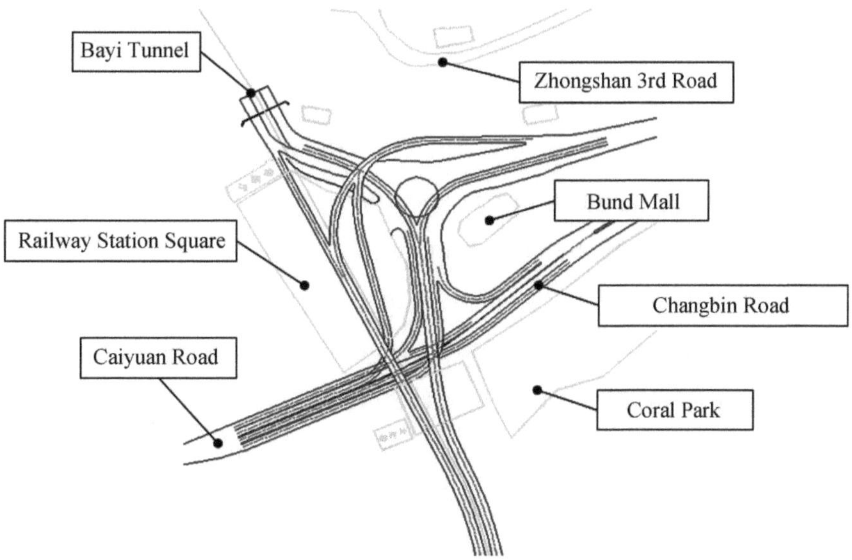


**Figure 7 - 25 Horizontal alignment of scheme 1**

(2)Scheme 2

Scheme 2 is a multi-storey semi-directional incomplete interchange (Figure 7 - 26). The interchange is compact, and the features of scheme 2 are as follows:

① It needs to remove part of the existing interchange ramps, unfavorable to traffic management during construction.

② It needs to dismantle Bund Mall with great demolition quantity.

③ The technical index of the main line is high, but the radii of other ramps are small; the utilization rate of existing ramps is low; there are very few ramps $R=30$ m; the technical index is low.

④ It can retain the railway station square and Coral Park.

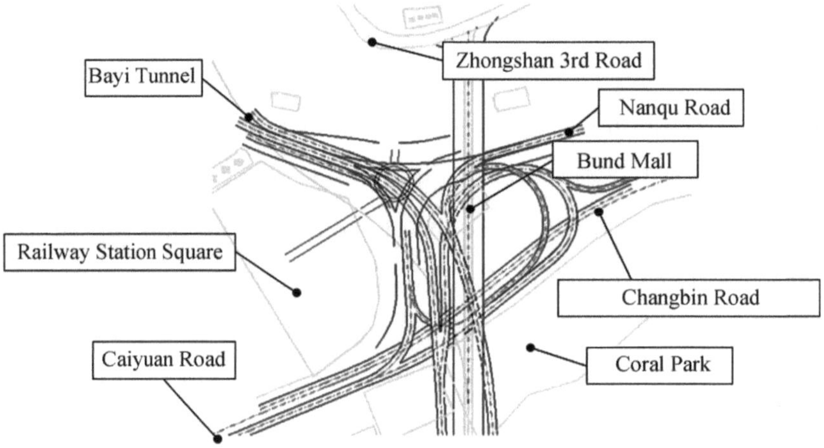


**Figure 7 - 26 Horizontal alignment of Scheme 2**

(3) Scheme 3

Scheme 3 is a multi-storey spiral incomplete interchange (Figure 7 - 27). The features of scheme 3 are as follows:

① It uses spiral ramps to solve the concern of height difference, extending the merging points of the main bridge and approach bridge, which is beneficial to the layout and elegance of the main bridge.

② It is favorable to the arrangement of rail-transit 3 in spatial.

③ It does not dismantle the existing interchange or change the original traffic organization, and is favorable to the traffic management during construction

④ It makes traffic flows in three-way merge into a loop ramp, and the weaving length is limited so that the ramp needs three lanes to accommodate weaving, which is difficult to connect with the main bridge.

⑤ The loop ramp covers a large area, the driving is uncomfortable, and the landscape is ugly.

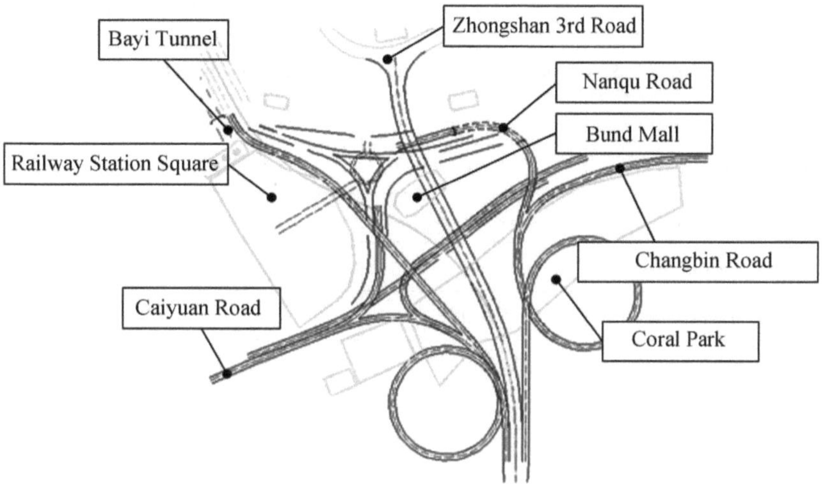


**Figure 7 - 27 Horizontal alignment of scheme 3**

(4) Scheme 4

Scheme 4 is a multi-storey semi-directional incomplete interchange (Figure 7 - 28). The interchange fully considers the utilization of existing interchange ramps and the nature of existing roads, buildings, and terrain. The features of scheme 4 are as follows:

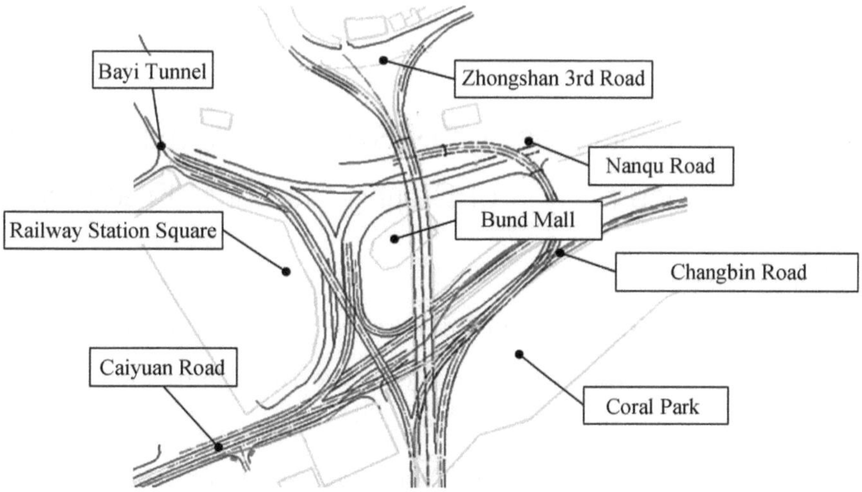


**Figure 7 - 28 Horizontal alignment of scheme 4**

① It retains Bund Mall and needs small land occupation and less demolition.

② The traffic flow direction is clear, the alignment index, minimum radius is 60 m, and the level of service is high.

③ The railway station square and Coral Park are preserved intact, which is favorable to the city landscape.

④ It connects Binjiang Road to Caiyuan Road to solve the inherited problem.

**Table 7 - 35 Comparison of Caiyuanba interchange schemes**

| Project | Scheme 2 | Scheme 3 | Scheme 4 |
|---|---|---|---|
| Ramp width/m | 9 | 9 | 9 |
| Ramp length/m | 3,907.84 | 3,748.93 | 3,005.48 |
| Minimum radius/m | 40 | 45 | 60 |
| Maximum grade/% | 5.00% | 5.00% | 5.00% |
| Bridge length/m | 3,355.37 | 3,250.68 | 2,592.57 |
| Land area/$m^2$ | 3,5307.6 | 33,740.40 | 27,049.3 |
| Quantity of demolition/$m^2$ | 51,288.2 | 8,156.07 | 8,144.56 |

Through comprehensive analysis and comparison of the above schemes, scheme 4 is relatively good, which better achieves the design objectives and satisfies the requirements, and the investment is more economical. The in-service Caiyuanba interchange adopts scheme 4 for construction. Scheme 4 reflects the characteristics of interchange design in the old city, and expounds on the problems that shall be paid special attention to in the design process. The implementation of this scheme provides an example for the interchange design and creativity in mountain cities.

### 7.4.2 Preliminary design of System Wangxing Interchange Reconstruction of Beijing-Shanghai Freeway

**1) Project Overview**

The existing Wangxing System Interchange is located in Wangxing Town, Huai'an City, and is a system interchange where the Beijing-Shanghai Freeway (G2) intersects with the Ning-Lian Freeway. It is a double-trumpet interchange, as shown in Figures 7 - 29 and 7 - 30. Wangxing West Interchange (directional type) is in service about 1.8 km to the west of Wangxing Interchange, and the traffic conversion with Beijing-Shanghai Freeway is realized through Wangxing Interchange. Auxiliary lanes are applied between the two interchanges to form a compound interchange. The present height clearance is 5.033 m where the main line of Beijing-Shanghai Freeway crosses the existing Ning-Lian Freeway, which cannot meet the expansion needs of Beijing-Shanghai Freeway and the long-term expansion needs of Ning-Lian Freeway. The vertical profile of the main line within the design area of Wangxing System Interchange is lifted. The influence range due to the lift is about 2.0 km, and the maximum lift height is 1.71 m.

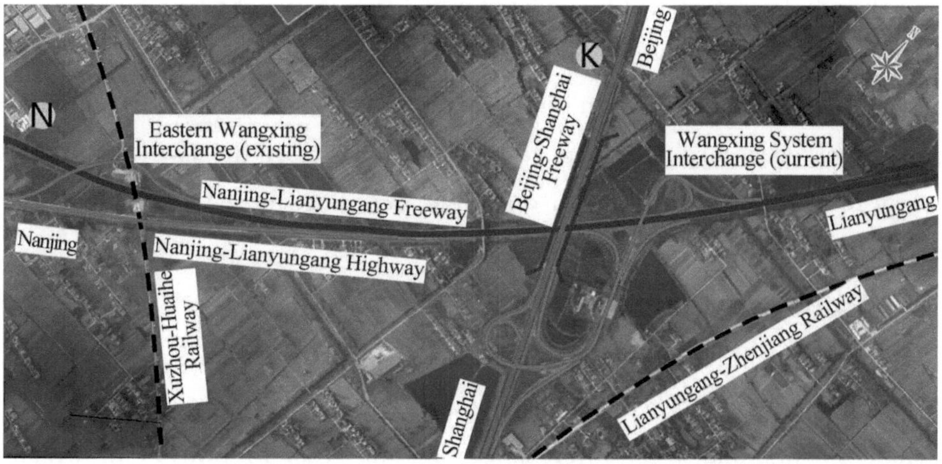


Figure 7 - 29 Sketch of Wangxing System Interchange

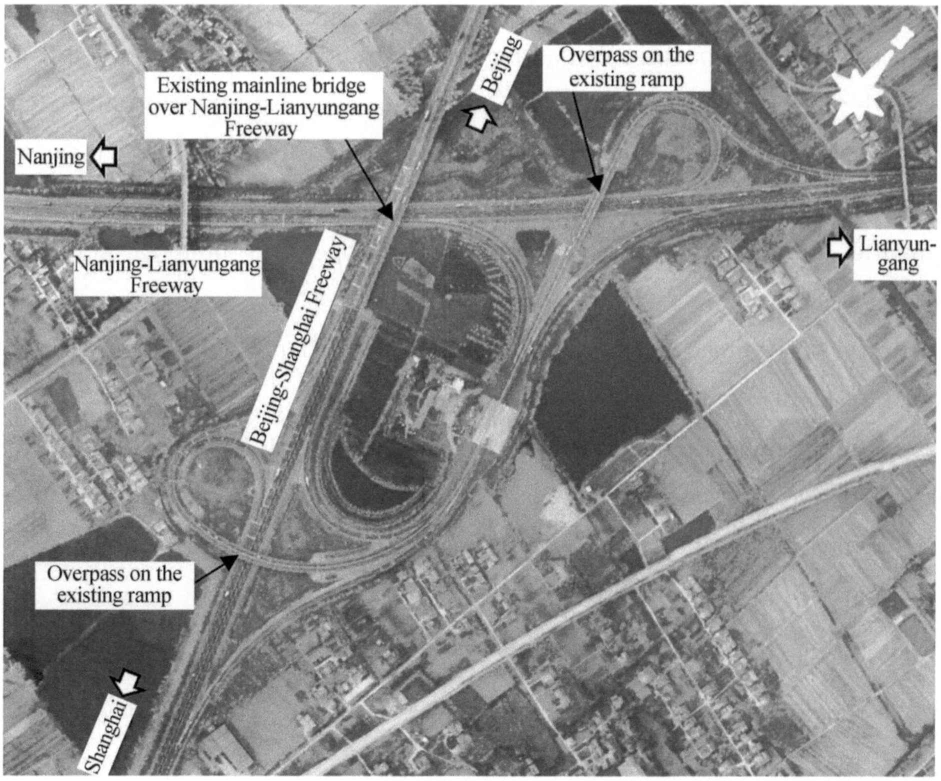


Figure 7 - 30 Present situation of Wangxing System Interchange (double-trumpet)

The existing Ning-Lian Freeway is a two-way four-lane road. The roadway width west of the merging end with the existing Ning-Lian Freeway is 28 m (medial separator width of 3 m) and the design speed is 120 km/h; the roadway width east of the merging end with the existing Ning-Lian Freeway is 24. 5 m (medial separator width of 2 m) and the design speed is 100 km/h. In the long term, the whole section will be updated into a two-way eight-lane highway with a design speed of 120 km/h and a roadway width of 42 m (medial separator width of 3 m). In the preliminary design of Wangxing System Interchange, the intersected road is reserved according to the technical standards of the two-way eight-lane freeway in the long term and is operated according to the two-way four-lane section in the short term.

**2) Traffic volume prediction**

According to the traffic volume predictions shown in Figure 7 - 31, the turning traffic volume of the interchange in the vision year of 2041 reaches 39,954 pcu/d (excluding Wangxing West Interchange of Ning-Huai Freeway), of which the mainstream direction is from Lianyungang to Shanghai, reaching 24,512 pcu/d. The existing inner-loop ramp cannot meet the needs of the traffic conversion in the vision year. The existing Wangxing System Interchange uses a double-trumpet type, which lacks security and operational efficiency. And it does not match the functional role of the transportation transfer hub for the proposed expansion project of Beijing-Shanghai Freeway and Huai-Lian Freeway. It is considered to dismantle the original double-trumpet interchange and build a new system interchange.

In order to reduce the expansion area of Ning-Lian Freeway and reduce the project scale of Wangxing System Interchange, the two ramps of Wangxing System Interchange to the direction of Lianyungang of Ning-Lian Freeway are implemented according to the two-lane section in the long term, and are operated with one-lane in the short term. According to the estimated traffic volume, it can meet the traffic demand before 2025.

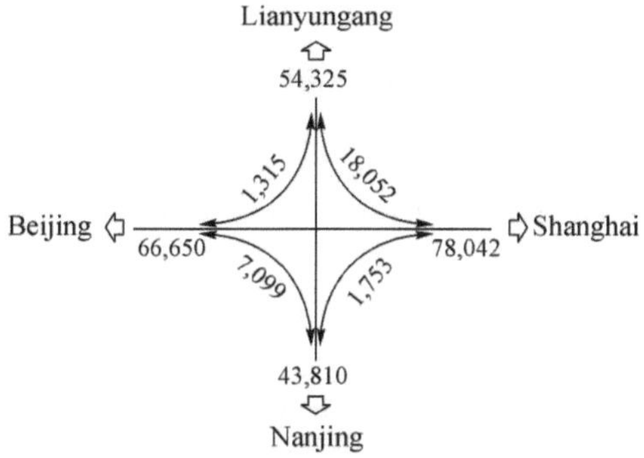

The total inbound and outbound traffic volume in 2025 is 28,219 pcu/d.

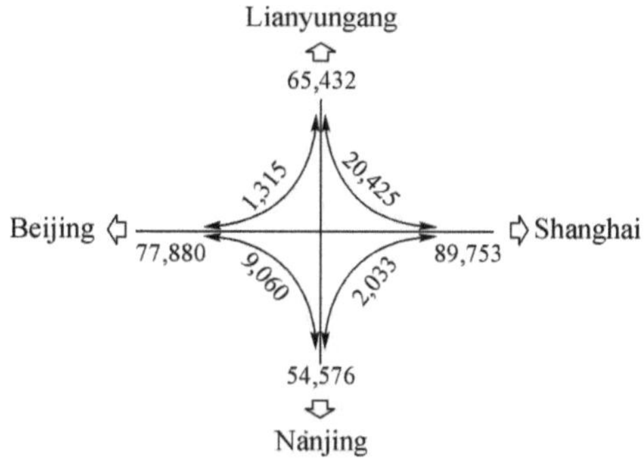

The total inbound and outbound traffic volume in 2030 is 33,043 pcu/d.

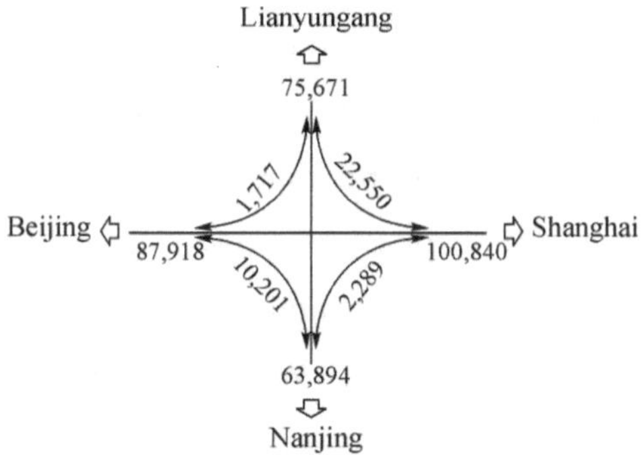

The total inbound and outbound traffic volume in 2035 is 36,757 pcu/d.

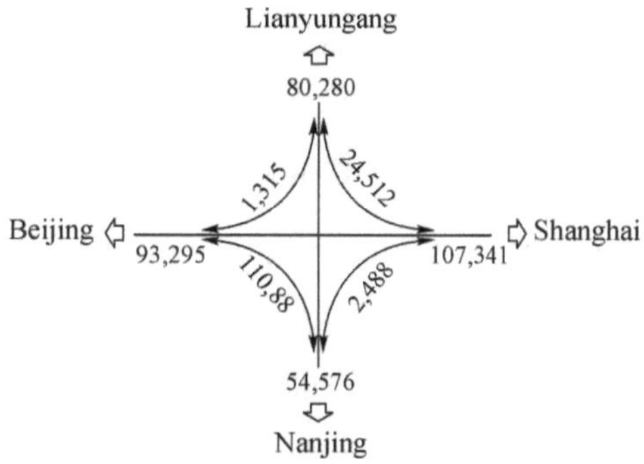

The total inbound and outbound traffic volume in 2041 is 39,954 pcu/d.

**Figure 7 - 31 Traffic volume forecast of Wangxing System Interchange**

**3) Scheme Design**

(1) Technical indicators

① Main line: The main lines of Beijing-Shanghai Freeway and Ning-Lian Freeway maintain the original design speeds.

② Ramps: The design speeds of interchange ramps are 60 km/h and 40 km/h.

Lane width: The width of the one-way one-lane ramp is 10.5 m, and the width of the one-way two-lane is 12.25 m and 10.5 m.

Ramp design line: For one-way one-lane and two-lane ramps, the centerline of the traveled lane is the design line.

③ Speed-change lane: The speed-change lane adopts both the tapered type and the parallel type. The entrance and exit of the two-lane ramp use the taper type. The length of the deceleration lane is not less than 225 m and the gradient rate is less than 1/22.5; the length of the acceleration lane is not less than 400 m and the gradient rate is less than 1/45. In order to maintain lane balance, the auxiliary lane is applied. The exit of the one-lane ramp uses the tapered type.

The length of the deceleration lane is not less than 145 m, and the gradient rate is less than 1/25; the acceleration lane uses the parallel type, and the length of the acceleration lane is not less than 230 m.

④ The grade of the vertical alignment shall be less than 4%, and is generally less than 3.5%.

(2) Control elements

① Main line design indicators: Within the design area of Wangxing System Interchange, the minimum radius of the horizontal curve of the main line is 7,000 m, the maximum grade of the vertical alignment is 1.290%, and the minimum radii of convex and sag vertical curves are 36,000 m and 16,500 m, which all meet the specification requirements. For the intersected road, the minimum radius of the horizontal curve is 3,600 m; the maximum grade of the vertical alignment is 0.328%, and the minimum radii of convex and sag vertical curves are 36,000 m and 66,000 m, which all meet the specification requirements.

② The existing Wangxing West Interchange and Wangxing System Interchange.

③ Water system and villages around the interchange.

④ Railways under construction: Xu-Huai Railway, Huai-Yang-Zhen Railway; the main line at the southern end of Wangxing System Interchange crosses Huai-Yang-Zhen Railway (under construction). The railway over-crosses the main line of Beijing-Shanghai Freeway, and railway bridge piers have been built on both sides of the Beijing-Shanghai Freeway at the intersection. The freeway sidelines on both sides are close to the bearing platform.

**4) Scheme Comparison**

(1) Scheme 1: Beijing-Shanghai and Ning-Lian Freeways are constructed with eight lanes (Ning-Lian Freeway is operated by four-lane delineation). The existing ramps are abolished and reconstructed in-situ using the partial cloverleaf system interchange. The main line bridge does not meet the clearance requirements of expansion and is demolished and reconstructed (Figure 7 - 32).

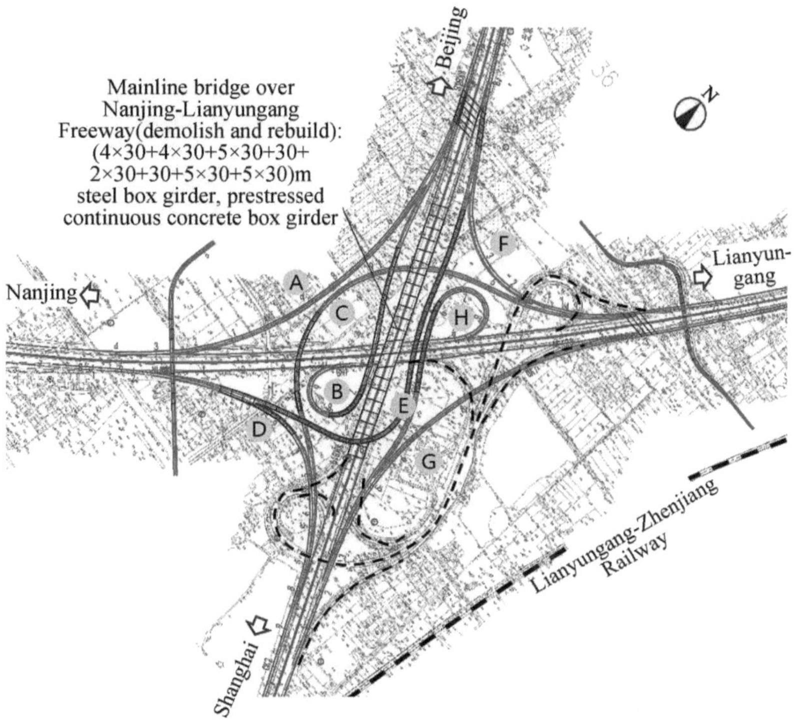


**Figure 7 - 32 Plan detail of reconstruction Scheme 1 of Wangxing System Interchange (blue line is bridge)**

For mainstreams of Shanghai ←→ Lianyungang direction and minor stream of Beijing ←→ Huai'an direction, the semi-directional ramp is adopted for a large traffic volume. For Shanghai ←→ Lianyungang direction, the one-way two-lane ramp is used with the design speed 60 km/h, the roadway width of 12.25 m, and the minimum circular curve radius of 280 m. The rest one-way two-lane ramps are with the design speed of 60 km/h and the roadway width of 10.5 m.

Ning-Lian Freeway within the design range of Wangxing System Interchange is constructed with eight lanes, thus Ramps I and J of the present Wangxing West Interchange need to be partially renovated.

The main line of the southern section within the Wangxing System interchange range crosses the railway. The nearest distance between the left edge of the main line and the edge of railway bearing platform is 0.28 m, and the right edge of the main line is 2.88 m from the edge of the bearing platform (Figure 7 - 33).

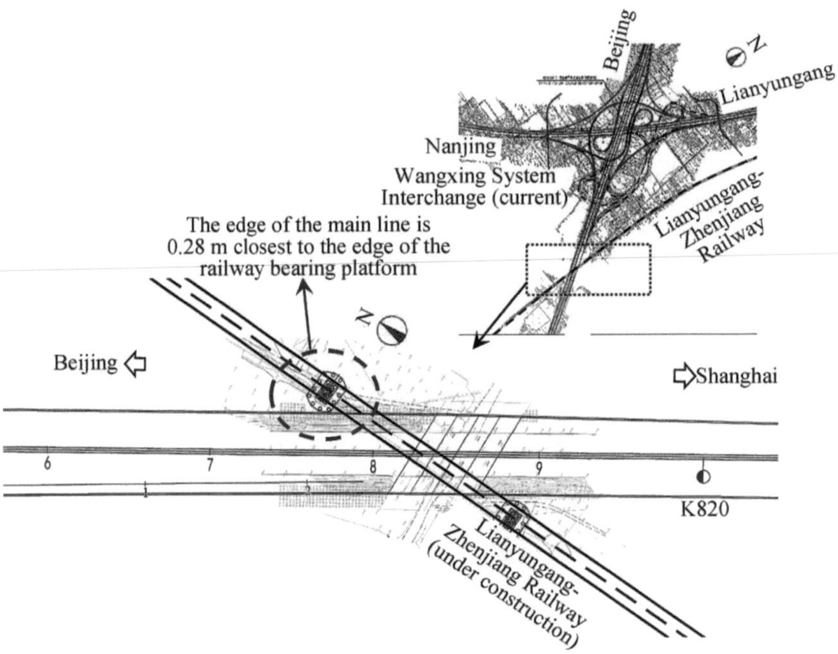


**Figure 7 - 33 Sketch of Wangxing System Interchange Scheme 1 and location of railway piers**

① Advantages of Scheme 1

• The overall layout of the interchange is well-proportioned and compact.

• Smooth alignment of the interchange ramp, good traffic adaptability and high level of service can be achieved.

• There is little conflict with the existing ramps, and the preliminary traffic organization scheme is convenient.

② Disadvantages of Scheme 1

The original interchange is less utilized and the land occupation is slightly larger. It is close to the Wangxing West Interchange, so it is necessary to build a composite interchange.

(2) Scheme 2: Beijing-Shanghai and Ning-Lian Freeways are constructed with eight lanes (Ning-Lian Freeway is operated by four-lane delineation). The existing ramps are abolished and reconstructed in-situ using the partial cloverleaf system interchange. The main line bridge does not meet the expansion clearance requirements and is demolished and reconstructed (Figure 7 - 34).

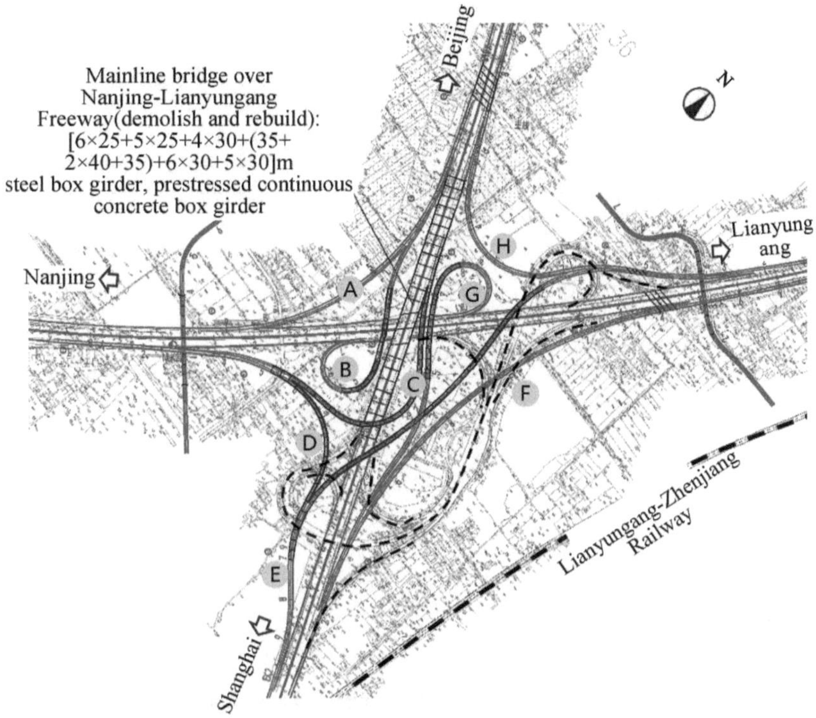


**Figure 7 - 34 Plan detail of reconstruction Scheme 2 of Wangxing System Interchange**

The ramp of Lianyungang to Shanghai direction is located in the southeast quadrant. It is a one-way two-lane ramp with the design speed of 60 km/h, the roadway width of 12.25 m, and the minimum circular curve radius of 300 m. The rest one-way two-lane ramp uses the design speed of 60km/h and the roadway width of 10.5 m.

Ning-Lian Freeway within the design range of Wangxing System Interchange is constructed with eight lanes, thus Ramps I and J of the present Wangxing West Interchange need to be partially renovated.

The main line of the southern section within the Wangxing System Interchange crosses the railway. The nearest distance between the left edge of the main line and the edge of railway bearing platform is 0.28 m, and the right edge line of the main line intrudes into the edge of the bearing platform by 0.20 m (Figure 7 - 35).

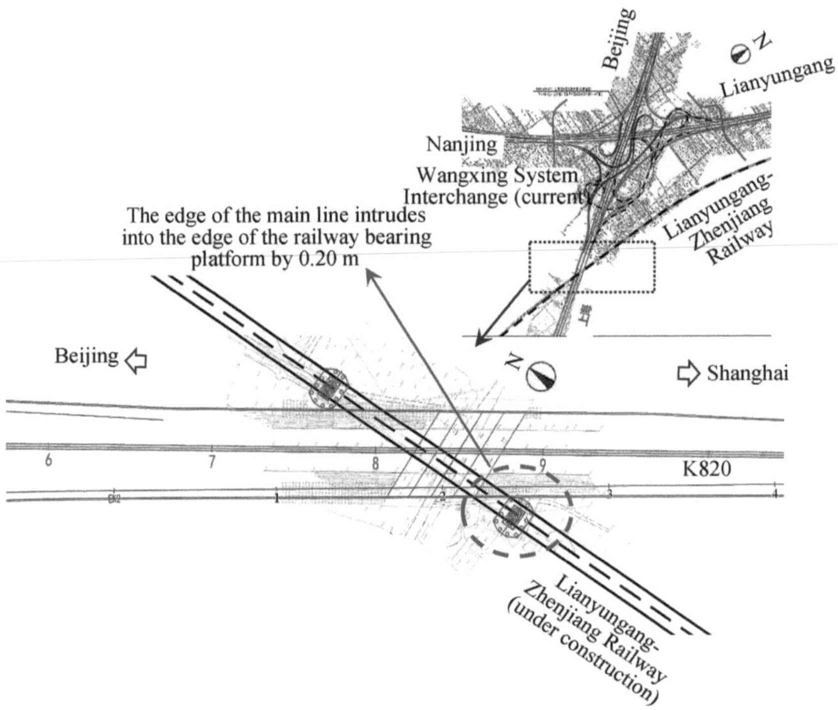


**Figure 7 - 35 Sketch of Wangxing System Interchange Scheme 2 and location of railway piers**

① Advantages of Scheme 2:

- The mainstream alignment is straight, with shorter distance and less detour.
- There are many overlaps with the existing interchanges which demand less new land.

② Disadvantages of Scheme 2:

- Ramp layout is not well-proportioned.
- The scale of interchange bridges is larger than that of Scheme 1.
- There are more conflicts with the existing ramps of Wangxing System Interchange, which is not convenient for traffic organization and requires high cost of traffic management during construction.
- Weaving length with Wangxing West Interchange is overlong.
- There is conflict with the railway control intersection, so it is difficult to coordinate.

(3) Scheme 3: Beijing-Shanghai and Ning-Lian Freeways are constructed with eight lanes (Ning-Lian Freeway is operated by four-lane delineation). The existing ramps are abolished and reconstructed in-situ using the partial cloverleaf

system interchange. The main line bridge does not meet the expansion clearance requirements and is demolished and reconstructed (Figure 7 - 36).

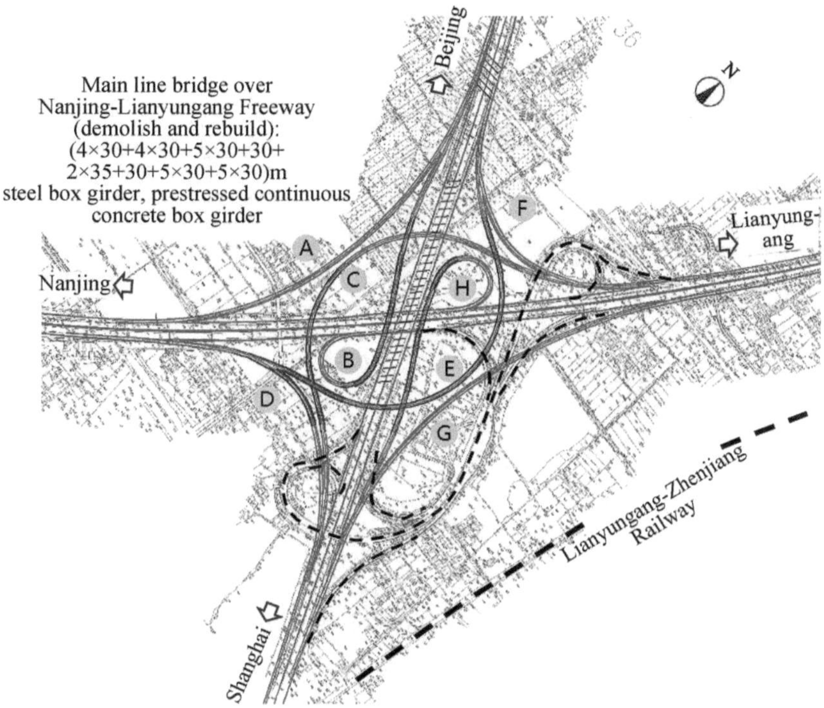


**Figure 7 - 36 Plan detail of reconstruction Scheme 3 of Wangxing System Interchange**

Ning-Lian Freeway within the design range of Wangxing System Interchange is constructed with eight lanes, thus Ramps I and J of the present Wangxing West Interchange need to be partially renovated .

Compared with Scheme 1, Ramp E crosses the main line of Beijing-Shanghai Freeway, Ramp H, and Ning-Lian Freeway continuously, and then is connected to the main line of Beijing-Shanghai Freeway once more.

**5) Scheme recommendation**

Compared with Scheme 1, the interchange configuration of Scheme 3 is basically the same. Ramp E in Scheme 3 needs to cross Ramp H one more time. Continuous overcrossing leads to a relatively large scale of Ramp E and a large scale of the total project. Therefore, Scheme 3 in the preliminary design stage will not be compared further.

The detailed comparisons of Schemes 1 and 2 are shown in Table 7 - 36.

Table 7 - 36 Scheme comparison of Wangxing System Interchange

| Project scheme | Scheme 1 | Scheme 2 |
|---|---|---|
| Comparison range of the main line/km | 3. 2 | 3. 2 |
| Reconstructed length of the intersected road/km | 3. 65 | 3. 73 |
| Subgrade earthwork/million $m^3$ | 44. 18 | 40. 60 |
| Soft foundation treatment/$m^2$ | 93,210 | 85,125 |
| Pavement works/thousand $m^2$ | 433. 028 | 426. 687 |
| Bridge works/m • $seat^{-1}$ | 3,356. 48/12 (new construction) | 3,522. 64/11 (new construction) |
| Bridge area/$m^2$ | 66,925. 04 | 71,450. 28 |
| Total land area/mu | 1,043. 04 | 1,067. 92 |
| House demolition/$m^2$ | 3,286. 52 | 3,433. 74 |
| Investment/million yuan | 69,173. 755 | 69,431. 940 (+258. 185) |
| Recommendation | Scheme 1 | |

General comparison: The investments of the two schemes are close. Scheme 1 is recommended in line with the interchange layout, project scale and local government opinions.

## 7.4.3 Preliminary design of Nanjing Pusi Interchange Reconstruction

### 1) Project Overview

Pusi Interchange is located at the intersection of Jiangbei Avenue and Puzhou Road in Nanjing, and is a system interchange, with G205 (Jiangbei Avenue) as the intersected road. Pusi Interchange has been reconstructed according to the requirements of traffic relief for the closure and reconstruction of Nanjing Yangtze River Bridge. Pusi Interchange has been transformed into a deformed single-trumpet interchange (Figure 7 - 37), which is an incomplete interchange. Traffic from north to east, from east to south and north are impassable, so vehicles need to bypass. The current traffic organization of Pusi Interchange is shown in Fugure 7 - 38.

**Figure 7 - 37 Present situation of Pusi Interchange**

**Figure 7 - 38 Sketch of present traffic organization of Pusi Interchange**

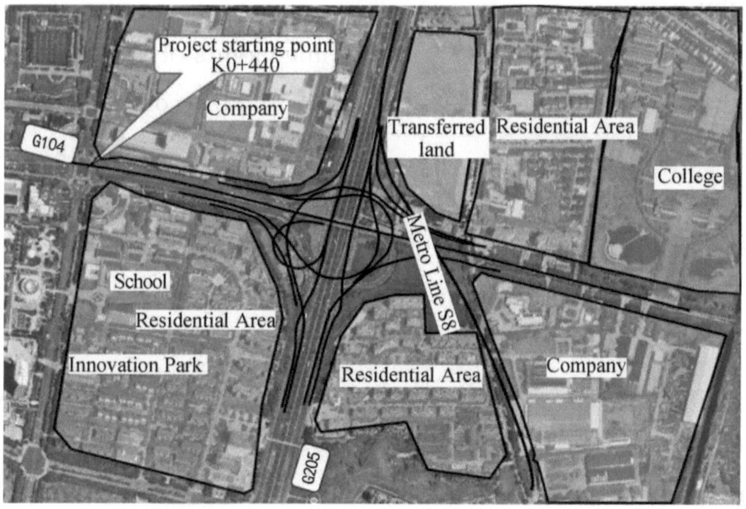


**Figure 7 - 39 Sketch of Pusi Interchange**

There are many control factors around Pusi Interchange, including two communities in the south, enterprises in the northwest, the transferred land and rail transit Ningtian Intercity Metro (Metro Line S8) in the northeast, etc. The interchange location is G104 (Pusi Road) to the west, as shown in Figure 7 - 39.

In the interchange design, in addition to paying attention to the bridge piers to avoid the impact on Metro Line S8, it is also necessary to consider that there is almost no room for expansion because the northwest quadrant enclosure is close to Jiangbei Avenue. The reconstruction of Pusi Interchange will involve the reconstruction of the enterprise enclosure in this section. According to the survey in the preliminary design stage, the status quo of the bridge over Jiangbei Avenue on Pusi road is a 4 m×16 m hollow slab (Figure 7 - 40), which mainly has the following problems:

**Figure 7 - 40 Present situation of the bridge over Jiangbei Avenue on Pusi Road**

(1) Insufficient clearance: The measured minimum clearance height of vehicles under the bridge is 4.9 m, which does not meet the requirements of the Class Ⅰ highway.

(2) Structural aging: Concretes in the hinge joints of girders fall off in many places, and some of the plate girders have been reinforced by pasting steel plates at the bottom.

(3) Insufficient width under the bridge: Widths of the non-motor vehicle lane and pedestrian lane on both sides of the side road of Jiangbei Avenue are 3.5 m and 2.5 m, respectively, leaving no space for pedestrians and non-motor vehicles under the bridge.

After comprehensive consideration and consultation with all parties, the bridge is considered to be demolished and rebuilt.

**2) Traffic volume prediction (2040)**

According to the traffic volume prediction, the annual turning traffic volume of Pusi Interchange in the vision year is 35,935 pcu/d, of which the southeast quadrant is the largest, reaching 58% of the total turning volume. In order to reduce detours of pedestrians and non-motor vehicles, a pedestrian and non-motor vehicle traffic system needs to be supplied inside the interchange to connect both sides of Jiangbei Avenue.

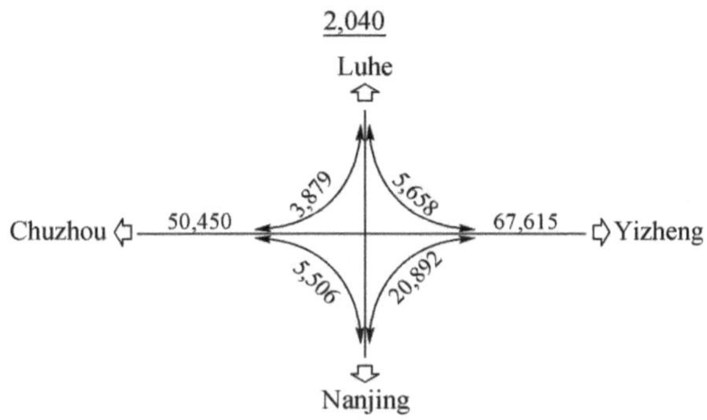


**Figure 7 - 41 Traffic volume prediction (pcu/d)**

**3) Selection of technical indicators**

The start point of the route is connected with the present Pusi Road, which is a two-way eight-lane Class Ⅰ highway with a design speed of 100 km/h on the main line and a subgrade width of 32 m. The speed limit of Pusi Road in this section is 60 km/h. Speed transition is conducted within the Pusi interchange range.

Center stake of Pusi Interchange is K1+000. 000. The main line within the interchange range is located in the sections of the tangent and horizontal curve with a radius of 2,300 m, and the maximum grade is 1. 50%.

Ramp design speed: 60 km/h is used for main-stream directional and semi-directional ramps with a minimum horizontal curve radius of 125 m, and 40 km/h is used for all other ramps with a minimum horizontal curve radius of 52 m.

Lane width: The one-way one-lane ramp width is 9. 0 m; the one-way two-lane ramp width is 10. 5 m.

Speed-change lane: The speed-change lane adopts the parallel type. The deceleration lane length of the one-lane ramp is not less than 125 m with the transition section of 90 m, and the acceleration lane length is not less than 200 m with the transition section of 80 m. The deceleration lane length of the two-lane ramp is not less than 190 m, the auxiliary lane length is not less than 250 m, and the transition section length is 80 m; the acceleration lane length is not less than 350 m, the auxiliary lane length is not less than 350 m, and the transition section length is 160 m.

The maximum grade of the vertical profile is less than 5.0%.

**4) Scheme Design**

Combining the above analysis and considering the conditions of interchange ramps, the following interchange schemes are proposed.

(1) Scheme 1: Deformed cloverleaf single-loop type

Except for the loop ramp used for left-turning in the northwest quadrant, semi-directional ramps are used for the other left-turnings. The advantage is the interchange land is small, but the engineering scale is larger compared to Scheme 2; affected by the existing buildings, some quadrants need to compress the width of pedestrian and non-motor vehicle lanes, with a minimum width of 3.0 m.

Relationship with communities: The southwest quadrant is 3.5 m from the fence of Gaoxin Huayuan Community, 14 m away from the community building, and 60 m away from the building of Tianhua Lügu Community in the southeast quadrant.

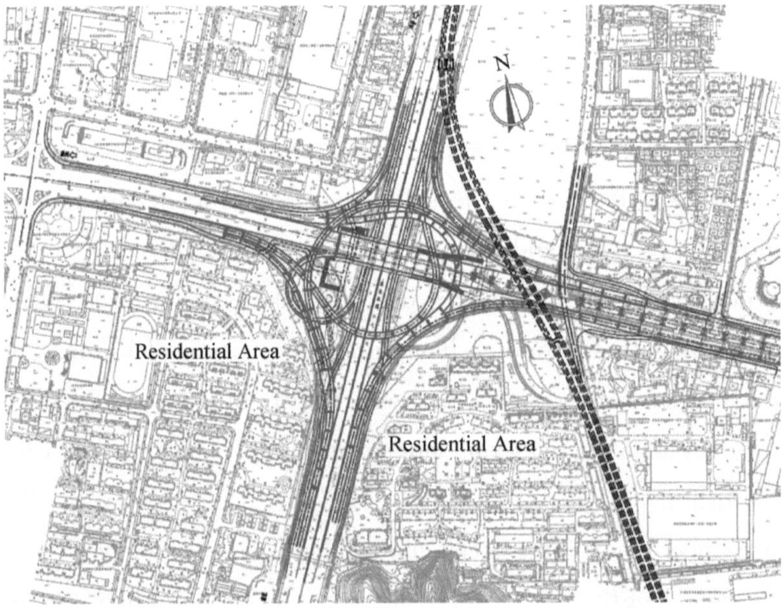


**Figure 7 - 42 Plan of Scheme 1 (deformed cloverleaf single-loop type)**

**Figure 7 - 43 Rendering of Scheme 1 (deformed cloverleaf single-loop type)**

(2) Scheme 2: Deformed cloverleaf symmetrical double-loop type

Loop ramps are used for the left-turning in both the northwest and northeast quadrants, and semi-directional ramps are used for the other left-turnings. The total length of ramps is short and the project scale is small, but the land area of Scheme 2 is larger than that of Scheme 1, so it is necessary to reduce the width of pedestrian and non-motor vehicle lanes.

Relationship with communities: The southwest quadrant is consistent with Scheme 1. The nearest distance from the community building is 14 m, and the nearest distance from the southeast quadrant to the building of Tianhua Lvgu Community is 30 m. Scheme 2 is closer to Tianhua Lvgu Community than Scheme 1, which has a relatively large impact.

**Table 7 - 37 Comparison of Pusi Interchange reconstruction schemes**

| Project scheme | | Scheme 1 | | Scheme 2 | |
|---|---|---|---|---|---|
| Comparison range | | K0+440 - K1+947.720 | | | |
| Ramp | Design speed/km • $h^{-1}$ | 40 | 60 | 40 | 60 |
| | Minimum horizontal curve radius/m | 52 | 125 | 52 | 125 |
| | Maximum grade/% | 4.974 | | 4.868 | |
| | Length/m | 3,892.659 | | 3,382.428 | |
| Bridge works/m • $seat^{-1}$ | | 3,754.5/8 | | 3,424.5/8 | |

**Continued Table 7 - 37**

| Project scheme | Scheme 1 | Scheme 2 |
|---|---|---|
| Bridge area/$m^2$ | 64,219 | 61,518 |
| Land area/mu | 295.4 | 299.3 |
| House demolition/$m^2$ | 6,322.7 | 6,322.7 |
| Recommendation | Scheme 1 | |

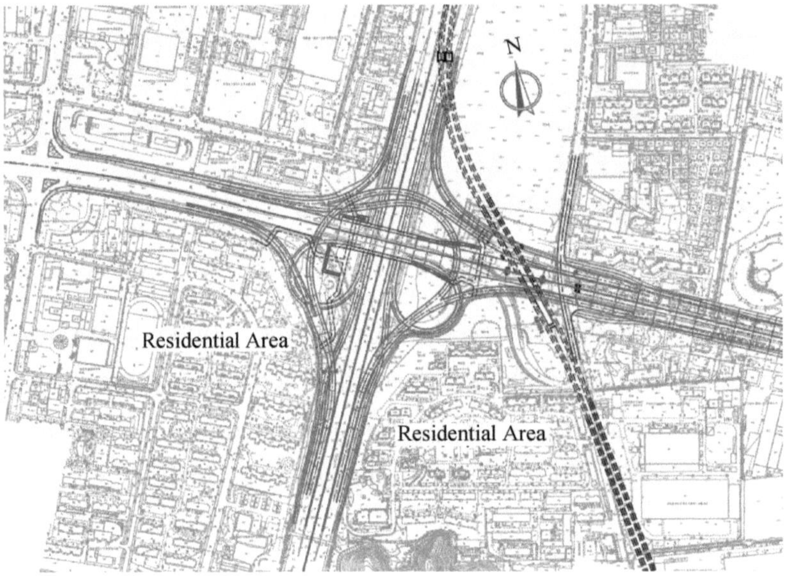


**Figure 7 - 44 Plan of Scheme 2 (deformed cloverleaf symmetrical double-ring type)**

Comprehensive comparison shows that although the project scale of Scheme 1 is slightly larger, compared with Scheme 2, it has less impact on Tianhua Lügu Community. When reporting to the planning department in the early stage, Scheme 1 is relatively good, so it is recommended at this stage.

**Questions:**

1. What are the components of interchanges and how to classify interchanges?
2. What are the common forms of interchanges and their applicable conditions?
3. What are the design contents and key points of ramp connections?
4. What are auxiliary lanes and collector-distributor lanes? What are their similarities and differences?
5. In view of the current toll method of freeways in China, what are the main

forms of three-way and four-way interchanges? What changes may occur in the future?

7. What is the transition form of the deceleration lane in front of the exit ramp? What transition form should be adopted for the one-lane ramp and the two-lane ramp?

# Chapter 8 Highway Traffic Engineering and Roadside Facilities

【**Objectives**: 1. Understand types, functions and principles of traffic engineering and roadside facilities; 2. Master the main categories of traffic safety facilities, and have the ability to design traffic safety facilities reasonably; 3. Master the main categories of service facilities, installation conditions and calculation of alignment indicators; 4. Master the main categories of management facilities, and have the ability to design management facilities reasonably.】

Traffic engineering and roadside facilities are an important part of the highway. Their construction scale and technical standards play a very important role in fulfilling highway functions, ensuring driving safety and improving levels of service and traffic capacity. It shall be designed based on comprehensive demonstrations such as the highway network planning, highway function, class, traffic volume, and operation conditions.

Traffic engineering and roadside facilities are organic components of the highway engineering and form a subordinate relationship with the main works of the highway, but they also complement and coordinate with each other. The overall design shall pay attention to the coordination of them, especially the layout, style, connection relationship of various facilities, etc. They shall be analyzed and demonstrated from the perspective of overall planning, and continuously optimized to finally form a unified organic whole.

The overall design of traffic engineering and roadside facilities shall meet the requirements of the overall design of the highway, match and coordinate with each other, and give full play to the overall benefits of the highway.

Traffic engineering and roadside facilities shall be designed in accordance with the principles of "ensuring safety, providing service, and facilitating management".

Traffic engineering and roadside facilities can be divided into three types:

© The Author(s), under exclusive license to Springer Nature Singapore Pte Ltd. 2025
B. Yu and J. Cheng, *Road Survey and Design*,
https://doi.org/10.1007/978-981-97-4836-5_8

traffic safety facilities, service facilities and management facilities. All facilities shall be applied in accordance with the principles of integrative coordination and overall design. It shall be gradually supplemented and improved in conjunction with the growth of traffic volume and technology development.

## 8.1 Traffic safety facilities

Traffic safety facilities belong to road infrastructure. They play an important role in reducing the severity of accidents, eliminating various vertical and horizontal disturbances, improving road level of service, providing sight guidance, and improving road landscapes. In particular, it is of special significance to give full play to the safe, fast, economical and comfortable functions of freeways.

Traffic safety facilities mainly include traffic signs, markings, guard rails, sight guidance facilities, isolation fences, anti-falling nets, anti-dazzling facilities, wind fences, snow (sand) fences, snow poles, etc.

Traffic safety facilities shall be applied according to the highway function, traffic composition, highway environment, and operation condition to support traffic safety management and services.

### 8.1.1 Traffic signs

Traffic signs are a kind of traffic management facility that uses colors, shapes, graphics, and texts to convey specific information to instruct, guide, warn, control or limit road traffic. It is generally set on the roadside or suspended above the road to provide traffic participants with clear traffic information.

**1) Classification of traffic signs**

According to their functions, traffic signs can be divided into two categories: main signs and auxiliary signs. The main signs are divided into:

(1) Warning signs: They are usually equilateral triangles. Colors of warning signs are with a yellow background, black border, and black pattern which are used to warn vehicles and pedestrians of dangers in the road ahead and the measures to be taken, such as intersections, sharp turns, railway crossings, slippery roads, uneven roads, dangerous roads near mountains, etc. Figure 8 - 1

shows examples of warning signs.

**Figure 8 - 1 Examples of warning signs**

(2) Prohibition signs: They are usually round, with white background, red circle, red bar, black pattern, or pattern bar. They are signs that prohibit or restrict the behavior of vehicles or pedestrians in order to ensure traffic safety based on road and traffic conditions, such as no passing, no parking, speed limit, etc. Figure 8 - 2 shows examples of prohibition signs.

**Figure 8 - 2 Examples of prohibition signs**

(3) Indication signs: They are usually round or rectangular, with blue background and white pattern. They are signs that instruct vehicles and pedestrians to proceed in a prescribed direction and place, such as going straight, turning left, turning right, and driving in one way. Figure 8 - 3 shows examples of indication signs.

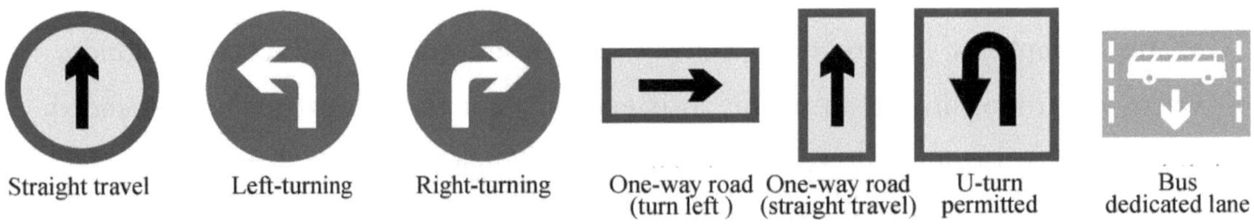


**Figure 8 - 3 Example of indication signs**

(4) Guide signs: They are usually rectangular with white characters on blue background (for general roads), or white characters on green background (for freeways). They are used to indicate the boundaries of cities, towns and villages, the directions and distances of destinations, highway entrances and exits, service areas, famous locations, etc. Figure 8 - 4 shows examples of guide signs.

Location | Famous location | Administrative division boundary | Intersection notice

**Figure 8 - 4 Examples of guide signs**

Auxiliary signs: They are attached to the main signs and serve as auxiliary explanations, which are rectangular, with black characters and black borders on white background. They can be used to indicate vehicle types, time, area or distance, prohibitions, warning reasons, etc. Auxiliary signs cannot be applied individually.

Small car | 200 m ahead | 100 m to the right | Combination sign | Accident | Landslide

**Figure 8 - 5 Examples of auxiliary signs**

**2) Three elements of traffic signs**

Traffic signs must enable drivers to recognize quickly and accurately within a certain distance, which requires clear visibility of traffic signs. There are three main factors that determine visibility: color, shape, and symbol. They are often referred to as the three elements of traffic signs.

(1) Color: Color selection of traffic signs takes the psychological effects of people into account. For example, red has a sense of danger, so red means stop and restraint, which is suitable for prohibition signs. Yellow is not as strong as red. It only produces vigilant psychological activities, so it is used to mean warning, attention, etc. Green has a sense of peace and safety. It means safety and passing in traffic and is generally used for guide signs. Blue has a sense of calm and quiet. It is generally used for guide and indication signs.

(2) Shape: Shape selection of traffic signs also requires clear visibility. The simplest shapes are selected, such as triangles, circles, rectangles, and squares. Generally, triangles indicate warning, circles indicate prohibition and restriction, and rectangles and squares indicate prompt.

(3) Symbol: Graphical symbol information is superior to text information in terms of recognition speed and recognition distance, and it is not restricted by

language. Therefore, traffic signs based on symbols have been adopted by most countries. Symbols indicate the specific meaning of traffic signs, and their meanings are required to be simple, clear and comply with international standards and conventions.

**3) General requirements of traffic sign design**

(1) Classification, color, shape, line, character, figure, size and installation of traffic signs shall comply with "Road Traffic Signs and Markings" (GB 5768—2017).

(2) Information provided by traffic signs shall all be related to the needs of traffic safety, service and management. The layout of traffic signs and supporting structures shall not be accompanied by commercial advertisements and other irrelevant information.

(3) Traffic signs shall be designed to facilitate clear identification, correct understanding and quick response of drivers. It shall consider factors such as the highway function, technical class, road network layout, traffic conditions, environmental conditions, road users and traffic management needs. It shall reasonably choose the parameters and scientifically determine the installation plan.

(4) Traffic sign design shall consider the information needs of road network, routes and sections at different levels, and adopt the methods of overall layout, level-by-level promotion and keypoint installation.

**4) Contents of traffic sign design**

Traffic sign design shall include the following contents:

(1) Location, type and content of traffic signs;

(2) Layout design;

(3) Support method;

(4) Material selection and design of the signboard, supporting structure, connecting piece and foundation;

(5) Strength and stability check;

(6) Construction process requirements, etc.

**5) Principles of traffic sign design**

(1) Highway traffic signs shall be designed for road users who are unfamiliar with the surrounding road network but have planned their travel routes, and

provide them with clear and concise information.

(2) Traffic signs shall be systematically arranged and installed according to the specific road section conditions on the basis of an in-depth analysis of traffic safety to match the actual traffic operation conditions of the road section. There should not be too many traffic signs at the same location and the traffic signs must not be contradictory.

(3) Warning signs shall be applied at the dangerous spots that affect driving safety and are not easy to be found along the highway and the roadside environment. It shall be used on the basis of sufficient argumentation and shall not be used excessively.

(4) Prohibition signs shall be applied in conspicuous locations near the start point of road sections where vehicles and pedestrians need to be clearly prohibited or restricted. Among them, speed limit signs shall consider factors such as the highway function, technical class, roadside development degree, route geometric characteristics, operating speed, traffic operation, traffic accidents and environment. On the basis of a thorough analysis of traffic safety, it is decided whether to set the speed limit value as well as the form of the speed limit sign. The speed limit sign is implemented after approval by the authority.

(5) Indication signs shall be applied in places where drivers and pedestrians are likely to be confused or where driving regulations must be obeyed according to the needs of traffic flow organization and traffic management.

(6) Guide signs shall be overall planned in accordance with the principle of road network integration, so that information is linked in an orderly manner, and there shall be no insufficient, improper or overloaded information. According to the functions of highways, the direction of traffic flows and the distribution of towns along the highway, priority shall be given to information instructions with greater traffic demand based on the distance, population, and the degree of social and economic development.

(7) Guide range shall be reasonably determined according to the class of tourist attractions and road network conditions when employing signs in tourist areas. When signs in tourist areas conflict with other traffic signs, the latter have priority.

(8) Installation of notice signs shall not affect the installation and visual

recognition of warning, prohibition, instruction and guide signs.

(9) Traffic signs at highway intersections shall be designed based on the complete consideration of the traffic management method, physical form, technical class of the intersection, traffic flow and other factors. It shall follow the principle of clear right of ways, reasonable channelization, explicit guidance, safety and order. Comprehensive schemes of different traffic signs shall be proposed to cooperate with traffic markings, and pilot vehicles to pass in an orderly manner.

(10) Except for special circumstances, traffic signs shall be placed above or on the right side of the traveled way in the forward direction, and traffic signs in other locations shall only be regarded as a supplement to the normal position. The specific location of traffic signs shall comply with "Road Traffic Signs and Markings" (GB 5768—2017). When the number of the one-way lanes is greater than or equal to three, or there are large traffic volumes or many large vehicles, or it is of poor visual recognition conditions, the specific installation of traffic signs shall be calculated and demonstrated based on the principles of traffic engineering.

## 8.1.2 Traffic markings

Traffic markings are traffic safety facilities composed of various road markings, arrows, text, elevation marks, raised road signs, and roadside outline signs. Their functions are to control and guide traffic, which can be used in conjunction with signs or alone.

Traffic markings include lane centerlines, lane dividing lines, lane edge lines, stop lines, yield lines, crosswalk lines, entrance and exit markings, guidances arrows, left-turning guidance line, road text markings, elevation marks, raised road signs, roadside outline signs, etc. Traffic markings are usually white or yellow. Road marking paint, plastic marking tape and other materials can be used.

**1) General stipulations of traffic markings design**

Classification, color, shape, characters, graphics, and size of traffic markings shall comply with "Road Traffic Signs and Markings" (GB 5768—2017) and "Specification for Layout of Highway Traffic Signs and Markings" (JTG

D82—2009). The chromaticity performance of road traffic markings shall comply with "Specification and Test Method for Road Traffic Markings" (GB/T 16311—2009).

Traffic markings shall be reflective. Within the normal service life, the retroreflective brightness coefficient of traffic markings shall meet the visibility requirements at night. When the raised road signs are used in conjunction with the painted marking, the directional reflective raised road signs shall be selected with their colors consistent with the marking color. For the raised road signs located on the boundary of the opposing traveled way and in the tunnel, they shall use the bidirectional reflective type.

Installation of highway traffic markings shall meet the following requirements:

(1) Installation of traffic markings shall match the traffic organization and traffic operation conditions.

(2) Traffic markings shall be coordinated with the geometric design of highways.

(3) Traffic markings shall be used in conjunction with other facilities such as traffic lights.

Traffic markings shall be designed according to the following critical procedures:

(1) Highway technical condition analysis, including the technical class, number of lanes, design speed, cross section change, intersections, etc.

(2) Determining the standard and scale of traffic markings, including colorful anti-slip markings as required.

(3) Traffic marking design for general road sections, including vertical markings, transverse markings, other markings, etc.

(4) Traffic marking design for special road sections, such as tunnel entrance and exit sections. Special road sections shall be deemed as an independent design unit, and the comprehensive design of traffic signs, markings, guardrails and other facilities shall be performed.

(5) Traffic marking design in complex areas, such as route crossings, toll plazas, etc.

**2) Principles of traffic marking application**

Traffic marking design of general road sections shall meet the following requirements:

(1) For the general road sections of the freeway and Class Ⅰ highways, there shall be edge lines for the traveled way and separation lines for lanes in the same way (as shown in Figure 8 - 6). For Class Ⅱ and lower-class highways, except for the single-lane road, separation lines for lanes in the opposing way shall be applied. And the following sections of Class Ⅱ and lower-class highways shall provide edge lines for the traveled way:

① The narrow bridge of highways and its upstream and downstream sections;

② The curve section with the lowest road design index and its upstream and downstream sections;

③ Road sections where the traffic flows merge or diverge;

④ Road sections where the road width changes;

⑤ Road sections where roadside obstacles are close to the traveled way;

⑥ Road sections that often experience heavy fog and other adverse weathers affecting safe driving;

⑦ Mixed traffic sections with many non-motor vehicles or pedestrians.

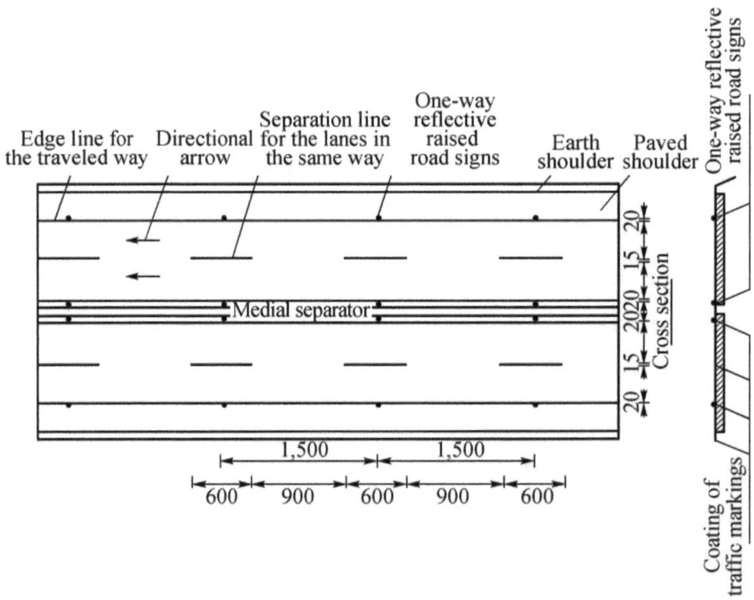


**Figure 8 - 6 Example of traffic marking design of general sections of the freeway with 100 km/h design speed (cm)**

(2) When there is a low-speed lane on a Class Ⅱ highway, separation lines for lanes in the opposing way, in the same way, and edge lines for the traveled way shall be applied.

Edge lines for the traveled way shall be applied within the paved shoulder on both sides, and they shall be applied on both sides of highways close to the outer edge of the traveled way if there is no paved shoulder. Separation lines for lanes in the same way shall be applied at the boundary of lanes in the same way.

The widths of traffic markings shall meet the requirements listed in Table 8-1.

**Table 8-1 Widths of traffic markings**

| Design speed/$km \cdot h^{-1}$ | | Edge line for the traveled way/cm | Separation line for lanes in the same way/cm | Separation line for lanes in the opposing way/cm |
|---|---|---|---|---|
| 120, 100 | | 20 | 15 | — |
| 80, 60 | Freeway, Class Ⅰ highway | 20 | 15 | — |
| | Class Ⅱ highway | 15 | 10 | 15 |
| 40, 30 | | 15 | 10 | 15 |
| 20 | Dual lane | 10 | — | 10 |
| | Single lane | 10 | — | — |

Traffic marking design for special road sections shall meet the following requirements:

(1) In bridge sections with frequent strong lateral winds, tunnel entrance and exit sections, sharp turning and steep grade sections, approaches of intersections, and sections close to the crosswalk line, it is necessary to set separation lines for lanes in the same way that prohibits crossing.

(2) Entrance and exit sections of tunnels should be deemed as independent design units, and the traffic marking design shall be integrated with traffic signs, guardrails, sight guidance signs, and other facilities.

(3) When there are piers of the overpass bridge in the center of highways or in the traveled way, ends of the medial separator, sign poles and other obstacles that may pose a threat to driving safety, markings that remind approaching obstacles shall be applied.

(4) On the pier and column facade of the overpass bridge and aqueduct, the end face of the side wall of tunnel openings and the facade of other obstacles that are close to the highway construction clearance, and on the central pier, toll island, physical refuge island or divisional island, lamp holder, sign base and other three-dimensional physical surfaces that may pose a threat to traffic safety, facade markings or physical markings shall be provided.

(5) Crosswalk lines should be drawn where there are no pedestrian crossing facilities on the road in front of schools, kindergartens, hospitals, and nursing homes.

(6) Transition markings shall be applied on road sections where the width of roads or the number of lanes changes.

(7) Longitudinal or transverse deceleration markings can be applied on road sections where vehicles need to be decelerated.

(8) For road sections where the speed hump is employed, the speed hump marking shall be applied in front of the speed hump.

(9) In addition to meeting the requirements of design specification for highway traffic safety facilities, traffic markings on highways across the city shall also consider the requirements for urban roads.

Traffic marking design at the entrances and exits of interchanges, service areas, and parking areas shall meet the following requirements:

(1) Traffic markings at the entrance and exit of interchanges, service areas, parking areas shall accurately reflect the principles of traffic flow organization. Separation lines for lanes in the same way that prohibit crossing shall be applied at appropriate locations of road entrance and exit sections (acceleration and deceleration lanes).

(2) At the entrances and exits of interchanges, service areas, and parking areas, directional arrows shall be provided. The specifications and repetition times of arrows shall comply with "Road Traffic Signs and Markings" (GB 5768—2017). The exit guidance arrow shall take the transition point of the deceleration lane as the reference point, and the entrance guidance arrow shall take the start point of the acceleration lane as the reference point.

(3) Within the service area and parking area, according to the traffic organization design and function planning of the area, traffic markings such as

parking space markings, roadway separation lines, and guidance arrows shall be applied respectively.

The design of at-grade intersection channelization markings shall meet the following requirements:

(1) Intersections between Class Ⅲ and above roads shall be channelized, and channelization markings are applied. Channelization islands should be applied when feasible, and the height of the curb should not exceed 10 cm. Intersections of other highways shall be provided with yield lines that are used in conjunction with stop or slow-down yield signs.

(2) Intersection channelization markings shall be designed in accordance with the actual situation of intersections and the actual characteristics of traffic flows.

Traffic markings design of the toll plaza shall meet the following requirements:

(1) Deceleration markings shall be applied when entering the toll plaza, and the spacing of each deceleration marking shall be calculated according to the entering speed and the length of the plaza. Toll island ground markings shall be provided in the direction of the approaching traffic flow, and physical markings shall be used on the toll island. At the exit of the toll plaza, partial separation lines of lanes in the same way can be applied.

(2) For toll plazas with ETC lanes, ETC road surface text and markings shall be provided in the ETC lanes, and relevant instructions and prohibition signs shall be provided.

(3) Toll plazas with more than five one-way toll lanes should be designed separately on the basis of traffic organization analysis.

Installation of raised road signs should meet the following requirements:

(1) Raised road signs should be installed on one side of pavement markings under the following conditions and they shall not intrude into lanes:

① Edge lines for the traveled way of freeways;

② Edge lines for the traveled way of Class Ⅰ and lower-class highway tunnels;

③ Edge lines for the traveled way of Class Ⅰ highway interchanges, service areas, and parking area sections;

④ Entrance sections of the interchange ramp.

(2) Raised road signs should be applied on the boundary of the traveled way of tunnels.

### 8.1.3 Guardrail

The guardrail is one of the traffic safety facilities, which is located on the side of the highway and medial separator. The function of the guardrail is to prevent vehicles from leaving the highway or entering the opposite lane, so as to minimize the injury to passengers and damage to vehicles. Also, it allows the vehicle to return to the normal driving direction, which is beneficial to induce drivers' sight. The anti-collision mechanism of the guardrail is to absorb the collision energy of vehicles through elastoplastic deformation, friction, and body displacement of the guardrail and vehicles, so as to protect people on vehicles. The significant difference between guardrails and other safety facilities is the self-destruction (deformation) of guardrails and vehicles to prevent more serious injury. In other words, if the accident severity of a certain vehicle colliding with a certain dangerous object under certain collision conditions is less than the accident severity of a vehicle colliding with a guardrail under the same conditions, the guardrail cannot be used to protect the dangerous object.

**1) Guardrail form**

Guardrail form selection shall depend on the specific conditions of each road, fully compare the performance of various guardrails, and analyze driving safety, pressure, sight guidance, and comfort of vision. It shall also consider the coordination with the surrounding environment of roads, combining factors such as economy, construction conditions, maintenance and repair, and is determined on the basis of comprehensive analysis.

(1) Corrugated beam guardrail

The corrugated beam guardrail is a semi-rigid structure, possessing a strong ability to absorb collision energy and a clear guidance of lines of sight. It can be well coordinated with the road alignment. Besides, it has an elegant appearance and can be used on small radius curves. The damaged part is easy to replace. Figure 8 - 7 is an example of the corrugated beam guardrail. For sections where vehicles rush out of roads (bridges) that may cause serious consequences, strengthened corrugated beam guardrails can be used.

**Figure 8 - 7 Example of the corrugated beam guardrail**

(2) Cable guardrail

The cable guardrail is a flexible structure and the cable functions within the elastic range when a vehicle collides. It can be reused and repaired easily. Figure 8 - 8 is an example of the cable guardrail. The column spacing is relatively flexible and less affected by uneven settlement. The cable guardrails is more elegant on scenic highways. In snowy areas, cable guardrails have fewer obstacles to snow sweeping. However, the construction of cable guardrail is complex, and it is difficult to repair the damaged end column. It is not suitable for the small radius curve section. Moreover, it has a poor line of sight and is not economical when the length is short.

**Figure 8 - 8 Example of the cable guardrail**

(3) Concrete guardrail

The concrete guardrail has a good effect on preventing vehicles from driving out of roads (bridges). It is suitable for narrow medial separators and sections with very dangerous roadsides. Because the concrete guardrail is hardly deformed, the maintenance cost is very low. Figure 8 - 9 is an example of the concrete guardrail.

**Figure 8 - 9 Example of the concrete guardrail**

(4) New Jersey guardrail

The New Jersey guardrail was developed by Stevens Institute of Technology in New Jersey, which is to separate the lanes of freeways. The New Jersey guardrail is a kind of wall-type guardrail with a certain section shape. When a vehicle collides with the guardrail, the guardrail does not move or deform (rigid state) under the action of the instantaneous moving load. The energy in the collision process is mainly absorbed by the vehicle climbing and turning along the guardrail surface in contact with the guardrail surface. At the same time, the crashed vehicle returns to the normal direction. In general, the New Jersey guardrail is more favorable to the steering of the crashed vehicle, so it is better to reduce the damage to the vehicle and protect the safety of passengers. But for the collision of high-speed heavy vehicles, it is easy to cross the guardrail and run to the opposite lane or out of the road.

**Figure 8 - 10 Example of the New Jersey guardrail**

**2) Roadside guardrail**

The roadside or medial separator of highways shall reduce the severity of accidents when vehicles drive out of the road or enter into the opposite lane by ensuring a reasonable clearance width. When the actual clearance width of highways is different from the calculated clearance width, it shall be determined whether to set guardrails according to the risks of accidents when driving out of the road or entering into the opposite lane by comprehensive analysis of traffic safety.

The risk of accidents when driving out of the road or entering into the opposite lane shall comprehensively consider the possibility and severity of accidents. Accident possibility when driving out of the road or entering into the opposite lane shall be determined according to factors such as the route alignment, traffic volume, traffic composition, and environmental conditions of road sections. Accident severity is related to the operating speed and roadside conditions. It can be divided into three levels: low, medium, and high.

The application principles of roadside guardrails and selection of protection levels are stipulated in Table 8 - 2.

**Table 8 - 2 Application principles of roadside guardrails and selection of protection levels**

| Accident severity and principle of guardrail installation | The following conditions exist within the width of roadside clearance | Highway class and design speed/km • $h^{-1}$ | Protection level / code |
|---|---|---|---|
| High, must set | High-speed railway, freeway, high voltage transmission tower, dangerous goods storage warehouse and other facilities | Freeway; 120 | Six (SS) |
| | | Freeway, Class Ⅰ; 100, 80 | Five (SA) |
| | | Class Ⅰ; 60 | Four (SB) |
| | | Class Ⅱ; 80, 60 | Four (SB) |
| | | Class Ⅲ; 40 | Three (A) |
| | | Class Ⅱ, Class Ⅳ; 30, 20 | Two (B) |
| Middle, shall set | (1) Sections where the slope and subgrade height of Class Ⅱ and above highways are within the shaded areas Ⅰ and Ⅱ in Figure 8 - 11, and road sections where there are cliffs, deep valleys, and deep ditch over 30 m on the side of Class Ⅲ and Ⅳ highways. (2) Road sections with rivers, lakes, seas, swamps and other water areas with a depth over 1.5 m; (3) Class Ⅰ railways, Class Ⅰ highways, etc.; (4) Road sections where there are facilities such as lights, cameras, traffic signs, sound barriers, bridge piers or abutments of overpass bridges, and access roads or portals at the entrance of tunnels outside the freeway and Class Ⅰ highway that cannot be safely crossed by vehicles | Freeway, Class Ⅰ; 120, 100, 80 | Four (SB) |
| | | Class Ⅰ; 60 | Three (A) |
| | | Class Ⅱ; 80, 60 | Three (A) |
| | | Class Ⅲ; 40 | Two (B) |
| | | Class Ⅲ, Class Ⅵ; 30, 20 | One (C) |

**Continued Table 8 - 2**

| Accident severity and principle of guardrail installation | The following conditions exist within the width of roadside clearance | Highway class and design speed/km • $h^{-1}$ | Protection level/ code |
|---|---|---|---|
| Low, can set | (1) Road sections where the slope and embankment height of Class Ⅱ and above highways are within the shaded area Ⅲ in Figure 8 - 11, and road sections where the slope and embankment heights of Class Ⅲ and Ⅳ highways are within the shaded area Ⅲ in Figure 8 - 10.<br>(2) Excavated road sections where the side ditch of Class Ⅱ and above roads have no slab and vehicles cannot pass safely.<br>(3) Road sections where there are obstacles such as concrete masonry or large boulders over 30 cm above the road surface or the excavated slope.<br>(4) Road sections where the triangle of the exit ramp has obstacles | Freeway Class Ⅰ; 120, 100, 80 | Three (A) |
| | | Class Ⅰ; 60 | Two (B) |
| | | Class Ⅱ; 80, 60 | Two (B) |
| | | ClassIII, Class Ⅵ; 40, 30, 20 | One (A) |

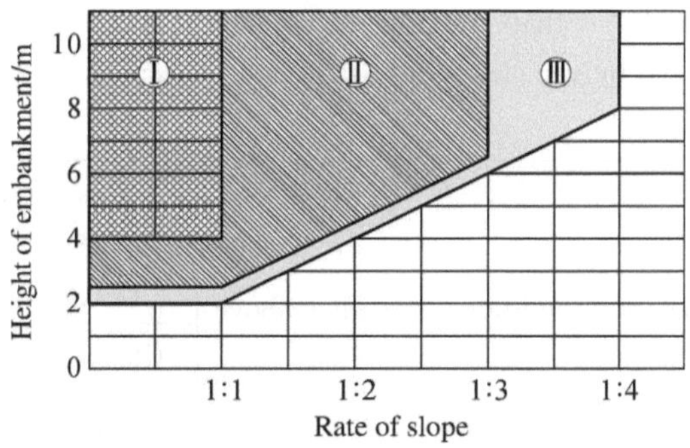


**Figure 8 - 11 Relationship between slope, height of embankment and installation of guardrails**

### 3) Medial separator guardrail

For freeways and Class Ⅰ highways as arterial roads, when the median width of the integral cross section is less than or equal to 12 m, or there are obstacles within the range of 12 m, the medial separator guardrail must be installed. Severity of the medial separator accident is determined and the protection level of the medial separator guardrail is selected according to Table 8 - 3.

**Table 8 - 3 Severity of the medial separator accident and protection level of the guardrail**

| Accident severity | Medial separator guardrail conditions | Highway class and design speed/$\text{km} \cdot \text{h}^{-1}$ | Protection level/ code |
|---|---|---|---|
| High | When the width of the medial separator is less than 2. 5 m and the integral guardrail is adopted | Freeway; 120 | Six (SSm) |
| | | Freeway, Class Ⅰ; 100, 80 | Five (SAm) |
| | | Class Ⅰ; 60 | Four (SBm) |
| Medium | For the two-way six-lane freeway, or two-way eight-lane and above highways without the left paved shoulder, the width of the medial separator is less than 2. 5 m and the separated guardrail is used, and there are sections with obstacles[1] that cannot be safely crossed by vehicles within the medial separator | Freeway; 120, 100, 80 | Four (SBm) |
| | For the Class Ⅰ highway with two-way six-lane and above, the width of the medial separator is less than 2. 5 m and the separated guardrail is used, and there are sections with obstacles[1] that cannot be safely crossed by vehicles within the medial separator | Class Ⅰ; 100, 80 | Four (SBm) |
| | | Class Ⅰ[2]; 60 | Three (Am) |
| Low | When the conditions specified above are not met | Freeway, Class Ⅰ; 120, 100, 80 | Three (Am) |
| | | Class Ⅰ[2]; 60 | Two (Bm) |
| | | Class Ⅱ[3]; 80, 60 | Two (Bm) |

*Notes*: (1) Obstacles refer to facilities such as lights, cameras, supporting structures of traffic signs, and piers of overpass bridges.

(2) Road sections of the Class Ⅰ highway with a design speed of 60 km/h serve as collector roads, which are restricted by topography, geology and other conditions. This table is suitable for the situation when it needs to set a medial separator guardrail.

(3) It is applicable to the Class Ⅱ highway with passing lanes but without isolation facilities, and with the possibility of entering into the opposite lane.

### 8.1.4 Sight guidance facilities

Sight guidance facilities refer to the general category of facilities installed on both sides of the lane to indicate the road direction, lane boundary, and spots of dangerous sections.

Sight guidance facilities actively inform the highway route, structures, potentially dangerous sections, small at-grade intersections, etc. Especially at night, it produces highly cost-effective benefits to actively guide driving through retroreflective materials or active luminescent systems. When feasible, line-of-sight guidance facilities can be appropriately added to realize its advantages of energy conservation and low cost.

Sight guidance facilities can be divided into contour marks, merging and diverging guidance marks alignment guidance marks, etc. According to their functions, alignment guidance marks can be further divided into the informative alignment guidance marks and warning alignment guidance marks.

Installation of contour marks shall meet the following requirements:

(1) Contour marks shall be applied continuously for the main lines of the freeway and Class Ⅰ highway, ramps of interchanges, service areas and parking areas, connecting lanes, medial separator openings, and refuge lanes.

(2) Contour marks should be applied on Class Ⅱ and below highway sections with poor sight distance, with changes in the lane number or lane width, and with continuous sharp turns and steep slopes. Contour marks can be applied on other sections if necessary.

(3) Contour marks shall be applied on the tunnel.

Contour marks are facilities installed at the edge of the traveled way, and their structure is related to roadside structures. When there is no structure on the roadside, contour marks are columns which are independently placed on the earth shoulder. When there are structures such as guardrails, bridge railings, and side walls on the roadside, contour marks are attached to the appropriate positions of these structures.

The installation of merging and diverging guide marks and alignment guide marks shall comply with the relevant standards and norms for the installation of highway traffic signs and markings.

### 8.1.5 Anti-glare facilities

Anti-glare facilities are artificial structures that prevent night-time traffic from being dazzled by the headlights of opposing vehicles. There are slatted anti-glare panels, fan-shaped anti-glare panels, anti-glare nets, and anti-glare sheds. In addition to beautifying the road, the plants in the medial separator also play a role in anti-glare, and can also be used as a type of anti-glare facilities.

For freeways and Class Ⅰ highways, if the medial separator width is less than 9m and one of the following conditions is met, anti-glare facilities should be installed:

(1) Road sections where the traffic volume at night is relatively large, and the sum of the original traffic volumes of trucks and buses in the design traffic volume is greater than or equal to 15%;

(2) Superelevated circular curve sections;

(3) Road sections with the radius of the sag vertical curve equal to or close to the minimum radius specified in the **Standard (2014)**;

(4) When the cross section of highways is separated, and the height difference between the traveled ways in both directions is less than or equal to 2 m;

(5) Road sections that are severely affected by glare from adjacent or crossed roads and railways;

(6) Road sections near the entrance and exit of the double-arch tunnel.

When choosing the form of anti-dazzle facilities, the performance of various anti-dazzle facilities shall be fully compared according to the horizontal and vertical alignments and climatic conditions of highways. Driving safety, pressure and landscape requirements shall be analyzed. Moreover, coordination with the surrounding environment of highways and factors such as economy, construction conditions, and maintenance shall be evaluated to choose anti-dazzle facilities. Table 8 – 4 shows the comparison of different anti-glare facilities.

**Table 8 - 4 Comparison of different anti-glare facilities**

| Characters | Plants (shrubs) | | Anti-glare board | Anti-glare net |
|---|---|---|---|---|
| | Dense type | Interval type | | |
| Appearance | Good | | Good | Relatively poor |
| Psychological impact on drivers | Low | High | Low | Relatively low |
| Wind resistance | High | | Low | High |
| Snow accumulation | Serious | | Good | Serious |
| Natural landscape coordination | Good | | Good | Poor |
| Anti-glare effect | Relatively good | | Good | Relatively poor |
| Economy | Poor | Good | Good | Relatively poor |
| Construction difficulty | Relatively difficult | | Easy | Difficult |
| Maintenance workload | Large | | Small | Small |
| Transverse visibility | Poor | Relatively good | Good | Good |
| Prevent pedestrians from crossing | Relatively good | Poor | Relatively good | Good |
| Landscape effect | Good | | Good | Poor |

## 8.1.6 Other traffic safety facilities

Installation of the highway isolation fence shall meet the following requirements:

(1) It should be applied continuously on both sides of the sections of freeways and Class Ⅰ highways that need to control access. Natural barriers can also be used at intervals.

(2) It can be applied on other roads if necessary.

Installation of the highway anti-falling net shall comply with the following requirements:

(1) When the highway crosses railways, navigable rivers or other highways with a large traffic volume.

(2) On the sections where there may be rockfall on the cut slope of highways and the traffic safety may be affected.

If the intersection angle between the crosswind on the roadside and the highway axis is greater than 30°, and one of the following conditions is met, the wind barrier can be applied on the windward side of the roadside:

(1) Road sections with design speed greater than or equal to 80 km/h and wind force greater than grade Ⅶ all year round;

(2) Road sections with design speed less than 80 km/h and wind force greater than grade Ⅷ all year round;

(3) When the instantaneous wind speed reaches the prescribed value in Table 8-5 at the tunnel entrance, pass, bridge, etc.

**Table 8-5 Safe driving wind speed**

| Design speed of highways/km • $h^{-1}$ | 100 | 80 | 60 | 40 | 20 |
|---|---|---|---|---|---|
| Wind speed/km • $h^{-1}$ | 15 | 17 | 19 | 20 | 20 |

Fixed snow fences can be installed on road sections where the amount of snow blown by wind is large and lasts for a long time, and with little change in wind direction. Mobile snow fences can be used in sections with variable wind direction, strong wind and heavy snow. Highway snow fence design shall meet the following requirements:

(1) Snow fence design shall effectively reduce the adverse effects of wind and snow on vehicle operation, while taking into account the protection of the highway subgrade.

(2) Snow fences shall be installed on the windward side of highways. When the terrain is open, the amount of snow is too large, and the wind is very strong, multiple rows of snow fences can be applied.

Highway sections with heavy, long-lasting snowfall and snow-covered lanes, snow poles can be applied. Highway snow pole design shall meet the following requirements:

(1) Snow poles should be applied on the earth shoulder of highways, and the installation position shall not invade highway construction clearance.

(2) The installation spacing of snow poles can refer to the one of outline marks.

## 8.2 Service facilities

Service facilities on the highway include service areas, parking areas and coach stations.

Location of service areas and parking areas shall be planned according to the regional road network, construction conditions, landscape and environmental protection requirements, etc. Location of coach stations should be planned in accordance with the regional road traffic planning, town distribution along the highway and travel needs.

### 8.2.1 Service area and parking area

Service areas and parking areas are mainly for vehicles and drivers. The main facilities for vehicles include the parking lot, gas station, vehicle maintenance station, etc. The main facilities for drivers and passengers include the restaurant, canteen, toilet, rest area, public telephone, square, garden, sidewalk, etc.

**1) Provisions of service areas and parking areas**

Service areas shall meet the following requirements:

(1) Service areas shall be provided on freeways and should be provided on Class Ⅰ and Class Ⅱ highways as arterial roads. The average interval should be 50 km. When towns along the highway are sparsely distributed and the supply of water and electricity is difficult, the interval can be elongated.

(2) The freeway service areas shall be equipped with facilities such as the parking lot, gas station, vehicle maintenance station, public toilet, indoor and outdoor rest area, restaurant, and retail outlet. According to the road environment and needs, facilities such as personnel accommodation and vehicle water filling can be provided.

(3) The parking lot, gas station, public toilet, outdoor rest area and other facilities should be provided in the service areas of Class Ⅰ and Class Ⅱ highways as arterial roads. The restaurant, retail store, vehicle water filling and other facilities can be provided when feasible.

Parking areas shall be applied in accordance with the following provisions:

(1) Parking areas should be provided on freeways, Class Ⅰ and Class Ⅱ highways as arterial roads. One or more parking areas can be offered between service areas, and the spacing between the parking area and service area or between parking areas should be 15 – 25 km.

(2) Parking areas shall be equipped with the parking lot, public toilet,

outdoor rest area, and other facilities.

The scale of service areas and parking areas shall be determined according to the traffic volume, traffic composition, town layout along the highway, land use condition and other factors. The corresponding land use and building areas shall comply with the relevant standards and specifications.

The spacing between the service area, parking area and interchange, tunnel should be greater than 2 km. It can refer to the relevant requirements of interchange spacing under constraint.

**(2) Type of service areas and parking areas**

The type of service areas and parking areas shall be determined according to the service demand, construction condition, natural environment, and other factors. Service facilities can be arranged on both sides, symmetrically and offset on both sides, asymmetrically on both sides or unilaterally.

When the service area is arranged bilaterally, it is divided into the following types according to the location of the main facilities in the service area. The type of parking areas can be applied according to the site conditions with reference to the service area.

(1) Separate outward type: Service areas are arranged on both sides of the main line of highways, and service facilities such as the parking lot and gas station are located between the restaurant and the main highway, as shown in Figure 8 - 12.

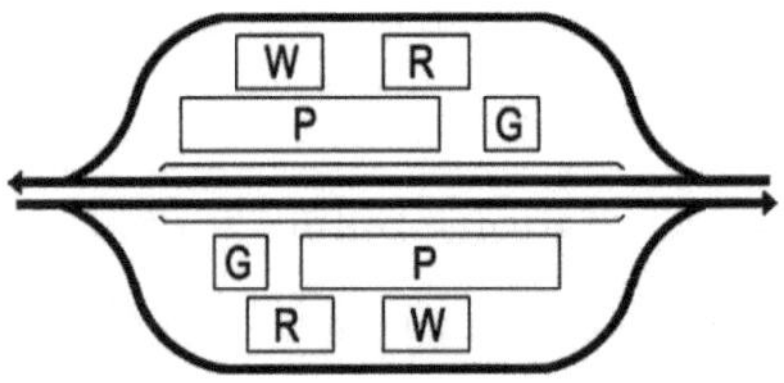


(P—parking lot; G—gas station; W—public toilet; R—restaurant)

**Figure 8 - 12 Separate outward type**

(2) Separate parallel type: Service areas are arranged on both sides of the main line of highways, and service facilities such as the restaurant, parking lot, and gas station are basically almost parallel to the main line of highways, as shown in Figure 8 - 13.

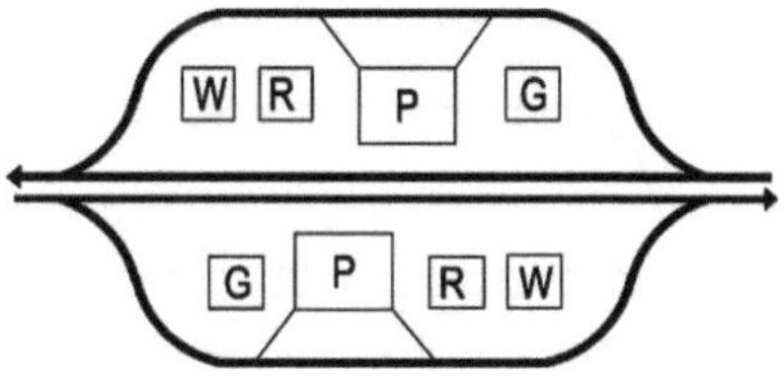


**Figure 8 - 13 Separate parallel type**

(3) Separate centralized type with the one-side restaurant: The service area is arranged on both sides of the main line of the highway, while the restaurants are only provided on one side of the service area. This form is suitable when the site on one side of the highway is relatively narrow, or the landscape on one side is beautiful which has a strong appeal for users, while the site conditions are limited, and the restaurant, lounge and other facilities have to be outward, as shown in Figure 8 - 14.

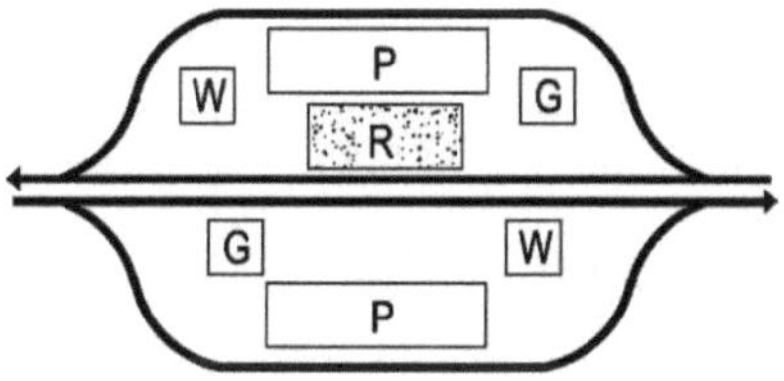


**Figure 8 - 14 Separate centralized type with the one-side restaurant**

(4) Separate centralized type with the loft restaurant: The service area is arranged on both sides of the main line of the highway, but the restaurant is built above the highway and for both sides to use to make full use of space. The restaurant is usually designed as the highway landscape, as shown in Figure 8 - 15.

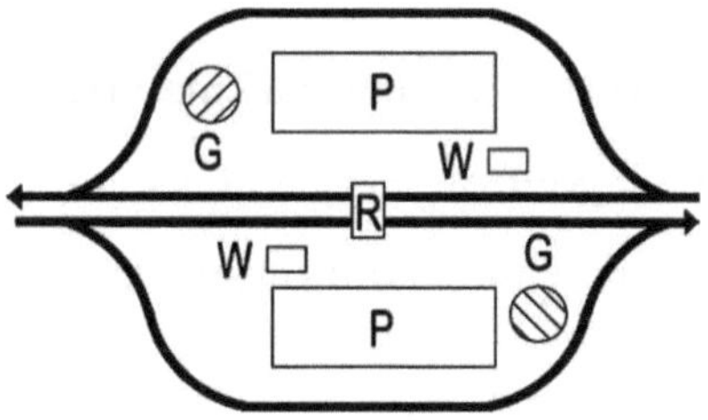


**Figure 8 - 15 Separate centralized type with the loft restaurant**

(5) Central or unilateral centralized type: The service area is located in the center or on one side of the main line of the highway, as shown in Figure 8 - 16. For closed toll roads, this form of service areas needs to completely separate the

parking lots on both sides, which is rarely used at present.

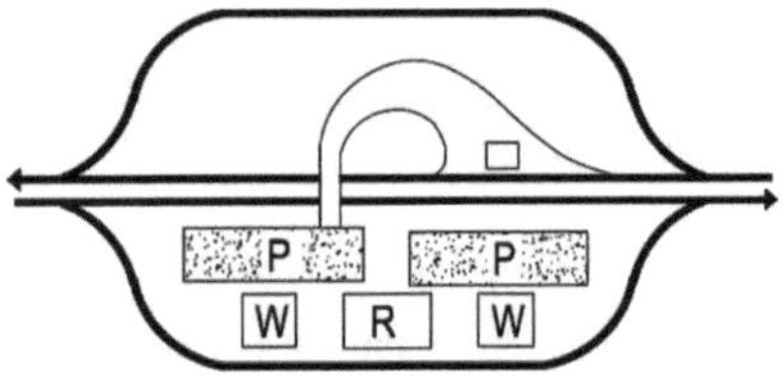


**Figure 8 - 16 Central or unilateral centralized type**

### 3) Main line alignment index within service areas and parking areas

Main line alignment index within service areas shall meet the requirements of main line alignment index within the interchange. Main line alignment index within parking areas shall meet the requirements specified in Table 8 - 6.

**Table 8 - 6 Main line alignment index within parking areas**

| Design speed/km · h$^{-1}$ | | 120 | 100 | 80 | 60 |
|---|---|---|---|---|---|
| Minimum circular curve radius/m | General value | 1,500 | 1,000 | 700 | 500 |
| | Limit value | 1,200 | 850 | 600 | 400 |
| Minimum crest vertical curve radius/m | General value | 45,000 | 25,000 | 12,000 | 6,000 |
| | Limit value | 23,000 | 15,000 | 6,000 | 3,000 |
| Minimum sag vertical curve radius/m | General value | 16,000 | 12,000 | 8,000 | 4,000 |
| | Limit value | 12,000 | 8,000 | 4,000 | 2,000 |
| Maximum grade/% | General value | 2 | 3 | 4 | 4.5 |
| | Maximum value | 3 | 4 | 5 | 5.5 |

*Notes*: (1) The grade shall be selected above the general value. In the limited terrain and special conditions, the maximum value can be used.
(2) This table refers to the **Specification (2017)**.

### 4) Overall layout of service areas and parking areas

The general geometric layout of service areas and parking areas shall include the acceleration (deceleration) lane, connecting lane, through lane parking lot, etc., as shown in Figure 8 - 17.

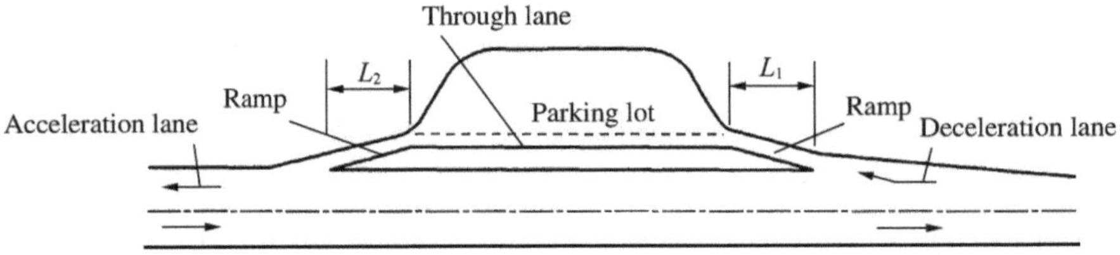


**Figure 8 - 17 Schematic diagram of lane layout in service areas and parking areas**

The design speed of the ramp in the service area and parking area should be 40 km/h, and shall not be less than 30 km/h under restriction. The minimum length of the ramp shall meet the requirements of Table 8 - 7.

**Table 8 - 7 Minimum length of the ramp**

| Design speed of the main line/km • $h^{-1}$ | | 120 | 100 | 80 |
|---|---|---|---|---|
| Side of the deceleration ramp/m | General value | 110 | 90 | 80 |
| | Limit value | 80 | 70 | 60 |
| Side of the acceleration ramp/m | General value | 80 | 70 | 60 |
| | Limit value | 60 | 60 | 60 |

*Note*: This table refers to the **Specification (2017)**.

Geometric design of the ramp and acceleration and deceleration lanes shall comply with the relevant regulations of the interchange.

Geometric design of the through lane form shall meet the following requirements:

(1) Design speed of the through lane should be 30 km/h.

(2) One-way one-lane shall be adopted for the through lane, with a lane width of 3. 50 m and a left and right marginal width of 0. 50 m.

(3) The vertical alignment of the through lane shall be designed considering the parking lot elevation and drainage needs.

The service area, parking area, and viewing platform of Class Ⅱ highways can be designed in a typical form by providing ramps, acceleration and deceleration lanes according to the function, traffic volume, site conditions, etc. Otherwise, the simple form without ramps that is integrated with the main line can also be adopted. The layout of simple service areas, parking areas, and viewing platforms shall meet the following requirements:

(1) The grade of the main line within the service area, parking area, and viewing platform shall not be greater than 2. 5%, and the traveled way of the main line and parking lot shall be separated by the side separator or road marking.

(2) Acceleration and deceleration zones shall be applied on both sides of the parking lot with equal length.

(3) The minimum grade length of the parking lot along the main line should be greater than 30 m.

### 8.2.2 Coach station

Coach stations shall not be applied on the main line side of freeways.

Facilities such as the gas station, public toilet and coach station can be applied as required for Class Ⅰ and Class Ⅱ highways as collector roads and Class Ⅲ and Class Ⅳ highways.

Coach stations shall be equipped with vehicle parking and passenger waiting facilities, and they can be integrated with the service area.

**1) Main line alignment index of coach stations**

The maximum grade of the main line within the range of coach stations shall not be greater than 2%, and shall not be greater than 3% when the terrain is particularly restrictive. Horizontal curve and vertical curve indexes of the main line shall conform to the provisions in Table 8 - 8.

**Table 8 - 8 Main line index within the range of coach stations**

| Design speed/km · h$^{-1}$ | 100 | 80 | 60 | ≤40 |
|---|---|---|---|---|
| Minimum radius of the circular curve/m | 800 | 500 | 250 | 150 |
| Minimum radius of the crest vertical curve/m | 10,000 | 4,500 | 2,000 | 1,000 |
| Minimum radius of the sag vertical curve/m | 4,500 | 3,000 | 1,500 | 1,000 |

*Note*: This table refers to the **Specification (2017)**.

**2) Coach station layout**

The layout of coach stations on the main line side of the Class Ⅰ highway shall include the transition section, acceleration (deceleration) lane, parking lane, etc., as shown in Figure 8 - 18, and it shall comply with the following provisions:

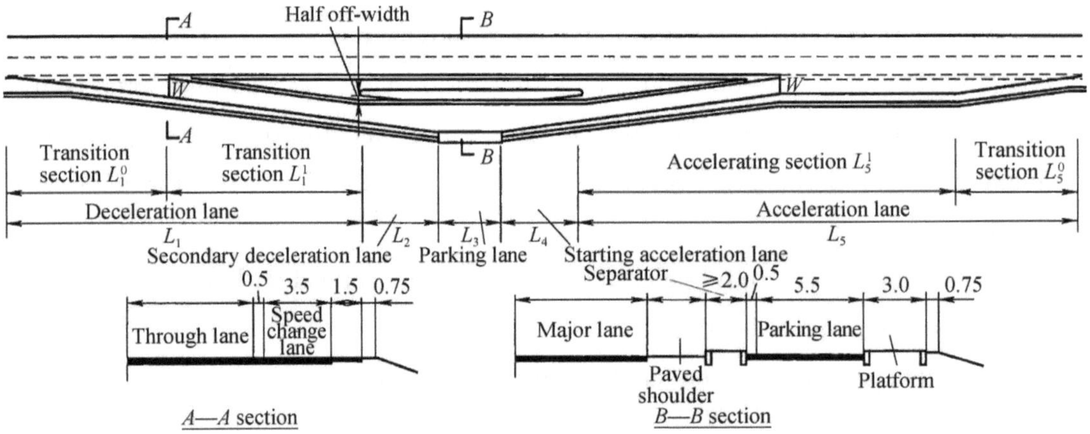


**Figure 8 - 18 Schematic diagram of coach stations on the Class Ⅰ highway (m)**

(1) A side separator or guardrail must be employed to separate the parking area from the right paved shoulder of the main line.

(2) The length of the speed-change lane and its transition section, and the length of the parking lane shall not be less than the values specified in Table 8 - 9.

(3) The width of the side separator shall not be less than 2.0 m, the width of the paved shoulder on the right side of the speed-change lane is 1.5 m, the width of the parking lane shall not be less than 5.5 m, and the width of the platform is 3.0 m.

**Table 8 - 9 Length of the speed-change lane and parking lane at the coach station of the Class Ⅰ highway**

| Main line design speed/km · $h^{-1}$ | | 120 | 100 | 80 |
|---|---|---|---|---|
| Deceleration lane $L_1$ | Transition section $L_1{}^0$ (1/20)/m | 70 | 70 | 70 |
| | Deceleration section $L_1{}^1$/m | 100 | 90 | 70 |
| Secondary deceleration lane $L_2$/m | | 50 | 50 | 40 |
| Parking lane $L_3$/m | | 30 | 30 | 20 |
| (Secondary) Initial acceleration lane $L_4$/m | | 40 | 40 | 30 |
| Acceleration lane $L_5$ | Acceleration section $L_5{}^1$/m | 130 | 110 | 80 |
| | Deceleration section $L_5{}^0$/m | 65 | 60 | 50 |

*Note*: This table refers to the **Specification (2017)**.

The layout of the coach station on the main line side of Class Ⅱ and lower highways shall include the acceleration (deceleration) section, parking lane, etc., as shown in Figure 8 - 19, and it shall meet the following requirements:

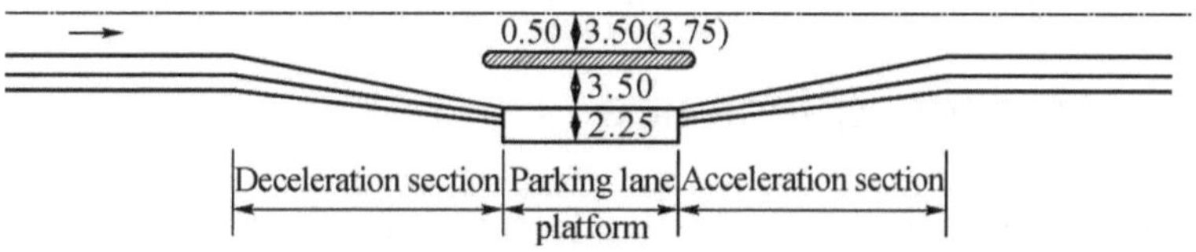


**Figure 8 - 19 Schematic diagram of the coach station of Class Ⅱ and lower highways (m)**

(1) The parking area and the traveled way shall be distinguished by road markings.

(2) Acceleration and deceleration sections with equal length shall be applied on both sides of the stop area in front of the coach station, and their lengths shall conform to the values specified in Table 8 - 10.

(3) Length of the parking lane is 15 m.

(4) Width of the separator (pavement marking) at the edge line of adjacent

lanes, width of the parking lane and width of the coach station platform are 0.5 m, 3.5 m and 2.25 m, respectively.

**Table 8 - 10 Length of the speed-change section of the coach station on the main line of Class Ⅱ and below highways**

| Main line design speed/km • $h^{-1}$ | 80 | 60 | 40 | 30 | 20 |
|---|---|---|---|---|---|
| Gradient | 1/15 | 1/12.5 | 1/10 | 1/7.5 | 1/5 |
| Length of the acceleration and deceleration section/m | 60 | 50 | 40 | 30 | 20 |

*Note*: This table refers to the **Specification (2017)**.

## 8.3 Management facilities

Management facilities include monitoring, tolling, communication, power supply, lighting, management and maintenance facilities, and they shall meet the following requirements:

(1) Freeways shall be equipped with facilities for monitoring, tolling, communication, power supply, lighting, management and maintenance. Other classes of roads can be equipped with facilities according to demands.

(2) Monitoring, tolling, communication, power supply, lighting, management, maintenance, and other facilities shall be designed and implemented in stages according to the traffic volume, and the foundation works, underground pipelines, reserved and embedded works, etc. shall be implemented accordingly.

### 8.3.1 Monitoring facilities

The monitoring facilities are divided into four levels: A, B, C, D.

Level A: Video monitoring, dynamic information release and traffic guidance shall be provided on the entire route, and traffic incident detection, traffic volume detection, environmental information detection and ramp control shall also be provided for toll stations, super-large bridges, tunnels, interchanges, service areas, and other key or special sections. Full monitoring, dynamic information release, and traffic guidance on the entire route shall be accomplished.

Level B: Video surveillance, traffic incident detection, traffic volume detection, environmental information detection, ramp control, dynamic information release and traffic guidance shall be applied in key or special road sections such as toll gates, long-span bridges, interchanges, service areas etc. Key monitoring, dynamic information release, and traffic guidance on the entire route shall be accomplished.

Level C: Video surveillance, traffic incident detection, traffic volume detection, dynamic information release and traffic guidance should be provided in key or special sections such as long span bridges, service areas, coach stations, and highway intersections.

Level D: Traffic volume detection, on-site traffic information prompts and traffic guidance may be provided in key or special sections such as long-span bridges, gas stations, coach stations and major highway intersections.

The applicable scope of each level of monitoring facilities can be determined according to Table 8 - 11.

**Table 8 - 11 Applicable scope of each level of monitoring facilities**

| Level of monitoring facilities | Applicable scope |
|---|---|
| A | Freeway (full monitoring) |
| B | Freeway (section monitoring) |
| C | Class Ⅰ and Class Ⅱ arterial highways |
| D | Collector and local highways |

*Note*: This table refers to the **Standard (2014)**.

When bridges and tunnels are equipped with structural monitoring, maintenance monitoring, and other facilities, they shall be planned, designed and managed in a unified way with the monitoring facilities on road sections.

### 8.3.2 Toll facilities

Toll facilities shall meet the following requirements:

(1) Toll facilities shall be coordinated with the level of service adopted in highway design. The number of toll lanes at the exit and entrance of toll plazas shall not be less than 2. The newly built toll facilities shall simultaneously construct ETC lanes.

(2) The electrical and mechanical equipment of the toll system can be configured according to the traffic volume in the 15th year after opening. The toll island, toll plaza, underground passage, toll shed, and other facilities should be

configured according to the traffic volume in the 15th year after opening. Toll plaza land, station building land, construction and earthwork land shall be implemented according to the traffic volume in the 20th year after opening.

(3) Passenger cars shall be charged by the vehicle type, and trucks should be charged by weight.

Toll plaza design shall meet the following requirements:

**1) Geometric indexes of the toll plaza**

(1) Main line toll plaza: The horizontal curve indexes shall comply with the provisions of the main line alignment indexes within the interchange range, and the vertical curve index shall not be less than the general value of the main line vertical alignment. Within a minimum of 100 m on both sides of the toll plaza centerline, the grade shall not be greater than 2%.

(2) Ramp toll plaza: The radius of the horizontal curve shall not be less than 200 m, and the radius of the vertical curve shall not be less than 800 m. Within the range of cement concrete pavement on both sides of the toll plaza centerline, the grade should not be greater than 2%, and shall not be greater than 3% under restriction.

(3) The cross slope of the toll plaza should be 1. 5%, and 2. 0% when drainage is needed.

**2) Toll plaza layout**

Design of the toll plaza shall meet the following requirements:

(1) The highway toll plaza shall be avoided at the bottom of the sag vertical curve.

(2) The geometric layout of the toll plaza is shown in Figure 8 - 20. The length $L_0$ of cement concrete pavement before and after the toll island shall meet the requirements of Table 8 - 12.

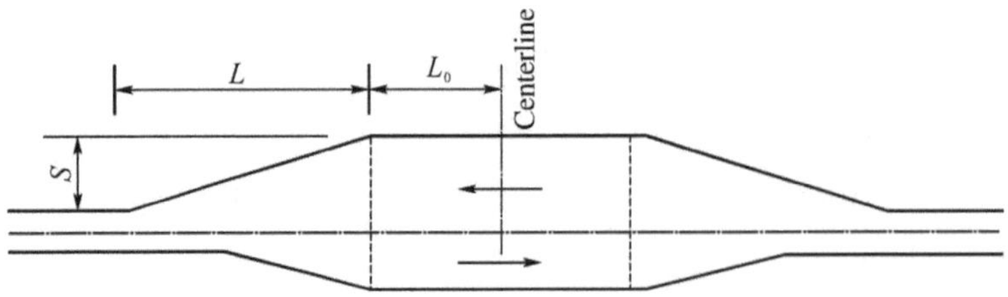


$L_0$—cement concrete pavement length at the plaza; $L$—plaza transition length; $S$—plaza transition width.

**Figure 8 - 20 Schematic diagram of toll plaza layout and transition at both ends**

**Table 8 - 12 Minimum length of cement concrete pavement before and after toll island $L_0$**

| Location of the toll plaza | | On the ramp | On the main line |
|---|---|---|---|
| Toll manner/m | One-way | 30 | 50 |
| | Two-way | 25 | 40 |

*Note*: This table refers to the **Specification (2017)**.

(3) The transition section at both ends of the toll plaza shall meet the requirements of Table 8 - 13.

**Table 8 - 13 Transition rate of traffic lanes at both ends of the toll plaza**

| Location of the toll plaza | On the ramp | On the main line |
|---|---|---|
| Plaza convergence rate/L • $S^{-1}$ | 4 - 6, limit value is 3 | 6 - 8 |

*Note*: This table refers to the **Specification (2017)**.

(4) The distance from the centerline of the ramp toll plaza to the ramp terminal should not be less than 100 m, and shall not be less than 75 m. The distance from the centerline of the ramp toll plaza to the intersecting point of the crossed road should not be less than 150 m, and if it cannot be satisfied, a waiting lane shall be added.

(5) Toll plaza width shall include the width of toll lanes, toll islands, and road shoulders (or curbs). Lane width between toll islands should be 3. 20 m. ETC lane width shall be 3. 50 m. The width of the super-wide lane should be 4. 5 m. Toll island width should be 2. 20 m. Paved shoulder width shall not be less than 0. 50 m. The cross section composition of the toll plaza is shown in Figure 8 - 21.

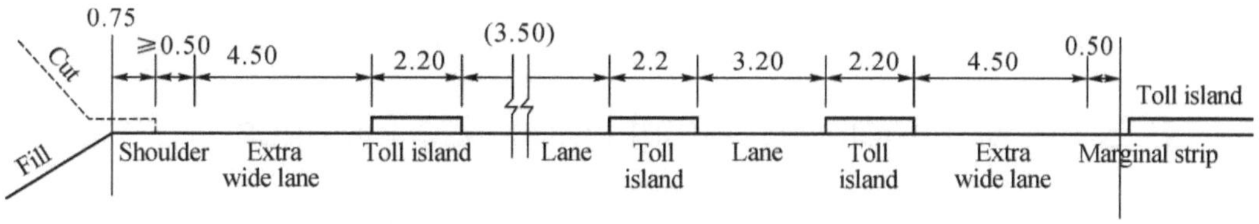


**Figure 8 - 21 Schematic diagram of cross section composition of the toll plaza centerline (m)**

## 8.3.3 Communication facilities

Communication facilities shall meet the following requirements:

(1) Communication facilities shall meet the business needs of monitoring, tolling and management, and provide voice, data, and image information service platforms in combination with unified road network planning, unified standards,

and unified systems.

(2) Communication pipelines of freeways shall be designed according to long-term planning. The laying capacity of communication pipelines shall be determined by comprehensive consideration of the demand for traffic network demand, social rental demand and capacity expansion requirements. A communication pipeline for arterial line networking shall be ensured between provinces.

### 8.3.4 Power supply and lighting facilities

Power supply and lighting facilities shall comply with the following provisions:

(1) According to the highway characteristics, system scale, load nature, electricity consumption, power supply conditions and power grid development planning, the power supply system scheme of external power supply and self-provided emergency power supply shall be reasonably determined, which shall meet the short-term requirements and take into account the long-term development needs.

(2) The high-voltage transmission line project shall be implemented based on the temporary power use for construction and the permanent power use for operation, as well as the characteristics, scale and long-term development of the project.

(3) Lighting facilities shall be provided for the toll plaza, service area plaza, escape lane, inspection point (station), etc. Lighting facilities can be provided for the interchange, long-span bridge, airport freeway and ring freeway located at urban entrances and exits.

### 8.3.5 Management and maintenance facilities

(1) The management center, management sub-center and management station should be applied according to the requirements of highway management.

(2) Maintenance work areas and road team houses shall be constructed according to the needs of highway maintenance business. Maintenance work areas should be constructed for freeways, and road team houses should be constructed for other classes of highways.

(3) Highway management and maintenance facilities should be jointly constructed in association with the terrain and business scope.

(4) Highway management buildings shall be reasonably arranged, economically applicable, environmentally friendly and energy-saving, and coordinated with the surrounding environment. The building scale should be determined according to the design traffic volume.

**Questions:**

1. Please describe the main categories of traffic safety facilities.
2. What are the types of service areas? How to choose them?
3. What is the anti-dazzle facility? Under what circumstances shall anti-dazzle facilities be used?
4. Please describe the level of monitoring facilities and their applicable scopes.

# References

[1] Ministry of Transport of the People's Republic of China, 2014. Technical Standard of Highway Engineering (JTG B01—2014). Beijing, China.

[2] Ministry of Transport of the People's Republic of China, 2017. Design Specification for Highway Alignment (JTG D20—2017). Beijing, China.

[3] Ministry of Transport of the People's Republic of China, 2014. Guidelines for Design of Highway Grade-separated Intersections (JTG/T D21—2014). Beijing, China.

[4] Ministry of Housing and Urban-Rural Development of the People's Republic of China, 2016. Code for Design of Urban Road Engineering (CJJ 37—2012). Beijing, China.

[5] Ministry of Housing and Urban-Rural Development of the People's Republic of China, 2016. Specification for Design of Intersections on Urban Roads (CJJ 152—2010). Beijing, China.

[6] Ministry of Housing and Urban-Rural Development of the People's Republic of China, 2012. Code for Design of Urban Road Alignment (CJJ 193—2012). Beijing, China.

[7] Ministry of Housing and Urban-Rural Development of the People's Republic of China, 2011. Code for Planning of Intersections on Urban Roads (GB 50647—2011). Beijing, China.

[8] Ministry of Transport of the People's Republic of China, 2015. Specifications for Highway Safety Audit (JTG B05—2015). Beijing, China.

[9] AASHTO. A Policy on Geometric Design of Highways and Streets [M]. Washington D. C.: American Association of State Highway and Transportation Officials, 2018.

[10] Ministry of Transport of the People's Republic of China. Highway Design Guidelines under Flexibility [M]. Beijing: China Communication Press, 2005.

[11] Qian, G. C. Innovative Designfor Expressway [M]. Beijing: China Communication Press, 2008.

© The Editor(s) (if applicable) and The Author(s), under exclusive license to Springer Nature Singapore Pte Ltd. 2025
B. Yu and J. Cheng, *Road Survey and Design*,
https://doi.org/10.1007/978-981-97-4836-5

[12] Xu, Y. C. Highway Geometric Design [M]. Taipei: Wunan Book Incorporation, 2010.

[13] Xu, J. L. Road Survey and Design[M]. Beijing: China Communication Press, 2018.

[14] Zhang, J. S.; Zhang, T. K. Road Survey and Design[M]. Shanghai: Tongji University Press, 2015.

[15] Pei, Y. L. ; Cheng, G. Z.; Zhang, Q. Road Survey and Design[M]. Beijing: China Communication Press, 2018.

[16] Sun, J. Y.; Li, S. Q.; Wang W. H. Road Survey and Design[M]. Beijing: China Communication Press, 2012.

[17] Zhang, J. S.; Zhang, T. K. Road Survey and Design[M]. Shanghai: Tongji University Press, 2012.

[18] Xu, Y. C. Theoretical Basis ofRoad Traffic Engineering Design[M]. Beijing: China Communication Press, 2020.

[19] Sichuan Communication Surveying & Design Institute Limited Company. Guideline of BIM Standard Establishment Application for Highway Engineering[M]. Beijing: China Machinery Press, 2020.

[20] Song, C. Z.; Li, X. L.; Fa, Z. Safety Analysis and Evaluation of Road Alignment[M]. Beijing: China Architecture & Building Press, 2017.

[21] Liu, L. M. Innovation in Route Computer Aided Design and Highway Alignment Design Practice[M]. Beijing: China Communication Press, 2020.

[22] Shanghai Planning and Land Resources Administration Bureau; Shanghai Municipal Transportation Commission; Shanghai Urban Planning & Design Research Institute. Shanghai Street Design Guidelines[M]. Shanghai: Tongji University Press, 2016.

[23] Fang, S. E.; Chen, Y. R. Road Planning and Geometric Design[M]. Beijing: China Communication Press, 2021.

MIX
Papier aus verantwortungsvollen Quellen
Paper from responsible sources
FSC® C105338

If you have any concerns about our products,
you can contact us on
**ProductSafety@springernature.com**

In case Publisher is established outside the EU,
the EU authorized representative is:
**Springer Nature Customer Service Center GmbH**
**Europaplatz 3, 69115 Heidelberg, Germany**

Printed by Libri Plureos GmbH
in Hamburg, Germany